T0171872

Birkhäuser

Applied and Numerical Harmonic Analysis

Lecture Notes in Applied and Numerical Harmonic Analysis

More information about this subseries at http://www.springer.com/series/13412

Vasily N. Malozemov · Sergey M. Masharsky

Foundations of Discrete Harmonic Analysis

Birkhäuser

Vasily N. Malozemov
Mathematics and Mechanics Faculty
Saint Petersburg State University
Saint Petersburg, Russia

Sergey M. Masharsky
Mathematics and Mechanics Faculty
Saint Petersburg State University
Saint Petersburg, Russia

ISSN 2296-5009 ISSN 2296-5017 (electronic)
Applied and Numerical Harmonic Analysis
ISSN 2512-6482 ISSN 2512-7209 (electronic)
Lecture Notes in Applied and Numerical Harmonic Analysis
ISBN 978-3-030-47047-0 ISBN 978-3-030-47048-7 (eBook)
https://doi.org/10.1007/978-3-030-47048-7

Mathematics Subject Classification (2010): 42C10, 42C20, 65D07, 65T50, 65T60

This book is published under the imprint Birkhäuser, www.birkhauser-science.com by the registered company Springer Nature Switzerland AG
The registered company address is: Gewerbestrasse 11, 6330 Cham, Switzerland

LN-ANHA Series Preface

The *Lecture Notes in Applied and Numerical Harmonic Analysis (LN-ANHA)* book series is a subseries of the widely known *Applied and Numerical Harmonic Analysis (ANHA)* series. The Lecture Notes series publishes paperback volumes, ranging from 80 to 200 pages in harmonic analysis as well as in engineering and scientific subjects having a significant harmonic analysis component. *LN-ANHA* provides a means of distributing brief-yet-rigorous works on similar subjects as the *ANHA* series in a timely fashion, reflecting the most current research in this rapidly evolving field.

The *ANHA* book series aims to provide the engineering, mathematical, and scientific communities with significant developments in harmonic analysis, ranging from abstract harmonic analysis to basic applications. The title of the series reflects the importance of applications and numerical implementation, but richness and relevance of applications and implementation depend fundamentally on the structure and depth of theoretical underpinnings. Thus, from our point of view, the interleaving of theory and applications and their creative symbiotic evolution is axiomatic.

Harmonic analysis is a wellspring of ideas and applicability that has flourished, developed, and deepened over time within many disciplines and by means of creative cross-fertilization with diverse areas. The intricate and fundamental relationship between harmonic analysis and fields such as signal processing, partial differential equations (PDEs), and image processing is reflected in our state-of-the-art *ANHA* series.

Our vision of modem harmonic analysis includes mathematical areas such as wavelet theory, Banach algebras, classical Fourier analysis, time-frequency analysis, and fractal geometry, as well as the diverse topics that impinge on them.

For example, wavelet theory can be considered an appropriate tool to deal with some basic problems in digital signal processing, speech and image processing, geophysics, pattern recognition, bio-medical engineering, and turbulence. These areas implement the latest technology from sampling methods on surfaces to fast algorithms and computer vision methods. The underlying mathematics of wavelet theory depends not only on classical Fourier analysis but also on ideas from abstract harmonic analysis, including von Neumann algebras and the affine group. This leads to a study of the Heisenberg group and its relationship to Gabor systems and of the metaplectic group for a meaningful interaction of signal decomposition methods.

The unifying influence of wavelet theory in the aforementioned topics illustrates the justification for providing a means for centralizing and disseminating information from the broader, but still focused, area of harmonic analysis. This will be a key role of *ANHA*. We intend to publish with the scope and interaction that such a host of issues demands.

Along with our commitment to publish mathematically significant works at the frontiers of harmonic analysis, we have a comparably strong commitment to publish major advances in applicable topics such as the following, where harmonic analysis plays a substantial role:

Bio-mathematics, bio-engineering, and bio-medical signal processing;
Communications and RADAR;
Compressive sensing (sampling) and sparse representations;
Data science, data mining and dimension reduction;
Fast algorithms;
Frame theory and noise reduction;
Image processing and super-resolution;
Machine learning;
Phaseless reconstruction;
Quantum informatics;
Remote sensing;
Sampling theory;
Spectral estimation;
Time-frequency and time-scale analysis–Gabor theory and wavelet theory

The above point of view for the *ANHA* book series is inspired by the history of Fourier analysis itself, whose tentacles reach into so many fields.

In the last two centuries Fourier analysis has had a major impact on the development of mathematics, on the understanding of many engineering and scientific phenomena, and on the solution of some of the most important problems in mathematics and the sciences. Historically, Fourier series were developed in the analysis of some of the classical PDEs of mathematical physics; these series were used to solve such equations. In order to understand Fourier series and the kinds of solutions they could represent, some of the most basic notions of analysis were defined, for example, the concept of "function." Since the coefficients of Fourier series are integrals, it is no surprise that Riemann integrals were conceived to deal with uniqueness properties of trigonometric series. Cantor's set theory was also developed because of such uniqueness questions.

A basic problem in Fourier analysis is to show how complicated phenomena, such as sound waves, can be described in terms of elementary harmonics. There are two aspects of this problem: first, to find, or even define properly, the harmonics or spectrum of a given phenomenon, e.g., the spectroscopy problem in optics; second, to determine which phenomena can be constructed from given classes of harmonics, as done, for example, by the mechanical synthesizers in tidal analysis.

Fourier analysis is also the natural setting for many other problems in engineering, mathematics, and the sciences. For example, Wiener's Tauberian theorem in Fourier analysis not only characterizes the behavior of the prime numbers but also provides the proper notion of spectrum for phenomena such as white light; this latter process leads to the Fourier analysis associated with correlation functions in filtering and prediction problems, and these problems, in turn, deal naturally with Hardy spaces in the theory of complex variables.

Nowadays, some of the theory of PDEs has given way to the study of Fourier integral operators. Problems in antenna theory are studied in terms of unimodular trigonometric polynomials. Applications of Fourier analysis abound in signal processing, whether with the fast Fourier transform (FFT), or filter design, or the adaptive modeling inherent in time-frequency-scale methods such as wavelet theory.

The coherent states of mathematical physics are translated and modulated Fourier transforms, and these are used, in conjunction with the uncertainty principle, for dealing with signal reconstruction in communications theory. We are back to the raison d'être of the *ANHA* series!

University of Maryland
College Park, MD, USA

John J. Benedetto
Series Editor

Preface

Discrete harmonic analysis is a mathematical discipline predominately targeted to advanced applications of digital signal processing. A notion of a *signal* requires a closer definition. A signal in discrete harmonic analysis is defined as a complex-valued periodic function of an integer argument.

In this book we study transforms of signals. One of the fundamental transforms is the *discrete Fourier transform* (DFT). In 1965, Cooley and Tukey in their paper [8] proposed the *fast Fourier transform* (FFT), a fast method of calculation of the DFT. Essentially, this discovery set the stage for development of discrete harmonic analysis as a self-consistent discipline.

The DFT inversion formula causes a signal to be expanded over the exponential basis. Expanding a signal over various bases is the main technique of digital signal processing.

An argument of a signal is interpreted as *time*. Components of the discrete Fourier transform comprise a *frequency spectrum* of a signal. Analysis in the time and frequency domain lets us uncover the structure of a signal and determine ways of transforming a signal to obtain required properties.

In practice, we are faced with a necessity to process signals of various natures such as acoustic, television, seismic, radio signals, or signals coming from the outer space. These signals are received by physical devices. When we take a reading of a device at regular intervals we obtain a discrete signal. It is this signal that is a subject of further digital processing. To start with, we calculate a frequency spectrum of the discrete signal. It corresponds to representing a signal in a form of a sum of simple summands being its frequency components. By manipulating with frequency components we achieve an improvement of specific features of a signal.

This book is aimed at an initial acquaintance with the subject. It is written on the basis of the lecture course that the first author has been delivering since 1995 on the Faculty of Mathematics and Mechanics of St. Petersburg State University.

The book consists of four chapters. The first chapter briefly exposes the facts that are being used in the main text. These facts are well known and are related to residuals, permutations, complex numbers, and finite differences.

In the second chapter we consider basic transforms of signals. The centerpieces are discrete Fourier transform, cyclic convolution, and cyclic correlation. We study the properties of these transforms. As an application, we provide solutions to the problems of optimal interpolation and optimal signal–filter pair. Separate sections are devoted to ensembles of signals and to the uncertainty principle in discrete harmonic analysis.

In the third chapter we introduce discrete periodic splines and study their fundamental properties. We establish an extremal property of the interpolation splines. In terms of splines, we offer an elegant solution to the problem of smoothing of discrete periodic data. We construct a system of orthogonal splines. With the aid of dual splines, we solve the problem of spline processing of discrete data with the least squares method.

We obtain a wavelet expansion of an arbitrary spline. We prove two limit theorems related to interpolation splines.

The focus of the fourth chapter is on fast algorithms: the fast Fourier transform, the fast Haar transform, and the fast Walsh transform. To build a fast algorithm we use an original approach stemming from introduction of a recurrent sequence of orthogonal bases in the space of discrete periodic signals. In this way we manage to form wavelet bases which altogether constitute a *wavelet packet*. In particular, Haar basis is a wavelet one. We pay a lot of attention to it in the book.

We investigate an important question of ordering of Walsh functions. We analyze in detail Ahmed–Rao bases that fall in between Walsh basis and the exponential basis.

The main version of the fast Fourier transform (it is called the *Cooley–Tukey algorithm*) is targeted to calculate the DFT whose order is a power of two. At the end of the fourth chapter, we show how to use the Cooley–Tukey algorithm to calculate a DFT of any order.

A specific feature of the book is a big number of exercises. They allow us to lessen the burden of the main text. Many special and auxiliary facts are formalized as exercises. All the exercises are endowed with solutions. Separate exercises or exercise groups are independent, so you as a reader can select only those that seem interesting to you. The most efficient way is solving an exercise and then checking your solution against the one presented in the book. It will let you actively master the matter.

At the end of the book we put the list of references. We lay emphasis on the books [5, 41, 49] that we used to study up the fundamentals of discrete harmonic analysis back in the day.

The first version of the book was published in 2003 as a preprint. In 2012 *Lan' publishers* published the book in Russian. This English edition is an extended and improved version of the Russian edition.

The seminar on discrete harmonic analysis and computer aided geometric design (shortly, DHA&CAGD) was held in St. Petersburg University from 2004 to 2014. The seminar's website is `http://dha.spb.ru`. The site was used to publish the proceedings of the seminar's members; these proceedings served as a basis for the books [46, 47, 34, 44, 7] published later on. Contents of the proceedings and the mentioned books can be considered as an addendum to this book.

St. Petersburg, Russia
July 2019

Vasily N. Malozemov
Sergey M. Masharsky

Acknowledgements First of all, the authors are thankful to the students and postgraduates who, over the years, attended the course of lectures on discrete harmonic analysis and offered beautiful solutions to some exercises.

The first author separately expresses his gratitude to his permanent co-author Prof. A. B. Pevnyi and to his former postgraduate students M. G. Ber and A. A. Tret'yakov. It is with these people that we accomplished our first works in the field of discrete harmonic analysis. We also give thanks to O. V. Prosekov, M. I. Grigoriev, and N. V. Chashnikov. By turn, they administered the website of DHA&CAGD over 10 years.

Contents

Acronyms

DFT	Discrete Fourier transform
DHT	Discrete Haar transform
DWT	Discrete Walsh transform
FFT	Fast Fourier transform
SLBF	Side-lobe blanking filter

Chapter 1
Preliminaries

The following notations are used throughout the book:

$\mathbb{Z}$, $\mathbb{R}$, $\mathbb{C}$ sets of integer, real, and complex numbers, respectively;
$m : n$ a set of consequent integer numbers $\{m, m+1, \dots, n\}$.

The notation $A := B$ or $B =: A$ means that A equals to B by a definition.

1.1 Residuals

Consider $j \in \mathbb{Z}$ and N being a natural number. There exists a unique integer p such that

$$p \leq j/N < p + 1. \tag{1.1.1}$$

It is referred to as an *integral part* of the fraction j/N and is noted as $p = \lfloor j/N \rfloor$. The difference $r = j - pN$ is called a *remainder after division of* j *by* N or a *modulo* N *residual of* j. It is noted as $r = \langle j \rangle_N$. For a given j we get a representation $j = pN + r$, where $p = \lfloor j/N \rfloor$ and $r = \langle j \rangle_N$.

It is not difficult to show that

$$\langle j \rangle_N \in 0 : N - 1. \tag{1.1.2}$$

Indeed, multiply the inequalities (1.1.1) by N and subtract pN. We obtain $0 \leq j - pN < N$, which is equivalent to (1.1.2).

It follows from the definitions that the equalities

$$\lfloor (j + kN)/N \rfloor = \lfloor j/N \rfloor + k, \tag{1.1.3}$$

$$\langle j + kN \rangle_N = \langle j \rangle_N \tag{1.1.4}$$

V. N. Malozemov and S. M. Masharsky, *Foundations of Discrete Harmonic Analysis*, Applied and Numerical Harmonic Analysis,
https://doi.org/10.1007/978-3-030-47048-7_1

hold for any integer k. The formal proof is carried out in this way. As long as $\lfloor j/N \rfloor \le j/N < \lfloor j/N \rfloor + 1$, after addition of k we obtain

$$\lfloor j/N \rfloor + k \le (j + kN)/N < \lfloor j/N \rfloor + k + 1.$$

This is equivalent to (1.1.3). Equality (1.1.4) is a direct consequence of (1.1.3). Indeed,

$$\langle j + kN \rangle_N = j + kN - \lfloor (j + kN)/N \rfloor N = j - \lfloor j/N \rfloor N = \langle j \rangle_N.$$

We mention two other properties of residuals that are simple yet important: for any integers j and k

$$\langle j + k \rangle_N = \big\langle \langle j \rangle_N + k \big\rangle_N = \big\langle \langle j \rangle_N + \langle k \rangle_N \big\rangle_N,$$

$$\langle jk \rangle_N = \big\langle \langle j \rangle_N \, k \big\rangle_N = \big\langle \langle j \rangle_N \, \langle k \rangle_N \big\rangle_N.$$

The proof of these equalities relies on formula (1.1.4).

1.2 Greatest Common Divisor

Take nonzero integers j and k. The largest natural number that divides both j and k is called the *greatest common divisor* of these numbers and is denoted by $\gcd(j, k)$.

Designate by M the set of linear combinations of the numbers j and k with integer coefficients:

$$M = \{aj + bk \mid a \in \mathbb{Z},\ b \in \mathbb{Z}\}.$$

Theorem 1.2.1 *The smallest natural number in M equals to* $\gcd(j, k)$.

Proof Let $d = a_0 j + b_0 k$ be the smallest natural number in M. We will show that j is divisible by d. Using the representation $j = pd + r$, where $r \in 0 : d - 1$, we write

$$r = j - pd = j - p(a_0 j + b_0 k) = (1 - pa_0) j - pb_0 k.$$

We see that $r \in M$ and $r < d$. It is possible only when $r = 0$, i.e. when j is divisible by d. Similarly we ascertain that k is divisible by d as well.

Now let j and k be divisible by a natural number d'. Then d is also divisible by d'. Hence $d = \gcd(j, k)$. The theorem is proved. □

According to Theorem 1.2.1, there exist integers a_0 and b_0 such that

$$\gcd(j, k) = a_0 j + b_0 k. \tag{1.2.1}$$

Formula (1.2.1) is referred to as a *linear representation* of the greatest common divisor.

Note that

$$\gcd(j, k) = \gcd(|j|, |k|)$$

since the integers j and $-j$ have the same divisors.

1.3 Relative Primes

Natural numbers n and N are referred to as *relative primes* if $\gcd(n, N) = 1$. For relative primes n and N equality (1.2.1) takes the form

$$a_0 n + b_0 N = 1. \tag{1.3.1}$$

Thus, for relative primes n and N there exist integers a_0 and b_0 such that equality (1.3.1) holds.

The inverse assertion is also valid: equality (1.3.1) guarantees relative primality of n and N. It follows from Theorem 1.2.1, since the unity is absolutely the smallest natural number.

Theorem 1.3.1 *If the product jn for some $j \in \mathbb{Z}$ is divisible by N, and the integers n and N are relative primes, then j is divisible by N.*

Proof Multiply both sides of equality (1.3.1) by j and take modulo N residuals. We get $\langle a_0 jn\rangle_N = \langle j\rangle_N$ and

$$\big\langle a_0 \langle jn\rangle_N \big\rangle_N = \langle j\rangle_N.$$

It is clear that $\langle j\rangle_N = 0$ if $\langle jn\rangle_N = 0$. This is a symbolic equivalent of the theorem's statement. □

Now multiply both sides of equality (1.3.1) by a number $k \in 0 : N-1$. Using modulo N residuals we come to the relation

$$\big\langle \langle a_0 k\rangle_N \, n \big\rangle_N = k.$$

This result can be interpreted as follows: an equation $\langle xn\rangle_N = k$ for any $k \in 0 : N-1$ has a solution $x_0 = \langle a_0 k\rangle_N$ on the set $0 : N-1$. Let's show that the solution is unique. Assume that $\langle x'n\rangle_N = k$ for some $x' \in 0 : N-1$. Then

$$\langle (x_0 - x')n\rangle_N = \big\langle \langle x_0 n\rangle_N - \langle x'n\rangle_N \big\rangle_N = 0.$$

Since n and N are the relative primes, Theorem 1.3.1 yields that $x_0 - x'$ is divisible by N. Taking into account the inequality $|x_0 - x'| \leq N - 1$ we conclude that $x' = x_0$.

Let's summarize.

Theorem 1.3.2 *If* $\gcd(n, N) = 1$ *then the equation* $\langle xn \rangle_N = k$ *has a unique solution on the set* $0 : N-1$ *for any* $k \in 0 : N-1$.

1.4 Permutations

Denote $f(j) = \langle jn \rangle_N$. By virtue of Theorem 1.3.2, provided that $\gcd(n, N) = 1$, the function $f(j)$ bijectively maps the set $J_N = \{0, 1, \ldots, N-1\}$ onto itself. Essentially, f performs a permutation of the elements of J_N. This is called an *Euler permutation*.

We describe a simple way of calculating the values $f(j)$. It is obvious that $f(0) = 0$ and

$$f(j+1) = \langle (j+1)n \rangle_N = \big\langle \langle jn \rangle_N + n \big\rangle_N = \langle f(j) + n \rangle_N.$$

We come to the recurrent relation

$$f(0) = 0; \quad f(j+1) = \langle f(j) + n \rangle_N, \; j = 0, 1, \ldots, N-2 \tag{1.4.1}$$

that makes it possible to consequently discover the values of the Euler permutation. The results of calculations with formula (1.4.1) for $n = 3$ and $N = 8$ are presented in Table 1.1.

Later on we will need two other permutations, rev_ν and grey_ν. They are defined on a set $\{0, 1, \ldots, 2^\nu - 1\}$ for a natural ν.

Recall that with the use of consequent bisections we can uniquely represent any integer $j \in 0 : 2^\nu - 1$ in the form

$$j = j_{\nu-1}2^{\nu-1} + j_{\nu-2}2^{\nu-2} + \cdots + j_1 2 + j_0, \tag{1.4.2}$$

where every coefficient j_k is equal to either zero or unity. Instead of (1.4.2), more compact notation is used: $j = (j_{\nu-1}, j_{\nu-2}, \ldots, j_0)_2$. The right side of the latter equality is referred to as a *binary code* of the number j.

Introduce a notation

$$\mathrm{rev}_\nu(j) = (j_0, j_1, \ldots, j_{\nu-1})_2.$$

Table 1.1 Euler permutation for $n = 3$ and $N = 8$

j	0	1	2	3	4	5	6	7
$\langle j3 \rangle_8$	0	3	6	1	4	7	2	5

Table 1.2 Permutation rev_ν for $\nu = 3$

j	$(j_2, j_1, j_0)_2$	$(j_0, j_1, j_2)_2$	$\text{rev}_3(j)$
0	$(0, 0, 0)_2$	$(0, 0, 0)_2$	0
1	$(0, 0, 1)_2$	$(1, 0, 0)_2$	4
2	$(0, 1, 0)_2$	$(0, 1, 0)_2$	2
3	$(0, 1, 1)_2$	$(1, 1, 0)_2$	6
4	$(1, 0, 0)_2$	$(0, 0, 1)_2$	1
5	$(1, 0, 1)_2$	$(1, 0, 1)_2$	5
6	$(1, 1, 0)_2$	$(0, 1, 1)_2$	3
7	$(1, 1, 1)_2$	$(1, 1, 1)_2$	7

A number $\text{rev}_\nu(j)$ belongs to a set $0 : 2^\nu - 1$, and its binary code equals to the reverted binary code of a number j. Identifier "rev" corresponds to a word *reverse*. Subscript ν determines the amount of reverted binary digits.

It is clear that $\text{rev}_\nu\big(\text{rev}_\nu(j)\big) = j$ for $j \in 0 : 2^\nu - 1$. Hence, in particular, it follows that the mapping $j \to \text{rev}_\nu(j)$ is a permutation of a set $\{0,\ 1,\ \ldots,\ 2^\nu - 1\}$.

By a definition, $\text{rev}_1(j) = j$ for $j \in 0 : 1$. It is reckoned that $\text{rev}_0(0) = 0$.

Table 1.2 shows how to form a permutation rev_ν for $\nu = 3$.

We continue with an investigation of a permutation rev_ν.

Theorem 1.4.1 *The following recurrent relation holds:*

$$\text{rev}_0(0) = 0;$$

$$\begin{aligned} \text{rev}_\nu(2k) &= \text{rev}_{\nu-1}(k), \\ \text{rev}_\nu(2k+1) &= 2^{\nu-1} + \text{rev}_{\nu-1}(k), \end{aligned} \tag{1.4.3}$$

$$k \in 0 : 2^{\nu-1} - 1, \quad \nu = 1,\ 2,\ \ldots$$

Proof Replace the second and the third lines in (1.4.3) with a single line

$$\begin{gathered} \text{rev}_\nu(2k + \sigma) = \sigma 2^{\nu-1} + \text{rev}_{\nu-1}(k), \\ \sigma \in 0 : 1, \quad k \in 0 : 2^{\nu-1} - 1, \quad \nu = 1, 2, \ldots \end{gathered} \tag{1.4.4}$$

When $\nu = 1$, formula (1.4.4) becomes of a known form $\text{rev}_1(\sigma) = \sigma, \sigma \in 0 : 1$. Let it be $\nu \geq 2$. For any $k \in 0 : 2^{\nu-1} - 1$ and $\sigma \in 0 : 1$ we have

$$k = k_{\nu-2}2^{\nu-2} + \cdots + k_1 2 + k_0, \quad 2k + \sigma = k_{\nu-2}2^{\nu-1} + \cdots + k_0 2 + \sigma,$$

$$\text{rev}_\nu(2k + \sigma) = \sigma 2^{\nu-1} + k_0 2^{\nu-2} + \cdots + k_{\nu-2} = \sigma 2^{\nu-1} + \text{rev}_{\nu-1}(k).$$

The theorem is proved. □

Table 1.3 Consequent calculation of permutations rev_ν

ν	$\mathrm{rev}_\nu(j)$ for $j = 0, 1, \ldots, 2^\nu - 1$
1	0 1
2	0 2 1 3
3	0 4 2 6 1 5 3 7

Theorem 1.4.1 makes it possible to consequently calculate the values $\mathrm{rev}_\nu(j)$ for $\nu = 1, 2, \ldots$ for all $j \in \{0, 1, \ldots, 2^\nu - 1\}$ at once. Table 1.3 presents the results of calculations of $\mathrm{rev}_1(j)$, $\mathrm{rev}_2(j)$, and $\mathrm{rev}_3(j)$. A transition between the $(\nu - 1)$-th and the ν-th rows was performed with accordance to formula (1.4.3). It was also taken into account that

$$\mathrm{rev}_\nu(2k + 1) = \mathrm{rev}_\nu(2k) + 2^{\nu-1}, \quad k = 0, 1, \ldots, 2^{\nu-1} - 1.$$

Now we turn to a permutation grey_ν. It is defined recursively:

$$\mathrm{grey}_0(0) = 0;$$

$$\begin{aligned} \mathrm{grey}_\nu(k) &= \mathrm{grey}_{\nu-1}(k), \\ \mathrm{grey}_\nu(2^\nu - 1 - k) &= 2^{\nu-1} + \mathrm{grey}_{\nu-1}(k), \end{aligned} \tag{1.4.5}$$

$$k \in 0 : 2^{\nu-1} - 1, \quad \nu = 1, 2, \ldots$$

Let's assure that the mapping $j \to \mathrm{grey}_\nu(j)$ is indeed a permutation of a set $\{0, 1, \ldots, 2^\nu - 1\}$.

When $\nu = 1$, this is obvious since, by a definition, $\mathrm{grey}_1(j) = j$ for $j \in 0 : 1$. Assume that the assertion is true for $\mathrm{grey}_{\nu-1}$. According to the second line of (1.4.5) the function $\mathrm{grey}_\nu(j)$ bijectively maps the set $\{0, 1, \ldots, 2^{\nu-1} - 1\}$ onto itself. The third line of (1.4.5) after argument replacement $k' = 2^{\nu-1} - 1 - k$ takes the form

$$\mathrm{grey}_\nu(2^{\nu-1} + k') = 2^{\nu-1} + \mathrm{grey}_{\nu-1}(2^{\nu-1} - 1 - k'), \tag{1.4.6}$$

$$k' \in 0 : 2^{\nu-1} - 1.$$

Hence the function $\mathrm{grey}_\nu(j)$ as well bijectively maps onto itself the set $\{2^{\nu-1}, \ldots, 2^\nu - 1\}$. Joining these two facts we conclude that the function $\mathrm{grey}_\nu(j)$ bijectively maps the set $\{0, 1, \ldots, 2^\nu - 1\}$ onto itself. In the other words, the mapping $j \to \mathrm{grey}_\nu(j)$ is a permutation of the set $\{0, 1, \ldots, 2^\nu - 1\}$.

Table 1.4 Consequent calculation of permutations grey_ν

ν	$\mathrm{grey}_\nu(j)$ for $j = 0, 1, \ldots, 2^\nu - 1$
1	0 1
2	0 1 3 2
3	0 1 3 2 6 7 5 4

Formula (1.4.5) make it possible to consequently calculate the values $\mathrm{grey}_\nu(j)$ for $\nu = 1, 2, \ldots$ for all $j \in \{0, 1, \ldots, 2^\nu - 1\}$ at once. Table 1.4 contains the results of calculations of $\mathrm{grey}_1(j)$, $\mathrm{grey}_2(j)$, and $\mathrm{grey}_3(j)$.

We adduce a characteristic property of a permutation grey_ν.

Theorem 1.4.2 *For* $\nu \geq 1$, *the binary codes of two adjacent elements* $\mathrm{grey}_\nu(k)$ *and* $\mathrm{grey}_\nu(k+1)$, $k \in 0 : 2^\nu - 2$, *differ in a single digit only.*

Proof When $\nu = 1$, the assertion is obvious since $\mathrm{grey}_1(0) = (0)_2$ and $\mathrm{grey}_1(1) = (1)_2$. We perform an induction step from $\nu - 1$ to ν, $\nu \geq 2$.

Let $k \in 0 : 2^{\nu-1} - 1$ and $\mathrm{grey}_{\nu-1}(k) = (p_{\nu-2}, \ldots, p_0)_2$. According to (1.4.5)

$$\begin{aligned} \mathrm{grey}_\nu(k) &= (0, p_{\nu-2}, \ldots, p_0)_2, \\ \mathrm{grey}_\nu(2^\nu - 1 - k) &= (1, p_{\nu-2}, \ldots, p_0)_2. \end{aligned} \tag{1.4.7}$$

If $k \in 0 : 2^{\nu-1} - 2$ and $\mathrm{grey}_{\nu-1}(k+1) = (q_{\nu-2}, \ldots, q_0)_2$ then by an inductive hypothesis the binary codes $(p_{\nu-2}, \ldots, p_0)_2$ and $(q_{\nu-2}, \ldots, q_0)_2$ differ only in one digit. Since $\mathrm{grey}_\nu(k+1) = (0, q_{\nu-2}, \ldots, q_0)_2$, the same is true for the binary codes of numbers $\mathrm{grey}_\nu(k)$ and $\mathrm{grey}_\nu(k+1)$.

In case of $k = 2^{\nu-1} - 1$ we have $k + 1 = 2^\nu - 1 - k$ so that the binary codes of numbers $\mathrm{grey}_\nu(k)$ and $\mathrm{grey}_\nu(k+1)$ differ in the most significant digit only, as can be seen from (1.4.7).

It is remaining to consider the indices k from a set $2^{\nu-1} : 2^\nu - 2$. Put $k' = 2^\nu - 1 - k$. It is clear that $k' \in 1 : 2^{\nu-1} - 1$. Let

$$\begin{aligned} \mathrm{grey}_{\nu-1}(k') &= (p'_{\nu-2}, \ldots, p'_0)_2, \\ \mathrm{grey}_{\nu-1}(k' - 1) &= (q'_{\nu-2}, \ldots, q'_0)_2. \end{aligned}$$

By an inductive hypothesis, the binary codes $(p'_{\nu-2}, \ldots, p'_0)_2$ and $(q'_{\nu-2}, \ldots, q'_0)_2$ differ only in a single digit. According to (1.4.5)

$$\begin{aligned} \mathrm{grey}_\nu(k) &= \mathrm{grey}_\nu(2^\nu - 1 - k') = (1, p'_{\nu-2}, \ldots, p'_0)_2, \\ \mathrm{grey}_\nu(k+1) &= \mathrm{grey}_\nu\big(2^\nu - 1 - (k' - 1)\big) = (1, q'_{\nu-2}, \ldots, q'_0)_2. \end{aligned}$$

It is evident that the binary codes of numbers $\text{grey}_\nu(k)$ and $\text{grey}_\nu(k+1)$ also differ in a single digit only. The theorem is proved. □

1.5 Bitwise Summation

Take two integers $j = (j_{s-1}, j_{s-2}, \ldots, j_0)_2$ and $k = (k_{s-1}, k_{s-2}, \ldots, k_0)_2$ from the set $0 : 2^s - 1$. An operation of bitwise summation $\oplus$ associates the numbers j and k with an integer $p = (p_{s-1}, p_{s-2}, \ldots, p_0)_2$ that has $p_\nu = \langle j_\nu + k_\nu \rangle_2$ for $\nu = 0, 1, \ldots, s-1$. Thus,

$$p = j \oplus k \quad \Leftrightarrow \quad p_\nu = \langle j_\nu + k_\nu \rangle_2, \quad \nu = 0, 1, \ldots, s-1.$$

It follows from a definition that

$$j \oplus j = 0 \quad \text{for all } j \in 0 : 2^s - 1. \tag{1.5.1}$$

Bitwise summation operation is commutative and associative, i.e. $j \oplus k = k \oplus j$ and

$$(j \oplus k) \oplus l = j \oplus (k \oplus l). \tag{1.5.2}$$

Let's verify the associativity. For $\nu \in 0 : s-1$ we have

$$\begin{aligned} \langle (j \oplus k)_\nu + l_\nu \rangle_2 &= \big\langle \langle j_\nu + k_\nu \rangle_2 + l_\nu \big\rangle_2 = \langle (j_\nu + k_\nu) + l_\nu \rangle_2 \\ &= \langle j_\nu + (k_\nu + l_\nu) \rangle_2 = \big\langle j_\nu + \langle k_\nu + l_\nu \rangle_2 \big\rangle_2 = \langle j_\nu + (k \oplus l)_\nu \rangle_2. \end{aligned}$$

This corresponds to (1.5.2).

An equation $x \oplus k = p$ with fixed k and p from $0 : 2^s - 1$ has a unique solution $x = p \oplus k$ on the set $0 : 2^s - 1$. Indeed, according to (1.5.2) and (1.5.1)

$$(p \oplus k) \oplus k = p \oplus (k \oplus k) = p.$$

By virtue of the mentioned properties of bitwise summation we may affirm that the mapping $j \to j \oplus k$ with a fixed k is a permutation of the set $\{0, 1, \ldots, 2^s - 1\}$. Table 1.5 shows an example of such a permutation for $s = 3$ and $k = 5 = (1, 0, 1)_2$.

We could introduce an operation of bitwise subtraction $k \ominus j$ by setting

$$(k \ominus j)_\alpha = \langle k_\alpha - j_\alpha \rangle_2, \quad \alpha \in 0 : s-1.$$

But this operation is redundant because

$$k \ominus j = k \oplus j.$$

Table 1.5 Permutation $j \to j \oplus k$ for $s = 3$ and $k = 5 = (1, 0, 1)_2$

j	$(j_2, j_1, j_0)_2$	$j \oplus k$
0	$(0, 0, 0)_2$	5
1	$(0, 0, 1)_2$	4
2	$(0, 1, 0)_2$	7
3	$(0, 1, 1)_2$	6
4	$(1, 0, 0)_2$	1
5	$(1, 0, 1)_2$	0
6	$(1, 1, 0)_2$	3
7	$(1, 1, 1)_2$	2

Let's verify the last equality. We write it in an expanded form

$$\langle k_\alpha - j_\alpha \rangle_2 = \langle k_\alpha + j_\alpha \rangle_2 , \quad \alpha \in 0 : s - 1, \tag{1.5.3}$$

where $k_\alpha, j_\alpha \in 0 : 1$. We fix α. When $k_\alpha = j_\alpha$, equality (1.5.3) is valid ($0 = 0$). Let $k_\alpha \neq j_\alpha$. Then $\langle k_\alpha + j_\alpha \rangle_2 = 1$. At the same time, $\langle k_\alpha - j_\alpha \rangle_2 = 1$ for $k_\alpha = 1$, $j_\alpha = 0$; and for $k_\alpha = 0$, $j_\alpha = 1$ we have

$$\langle k_\alpha - j_\alpha \rangle_2 = \langle -1 \rangle_2 = \langle 2 - 1 \rangle_2 = 1.$$

1.6 Complex Numbers

It is assumed that the reader is familiar with the arithmetic operations on complex numbers. We remind some notations:

$z = u + iv$	a complex number,
$u = \operatorname{Re} z$	a real part of a complex number,
$v = \operatorname{Im} z$	an imaginary part of a complex number,
$\overline{z} = u - iv$	a conjugate complex number,
$\lvert z \rvert = \sqrt{u^2 + v^2}$	a modulus of a complex number.

It is obvious that $|z|^2 = z\overline{z}$. Also valid are the formulae

$$|z_1 + z_2|^2 = |z_1|^2 + |z_2|^2 + 2 \operatorname{Re}(z_1 \overline{z}_2),$$

$$|z_1 + i z_2|^2 = |z_1|^2 + |z_2|^2 + 2 \operatorname{Im}(z_1 \overline{z}_2).$$

Let's verify, for example, the latter one. We have

$$|z_1 + iz_2|^2 = (z_1 + iz_2)(\overline{z}_1 - i\overline{z}_2) = |z_1|^2 + |z_2|^2 - i(z_1\overline{z}_2 - \overline{z}_1 z_2)$$
$$= |z_1|^2 + |z_2|^2 + 2\,\mathrm{Im}(z_1\overline{z}_2).$$

The following two formulae are well known: for a natural n

$$(z_1 + z_2)^n = \sum_{k=0}^{n} \binom{n}{k} z_1^{n-k} z_2^k, \tag{1.6.1}$$

where $\binom{n}{k}$ is a binomial coefficient; for $z \neq 1$

$$\sum_{k=0}^{n-1} z^k = \frac{1 - z^n}{1 - z}.$$

Substituting the values $z_1 = 1$, $z_2 = 1$, and $z_1 = 1$, $z_2 = -1$ into (1.6.1) we obtain, in particular,

$$\sum_{k=0}^{n} \binom{n}{k} = 2^n, \quad \sum_{k=0}^{n} (-1)^k \binom{n}{k} = 0.$$

We will need one more formula: for $z \neq 1$

$$\sum_{k=1}^{n-1} kz^k = \frac{z}{(1 - z)^2}[1 - nz^{n-1} + (n - 1)z^n]. \tag{1.6.2}$$

In order to prove it we write

$$(1 - z)\sum_{k=1}^{n-1} kz^k = \sum_{k=1}^{n-1} kz^k - \sum_{k=2}^{n} (k - 1)z^k = z - (n - 1)z^n + \sum_{k=2}^{n-1} z^k$$
$$= z - (n - 1)z^n + z^2\,\frac{1 - z^{n-2}}{1 - z} = \frac{z}{1 - z}[1 - nz^{n-1} + (n - 1)z^n],$$

which is equivalent to (1.6.2) when $z \neq 1$.

1.7 Roots of Unity

Let N be a natural number, $N \geq 2$. We introduce a complex number

$$\omega_N = \cos\tfrac{2\pi}{N} + i\,\sin\tfrac{2\pi}{N}.$$

With respect to Moivre formula for a natural k we write

$$\omega_N^k = \cos\tfrac{2\pi k}{N} + i\,\sin\tfrac{2\pi k}{N}. \tag{1.7.1}$$

In particular, $\omega_N^N = 1$. The number ω_N is referred to as the *N-th degree root of unity*.

Formula (1.7.1) is valid for $k = 0$. It is also valid for negative integer powers of ω_N. Indeed,

$$\begin{aligned}\omega_N^{-k} &= \frac{1}{\cos(2\pi k/N) + i\,\sin(2\pi k/N)} = \cos\tfrac{2\pi k}{N} - i\,\sin\tfrac{2\pi k}{N} \\ &= \cos\tfrac{2\pi(-k)}{N} + i\,\sin\tfrac{2\pi(-k)}{N}.\end{aligned}$$

It means that formula (1.7.1) is valid for all $k \in \mathbb{Z}$.

We note that $\omega_N^{-1} = \overline{\omega_N}$ and $\omega_{nN}^n = \omega_N$ for a natural n. From (1.7.1) and from the properties of trigonometric functions it follows that for all integers j and k

$$(\omega_N^k)^j = \omega_N^{kj}, \quad \omega_N^k\,\omega_N^j = \omega_N^{k+j}.$$

With respect to Euler formula we write $\omega_N = \exp(2\pi i/N)$. It is this compact form of the number ω_N that will be used throughout the book.

1.8 Finite Differences

Take a complex-valued function of an integer argument $f(j)$, $j \in \mathbb{Z}$. The finite differences of the function f are defined recursively:

$$[\Delta(f)](j) = [\Delta^1(f)](j) = f(j+1) - f(j),$$

$$[\Delta^r(f)](j) = \big[\Delta\big(\Delta^{r-1}(f)\big)\big](j) = [\Delta^{r-1}(f)](j+1) - [\Delta^{r-1}(f)](j),$$

$$r = 2, 3, \ldots$$

Usually the notation $\Delta^r f(j)$ is used instead of $[\Delta^r(f)](j)$.

The finite difference of the r-th order $\Delta^r f(j)$ can be expressed by means of the values of the function $f(j)$ directly. The following formula is valid:

$$\Delta^r f(j) = \sum_{k=0}^{r} (-1)^{r-k} \binom{r}{k} f(j+k).$$

It can be easily proved by an induction on r.

It is obvious that a finite difference of any order of the function $f(j) \equiv \text{const}$ equals to zero identically.

Exercises

1.1 Let $j \in \mathbb{Z}$ and N be a natural number. Prove that

$$\left\lfloor -\frac{j}{N} \right\rfloor = -\left\lfloor \frac{j-1}{N} \right\rfloor - 1.$$

1.2 Prove that for $j \in \mathbb{Z}$ and natural n and N the following equality is valid:

$$\langle nj \rangle_{nN} = n \langle j \rangle_N.$$

1.3 Let $f(j) = \langle jn \rangle_N$. Prove that, provided $\langle n^2 \rangle_N = 1$, the equality $f\big(f(j)\big) = j$ is valid for $j \in 0 : N-1$.

1.4 We put $f(j) = \langle jn + l \rangle_N$, where n and N are relatively prime natural numbers and $l \in \mathbb{Z}$. Prove that the sequence $f(0),\ f(1),\ \ldots,\ f(N-1)$ is a permutation of numbers $0,\ 1,\ \ldots,\ N-1$.

1.5 Find the values of a function $f(j) = \langle jn \rangle_N$ for $j = 0,\ 1,\ \ldots,\ N-1$ in case of $\gcd(n, N) = d$.

1.6 Prove that $\gcd(j, k) = \gcd(j - k,\ k)$.

1.7 Let $n_1,\ n_2$ be relative primes, $N = n_1 n_2$ and $j \in 0 : N-1$. Prove that there exist unique integers $j_1 \in 0 : n_1 - 1$ and $j_2 \in 0 : n_2 - 1$ such that $j = \langle j_1 n_2 + j_2 n_1 \rangle_N$.

1.8 Assume that integers $n_1, n_2, \ldots, n_s$ are relatively prime with an integer m. Prove that the product of these numbers $N = n_1 n_2 \cdots n_s$ is also relatively prime with m.

1.9 We take pairwise relatively prime numbers $n_1, n_2, \ldots, n_s$. Prove that if a number $j \in \mathbb{Z}$ is divisible by each n_α, $\alpha \in 1 : s$, then j is also divisible by their product $N = n_1 n_2 \cdots n_s$.

1.10 Let $n_1,\ n_2,\ \ldots,\ n_s$ be pairwise relatively prime numbers unequal to unity. Let $N = n_1 n_2 \cdots n_s$. We denote $\widehat{N}_\alpha = N / n_\alpha$. Prove that there exist integers $b_1,\ b_2,\ \ldots,\ b_s$ such that

$$b_1 \widehat{N}_1 + b_2 \widehat{N}_2 + \cdots + b_s \widehat{N}_s = 1.$$

1.11 Under conditions of the previous exercise, prove that any integer $j \in 0 : N-1$ can be uniquely represented in a form

$$j = \left\langle \sum_{\alpha=1}^{s} j_\alpha \widehat{N}_\alpha \right\rangle_N,$$

where $j_\alpha \in 0 : n_\alpha - 1$. Find the explicit expression for the coefficients j_α.

1.12 Let conditions of the Exercise 1.10 hold. For each $\alpha \in 1:s$ the equation $\langle x\widehat{N}_\alpha\rangle_{n_\alpha} = 1$ has a unique solution on the set $0 : n_\alpha - 1$. We denote it by p_α. Prove that any integer $k \in 0 : N-1$ can be uniquely represented in a form

$$k = \left\langle \sum_{\alpha=1}^{s} k_\alpha p_\alpha \widehat{N}_\alpha \right\rangle_N ,$$

where $k_\alpha \in 0 : n_\alpha - 1$. Find the explicit expression for the coefficients k_α.

1.13 Let $j = (j_{\nu-1}, j_{\nu-2}, \ldots, j_0)_2$. Prove that

$$\text{grey}_\nu(j) = j_{\nu-1}2^{\nu-1} + \sum_{k=2}^{\nu} \langle j_{\nu-k+1} + j_{\nu-k}\rangle_2 \, 2^{\nu-k}.$$

1.14 We take $p = (p_{\nu-1}, p_{\nu-2}, \ldots, p_0)_2$. Prove that the unique solution of the equation $\text{grey}_\nu(j) = p$ is an integer $j = (j_{\nu-1}, j_{\nu-2}, \ldots, j_0)_2$ which has

$$j_{\nu-1} = p_{\nu-1},$$

$$j_{\nu-k} = \langle p_{\nu-1} + p_{\nu-2} + \cdots + p_{\nu-k}\rangle_2, \quad k = 2, \ldots, \nu.$$

1.15 Prove that

$$\sum_{k=1}^{n} k^2 = \frac{n(n+1)(2n+1)}{6}.$$

1.16 Prove that

$$\sum_{k=1}^{n} k \binom{n}{k} = n\, 2^{n-1}.$$

1.17 Let n and N be relatively prime natural numbers. We put $\varepsilon_n = \omega_N^n$, where $\omega_N = \exp(2\pi i/N)$. Prove that the sets $\{\varepsilon_n^k\}_{k=0}^{N-1}$ and $\{\omega_N^j\}_{j=0}^{N-1}$ are equal, i.e. that they consist of the same elements.

1.18 Prove that for relative primes m and n there exist unique integers $p \in 0 : m-1$ and $q \in 0 : n-1$ with the following properties: $\gcd(p, m) = 1$, $\gcd(q, n) = 1$, and $\omega_{mn} = \omega_m^p\, \omega_n^q$.

1.19 Prove that

$$\sum_{k=0}^{N-1} z^k = \prod_{j=1}^{N-1} (z - \omega_N^j).$$

1.20 Let P_r be an algebraic polynomial of the r-th degree. Prove that a finite difference of the $(r+1)$-th order of P_r equals to zero identically.

Chapter 2
Signal Transforms

2.1 Space of Signals

2.1.1 We fix a natural number N. The term *signal* is used to refer to an N-periodic complex-valued function of an integer argument $x = x(j)$, $j \in \mathbb{Z}$. We denote the set of all signals by $\mathbb{C}_N$. Two operations are introduced in $\mathbb{C}_N$ in a natural manner—the operation of addition of two signals and the operation of multiplication of a signal by a complex number:

$$y = x_1 + x_2 \Leftrightarrow y(j) = x_1(j) + x_2(j), \quad j \in \mathbb{Z};$$
$$y = c\,x \Leftrightarrow y(j) = c\,x(j), \quad j \in \mathbb{Z}.$$

As a result $\mathbb{C}_N$ becomes a linear complex space. Zero element of $\mathbb{C}_N$ is a signal $\mathbb{O}$ such that $\mathbb{O}(j) = 0$ for all $j \in \mathbb{Z}$.

2.1.2 A *unit N-periodic pulse* is a signal δ_N which is equal to unity if j is divisible by N and equal to zero for other $j \in \mathbb{Z}$. It is clear that $\delta_N(-j) = \delta_N(j)$.

Lemma 2.1.1 *Given $x \in \mathbb{C}_N$, valid is the equality*

$$x(j) = \sum_{k=0}^{N-1} x(k)\,\delta_N(j-k), \quad j \in \mathbb{Z}. \tag{2.1.1}$$

Proof Both sides of (2.1.1) contain N-periodic functions, therefore it is sufficient to verify the equality for $j \in 0 : N-1$. Since the inequalities $-(N-1) \le j-k \le N-1$ hold for $k,\ j \in 0 : N-1$, it follows that $\delta_N(j-k) = 0$ for $k \ne j$. Hence

$$\sum_{k=0}^{N-1} x(k)\,\delta_N(j-k) = x(j)\,\delta_N(0) = x(j).$$

The lemma is proved. □

V. N. Malozemov and S. M. Masharsky, *Foundations of Discrete Harmonic Analysis*, Applied and Numerical Harmonic Analysis,
https://doi.org/10.1007/978-3-030-47048-7_2

Formula (2.1.1) gives an analytic representation of a signal x through its values on the main period $J_N = 0 : N-1$.

Consider a system of shifts of the unit pulse

$$\delta_N(j),\ \delta_N(j-1),\ \ldots,\ \delta_N(j-N+1). \tag{2.1.2}$$

This system is linearly independent on $\mathbb{Z}$. Indeed, let

$$\sum_{k=0}^{N-1} c(k)\,\delta_N(j-k) = 0 \quad \text{for} \quad j \in 0 : N-1.$$

As it was mentioned, the left side of this equality equals to $c(j)$, thus $c(j) = 0$ for all $j \in 0 : N-1$.

According to Lemma 2.1.1 any signal x can be expanded over the linearly independent system (2.1.2). It means that the system (2.1.2) is a *basis* of the space $\mathbb{C}_N$. Moreover, the dimension of $\mathbb{C}_N$ is equal to N.

2.1.3 The following assertion will be frequently used later on.

Lemma 2.1.2 *Given a signal x, the following equality holds for all $l \in \mathbb{Z}$:*

$$\sum_{j=0}^{N-1} x(j+l) = \sum_{j=0}^{N-1} x(j). \tag{2.1.3}$$

Proof Let $l = pN + r$, where $p = \lfloor l/N \rfloor$ and $r = \langle l \rangle_N$ (see Sect. 1.1). Using the N-periodicity of a signal x and the fact that $r \in 0 : N-1$, we obtain

$$\begin{aligned}\sum_{j=0}^{N-1} x(j+l) &= \sum_{j=0}^{N-1} x(j+r) = \sum_{j=0}^{N-r-1} x(j+r) + \sum_{j=N-r}^{N-1} x(j+r-N)\\ &= \sum_{j'=r}^{N-1} x(j') + \sum_{j'=0}^{r-1} x(j') = \sum_{j=0}^{N-1} x(j).\end{aligned}$$

The lemma is proved. □

Corollary 2.1.1 *Under conditions of Lemma 2.1.2, valid is the equality*

$$\sum_{j=0}^{N-1} x(l-j) = \sum_{j=0}^{N-1} x(j). \tag{2.1.4}$$

Indeed,

$$\sum_{j=0}^{N-1} x(l-j) = x(l) + \sum_{j=1}^{N-1} x\big(l+(N-j)\big) = x(l) + \sum_{j'=1}^{N-1} x(l+j')$$
$$= \sum_{j'=0}^{N-1} x(j'+l) = \sum_{j=0}^{N-1} x(j).$$

The following result is related to Lemma 2.1.2.

Lemma 2.1.3 *Let $N = mn$, where m and n are natural numbers, $x \in \mathbb{C}_N$ and*

$$y(j) = \sum_{p=0}^{m-1} x(j-pn), \quad j \in \mathbb{Z}.$$

We assert that $y \in \mathbb{C}_n$.

Proof We need to verify that for any j and l from $\mathbb{Z}$ there holds the equality $y(j+ln) = y(j)$, or, equivalently,

$$\sum_{p=0}^{m-1} x\big(j-(p-ln)\big) = \sum_{p=0}^{m-1} x(j-pn). \tag{2.1.5}$$

We fix j and introduce a signal $z(p) = x(j-pn)$. This signal is m-periodic. According to (2.1.3),

$$\sum_{p=0}^{m-1} z(p-l) = \sum_{p=0}^{m-1} z(p).$$

It corresponds to (2.1.5). □

2.1.4 We introduce the inner (scalar) product and the norm in $\mathbb{C}_N$:

$$\langle x,\, y\rangle = \sum_{j=0}^{N-1} x(j)\,\overline{y(j)}, \quad \|x\| = \langle x,\, x\rangle^{1/2}.$$

Two signals x, y are called *orthogonal* if $\langle x,\, y\rangle = 0$. A signal x is called *normalized* if $\|x\| = 1$.

We denote a shift $x(j-k)$ of a signal $x(j)$ as an element of the space $\mathbb{C}_N$ by $x(\cdot - k)$.

Lemma 2.1.4 *For all $k,\, l \in \mathbb{Z}$ there holds an equality*

$$\langle \delta_N(\cdot - k),\, \delta_N(\cdot - l)\rangle = \delta_N(k-l).$$

Proof We fix an integer k and introduce a signal $x_k(j) = \delta_N(j-k)$. Recall that $\delta_N(j) = \delta_N(-j)$ for all $j \in \mathbb{Z}$. Taking into account formula (2.1.1) we write

$$\begin{aligned}\langle \delta_N(\cdot - k),\ \delta_N(\cdot - l)\rangle &= \sum_{j=0}^{N-1} \delta_N(j-k)\,\delta_N(j-l) = \sum_{j=0}^{N-1} x_k(j)\,\delta_N(l-j) \\ &= x_k(l) = \delta_N(l-k) = \delta_N(k-l).\end{aligned}$$

The lemma is proved. □

Corollary 2.1.2 *The system of signals* (2.1.2) *is orthonormal, i.e. it constitutes an orthonormal basis in the space* $\mathbb{C}_N$.

Lemma 2.1.5 *Given arbitrary signals x and y, there holds a Cauchy–Bunyakovskii inequality*

$$|\langle x,\ y\rangle| \le \|x\| \times \|y\|. \tag{2.1.6}$$

Provided $x \ne \mathbb{O}$, the inequality turns into an equality if and only if $y = cx$ for some $c \in \mathbb{C}$.

Proof Provided $x = \mathbb{O}$, the inequality (2.1.6) holds as an equality. Assume that $x \ne \mathbb{O}$. We take a signal y and denote $c = \langle y,\ x\rangle / \langle x,\ x\rangle$. For a signal $z = y - c\,x$ we have $\langle z,\ x\rangle = 0$. Taking into account that $\langle y,\ x\rangle = \overline{\langle x,\ y\rangle}$, we write

$$\begin{aligned}\|z\|^2 &= \langle z,\ y - c\,x\rangle = \langle z,\ y\rangle = \langle y - c\,x,\ y\rangle \\ &= \|y\|^2 - c\langle x,\ y\rangle = \|y\|^2 - |\langle x,\ y\rangle|^2 / \|x\|^2.\end{aligned}$$

We come to the equality

$$\|x\|^2 \times \|y\|^2 - |\langle x,\ y\rangle|^2 = \|x\|^2 \times \|z\|^2.$$

Hence follows both the inequality (2.1.6) and the condition of turning this inequality into an equality. The lemma is proved. □

2.1.5 We can introduce an operation of multiplication of two signals in the linear complex space $\mathbb{C}_N$:

$$y = x_1 x_2 \quad \Leftrightarrow \quad y(j) = x_1(j)\,x_2(j), \quad j \in \mathbb{Z}.$$

In this case $\mathbb{C}_N$ becomes a commutative algebra with unity. Unity element is a signal $\mathbb{I}$ that has $\mathbb{I}(j) = 1$ for all $j \in \mathbb{Z}$. Given a signal x, the inverse signal x^{-1} is determined from a condition $xx^{-1} = x^{-1}x = \mathbb{I}$. It exists if and only if all values $x(j)$ are nonzero. In this case $x^{-1}(j) = [x(j)]^{-1}$, $j \in \mathbb{Z}$.

2.1.6 Along with a signal x we will consider signals $\overline{x}$, $\operatorname{Re} x$, $\operatorname{Im} x$, $|x|$ with values $\overline{x}(j) = \overline{x(j)}$, $[\operatorname{Re} x](j) = \operatorname{Re} x(j)$, $[\operatorname{Im} x](j) = \operatorname{Im} x(j)$, $|x|(j) = |x(j)|$. Note that $x\overline{x} = |x|^2$.

A signal x is called even if $x(-j) = \overline{x}(j)$ and odd if $x(-j) = -\overline{x}(j)$ for all $j \in \mathbb{Z}$. A signal x is called real if $\operatorname{Im} x = \mathbb{O}$ and imaginary if $\operatorname{Re} x = \mathbb{O}$.

2.1.7 Later on we will be interested in a space $\mathbb{C}_N$ for $N \geq 2$. However, for $N = 1$ this space also has meaning: $\mathbb{C}_1$ consists of signals x with $x(j) \equiv c$, where c is a complex number. In this case $\delta_1 = \mathbb{I}$.

2.2 Discrete Fourier Transform

2.2.1 We take the N-th degree root of unity which we denote by $\omega_N = \exp(2\pi i/N)$.

Lemma 2.2.1 *Valid is the equality*

$$\frac{1}{N}\sum_{k=0}^{N-1} \omega_N^{kj} = \delta_N(j), \qquad j \in \mathbb{Z}. \tag{2.2.1}$$

Proof The left side of (2.2.1) contains an N-periodic function, it follows from the relation

$$\omega_N^{k(j+lN)} = \omega_N^{kj}\left(\omega_N^N\right)^{kl} = \omega_N^{kj} \quad \text{for} \quad l \in \mathbb{Z}$$

(see Sect. 1.7). A unit pulse $\delta_N(j)$ is N-periodic as well. Thus, it is sufficient to verify equality (2.2.1) for $j \in 0 : N-1$.

For $j = 0$ it is trivial. Let $j \in 1 : N-1$. We will use the geometric progression summation formula

$$\sum_{k=0}^{N-1} z^k = \frac{1-z^N}{1-z} \quad \text{for} \quad z \neq 1.$$

By putting $z = \omega_N^j$ we obtain

$$\frac{1}{N}\sum_{k=0}^{N-1} \omega_N^{kj} = \frac{1-\omega_N^{Nj}}{N(1-\omega_N^j)} = 0 = \delta_N(j) \quad \text{for} \quad j \in 1 : N-1.$$

The lemma is proved. □

2.2.2 *Discrete Fourier transform* (DFT) is a mapping $\mathcal{F}_N : \mathbb{C}_N \to \mathbb{C}_N$ that associates a signal x with a signal $X = \mathcal{F}_N(x)$ with the values

$$X(k) = \sum_{j=0}^{N-1} x(j)\, \omega_N^{-kj}, \qquad k \in \mathbb{Z}. \tag{2.2.2}$$

The signal X is referred to as a *Fourier spectrum* of the signal x, or just a *spectrum*. The values $X(k)$ are called *spectral components*.

Theorem 2.2.1 *Valid is the inversion formula*

$$x(j) = \frac{1}{N} \sum_{k=0}^{N-1} X(k)\, \omega_N^{kj}, \qquad j \in \mathbb{Z}. \tag{2.2.3}$$

Proof According to (2.2.2) and (2.2.1) we have

$$\frac{1}{N} \sum_{k=0}^{N-1} X(k)\, \omega_N^{kj} = \frac{1}{N} \sum_{k=0}^{N-1} \left(\sum_{l=0}^{N-1} x(l)\, \omega_N^{-kl} \right) \omega_N^{kj}$$
$$= \sum_{l=0}^{N-1} x(l) \left\{ \frac{1}{N} \sum_{k=0}^{N-1} \omega_N^{k(j-l)} \right\} = \sum_{l=0}^{N-1} x(l)\, \delta_N(j-l) = x(j).$$

The theorem is proved. □

Formula (2.2.3) can be written in a shorter way: $x = \mathcal{F}_N^{-1}(X)$. If we now substitute $\mathcal{F}_N(x)$ instead of X into the right side, we will get $x = \mathcal{F}_N^{-1}\big(\mathcal{F}_N(x)\big)$, so that $\mathcal{F}_N^{-1}\mathcal{F}_N$ is an identity operator. As far as $X = \mathcal{F}_N(x) = \mathcal{F}_N\big(\mathcal{F}_N^{-1}(X)\big)$, we conclude that $\mathcal{F}_N\mathcal{F}_N^{-1}$ is also an identity operator.

The mapping $\mathcal{F}_N^{-1} : \mathbb{C}_N \to \mathbb{C}_N$ is called an inverse DFT.

2.2.3 We introduce the notation $u_k(j) = \omega_N^{kj}$. This time an inversion formula for DFT takes a form

$$x(j) = \frac{1}{N} \sum_{k=0}^{N-1} X(k)\, u_k(j). \tag{2.2.4}$$

It means that a signal $x(j)$ is expanded over the system of signals

$$u_0(j),\ u_1(j),\ \ldots,\ u_{N-1}(j). \tag{2.2.5}$$

The coefficients of this expansion are the spectral components.

Lemma 2.2.2 *The system of signals* (2.2.5) *is orthogonal. In addition,* $\|u_k\|^2 = N$ *for all* $k \in 0 : N-1$.

Proof For $k,\ l \in 0 : N-1$ we have

$$\langle u_k,\ u_l\rangle = \sum_{j=0}^{N-1} u_k(j)\,\overline{u}_l(j) = \sum_{j=0}^{N-1} \omega_N^{(k-l)j} = N\delta_N(k-l).$$

Hence the lemma's statement follows evidently. □

It is ascertained that the system (2.2.5) forms an orthogonal basis in the space $\mathbb{C}_N$. This basis is called *exponential*.

The coefficients of the expansion (2.2.4) are determined uniquely. More to the point, if

$$x(j) = \frac{1}{N}\sum_{l=0}^{N-1} a(l)\,u_l(j), \quad j \in \mathbb{Z}, \tag{2.2.6}$$

then necessarily $a(k) = X(k)$ for all $k \in 0 : N-1$. Indeed, let us multiply both parts of equality (2.2.6) by $u_k(j)$ scalarly. According to Lemma 2.2.2 we obtain

$$\langle x,\ u_k\rangle = \frac{1}{N}\sum_{l=0}^{N-1} a(l)\,\langle u_l,\ u_k\rangle = a(k),$$

thus

$$a(k) = \langle x,\ u_k\rangle = \sum_{j=0}^{N-1} x(j)\,\omega_N^{-kj} = X(k).$$

We rewrite formula (2.2.1) in a form

$$\delta_N(j) = \frac{1}{N}\sum_{k=0}^{N-1} u_k(j).$$

We have the expansion of the unit pulse over the exponential basis. All the coefficients in this expansion are equal to unity. By virtue of the uniqueness of such an expansion, $\mathcal{F}_N(\delta_N) = \mathbb{I}$.

2.2.4 Here we provide frequently used properties of discrete Fourier transform.

Theorem 2.2.2 *A signal x is real if and only if its spectrum X is even.*

Proof
Necessity Let x be a real signal. We write

$$\overline{X}(-k) = \sum_{j=0}^{N-1} \overline{x}(j)\,\omega_N^{-kj} = \sum_{j=0}^{N-1} x(j)\,\omega_N^{-kj} = X(k).$$

Hence it follows that $X(-k) = \overline{X}(k)$ for all $k \in \mathbb{Z}$. We ascertained evenness of the spectrum X.

Sufficiency By virtue of evenness of the spectrum X, Theorem 2.2.1, and the corollary to Lemma 2.1.2 (for $l = 0$), we have

$$\begin{aligned}\overline{x}(j) &= \frac{1}{N}\sum_{k=0}^{N-1} \overline{X}\big(-(-k)\big)\,\omega_N^{(-k)j} = \frac{1}{N}\sum_{k=0}^{N-1} \overline{X}(-k)\,\omega_N^{kj} \\ &= \frac{1}{N}\sum_{k=0}^{N-1} X(k)\,\omega_N^{kj} = x(j).\end{aligned}$$

Thus, x is a real signal.

The theorem is proved. □

Theorem 2.2.3 *A signal x is even if and only if its spectrum X is real.*

Proof

Necessity By virtue of evenness of the signal x and the corollary to Lemma 2.1.2 (for $l = 0$) we get

$$\begin{aligned}\overline{X}(k) &= \sum_{j=0}^{N-1} \overline{x}\big(-(-j)\big)\,\omega_N^{-k(-j)} = \sum_{j=0}^{N-1} \overline{x}(-j)\,\omega_N^{-kj} \\ &= \sum_{j=0}^{N-1} x(j)\,\omega_N^{-kj} = X(k).\end{aligned}$$

Sufficiency We have

$$\overline{x}(-j) = \frac{1}{N}\sum_{k=0}^{N-1} \overline{X}(k)\,\omega_N^{kj} = \frac{1}{N}\sum_{k=0}^{N-1} X(k)\,\omega_N^{kj} = x(j).$$

The theorem is proved. □

As a consequence of Theorems 2.2.2 and 2.2.3 we get the following result: *a signal x is real and even if and only if its spectrum X is real and even.*

2.2.5 Below we present two examples of DFT calculation. Note that it is sufficient to define signals from $\mathbb{C}_N$ by their values on the main period $0 : N-1$.

Example 2.2.1 Let m be a natural number, $2m \le N$, and

$$x(j) = \begin{cases} 1 & \text{for } j \in 0 : m-1 \text{ and } j \in N-m+1 : N-1; \\ 0 & \text{for } j \in m : N-m. \end{cases}$$

(In case of $m = 1$ the signal $x(j)$ coincides with $\delta_N(j)$.)

We will show that

$$X(k) = \begin{cases} 2m-1 & \text{for } k = 0; \\ \dfrac{\sin\big((2m-1)k\pi/N\big)}{\sin(k\pi/N)} & \text{for } k \in 1 : N-1. \end{cases}$$

Indeed, by the definition of DFT

$$X(k) = \sum_{j=0}^{m-1} \omega_N^{-kj} + \sum_{j=N-m+1}^{N-1} \omega_N^{k(N-j)} = \sum_{j=-(m-1)}^{m-1} \omega_N^{kj}.$$

In particular, $X(0) = 2m - 1$. Further, by the geometric progression summation formula, for $k \in 1 : N - 1$ we have

$$\begin{aligned} X(k) &= \frac{\omega_N^{-k(m-1)} - \omega_N^{km}}{1 - \omega_N^k} \times \frac{1 - \omega_N^{-k}}{1 - \omega_N^{-k}} = \\ &= \frac{\omega_N^{-k(m-1)} - \omega_N^{km} - \omega_N^{-km} + \omega_N^{k(m-1)}}{2 - \omega_N^k - \omega_N^{-k}} \\ &= \frac{\cos\big(2(m-1)k\pi/N\big) - \cos(2mk\pi/N)}{1 - \cos(2k\pi/N)} \\ &= \frac{\sin\big((2m-1)k\pi/N\big)\,\sin(k\pi/N)}{\sin^2(k\pi/N)} = \frac{\sin\big((2m-1)k\pi/N\big)}{\sin(k\pi/N)}. \end{aligned}$$

Example 2.2.2 Let $N = 2n$ and

$$x(j) = \begin{cases} 1 & \text{for } j \in 0 : n-1, \\ -1 & \text{for } j \in n : N-1. \end{cases}$$

Let us show that

$$X(k) = \begin{cases} 0 & \text{for even } k, \\ 2\,(1 - i \cot \frac{\pi k}{N}) & \text{for odd } k. \end{cases}$$

By the definition of DFT

$$X(k) = \sum_{j=0}^{n-1} \omega_N^{-kj} - \sum_{j=n}^{2n-1} \omega_N^{-k(j-n)-kn} = (1 - \omega_N^{-kn}) \sum_{j=0}^{n-1} \omega_N^{-kj}.$$

Since $\omega_2 = -1$, we have $\omega_N^{-kn} = \omega_{2n}^{-kn} = \omega_2^{-k} = (-1)^k$, so that

$$X(k) = \begin{cases} 0 & \text{for even } k, \\ \dfrac{2\,(1 - \omega_N^{-kn})}{1 - \omega_N^{-k}} = \dfrac{4}{1 - \omega_N^{-k}} & \text{for odd } k. \end{cases}$$

Now it is remaining to mention that in a case when k is not divisible by N (in particular, when k is odd) the following equality holds:

$$\begin{aligned} \frac{1}{1 - \omega_N^{-k}} &= \frac{1}{\left(1 - \cos \frac{2\pi k}{N}\right) + i \sin \frac{2\pi k}{N}} \\ &= \frac{1}{2 \sin \frac{\pi k}{N} \left(\sin \frac{\pi k}{N} + i \cos \frac{\pi k}{N}\right)} \\ &= \frac{\sin \frac{\pi k}{N} - i \cos \frac{\pi k}{N}}{2 \sin \frac{\pi k}{N}} = \frac{1}{2}\left(1 - i \cot \frac{\pi k}{N}\right). \end{aligned} \tag{2.2.7}$$

2.3 Parseval Equality

2.3.1 The following statement is true.

Theorem 2.3.1 *Let* $X = \mathcal{F}_N(x)$, $Y = \mathcal{F}_N(y)$. *Then*

$$\langle x,\ y\rangle = N^{-1}\langle X,\ Y\rangle. \tag{2.3.1}$$

Proof According to (2.2.2) and (2.2.3) we obtain

$$\begin{aligned} \frac{1}{N}\sum_{k=0}^{N-1} X(k)\,\overline{Y}(k) &= \frac{1}{N}\sum_{k=0}^{N-1}\left(\sum_{j=0}^{N-1} x(j)\,\omega_N^{-kj}\right)\overline{Y}(k) \\ &= \sum_{j=0}^{N-1} x(j)\left\{\frac{1}{N}\sum_{k=0}^{N-1}\overline{Y}(k)\,\omega_N^{-kj}\right\} = \sum_{j=0}^{N-1} x(j)\,\overline{y}(j). \end{aligned}$$

The theorem is proved. □

Corollary 2.3.1 *The following equality is valid:*

$$\|x\|^2 = N^{-1}\|X\|^2. \tag{2.3.2}$$

Formula (2.3.2) is referred to as a *Parseval equality* and formula (2.3.1) as a *generalized Parseval equality*.

2.3.2 Parseval equality can be used for calculation of trigonometric sums in cases where explicit formulae for the spectral components of a signal are available. Let us revisit the Example 2.2.2 from the previous section. For the signal that was considered there we have

$$\|x\|^2 = N = 2n,$$

$$\|X\|^2 = 4\sum_{k=0}^{n-1}\left|1 - i\cot\frac{(2k+1)\pi}{2n}\right|^2 = 4\sum_{k=0}^{n-1}\frac{1}{\sin^2\frac{(2k+1)\pi}{2n}}.$$

By virtue of (2.3.2) we obtain

$$\sum_{k=0}^{n-1}\frac{1}{\sin^2\frac{(2k+1)\pi}{2n}} = n^2.$$

Consider one more example. Let

$$x(j) = j, \quad j \in 0 : N-1.$$

We will show that

$$X(k) = \begin{cases} \frac{1}{2}N(N-1) & \text{for } k = 0; \\ -\frac{1}{2}N(1 - i\cot\frac{\pi k}{N}) & \text{for } k \in 1 : N-1. \end{cases} \tag{2.3.3}$$

By the definition of DFT we have

$$X(k) = \sum_{j=0}^{N-1} j\,\omega_N^{-kj}.$$

In particular, $X(0) = \frac{1}{2}N(N-1)$. Let $k \in 1 : N-1$. We write

$$\sum_{j=0}^{N-1}(j+1)\,\omega_N^{-kj} = X(k) + N\delta_N(k) = X(k).$$

At the same time,

$$\begin{aligned}\sum_{j=0}^{N-1}(j+1)\,\omega_N^{-kj} &= \omega_N^k\sum_{j=0}^{N-1}(j+1)\,\omega_N^{-k(j+1)} \\ &= \omega_N^k\sum_{j'=1}^{N} j'\,\omega_N^{-kj'} = \omega_N^k\big(X(k) + N\big).\end{aligned}$$

We come to the equation $X(k) = \omega_N^k\big(X(k) + N\big)$, from which, by virtue of (2.2.7), it follows that

$$X(k) = \frac{N\omega_N^k}{1-\omega_N^k} = -\frac{N}{1-\omega_N^{-k}} = -\frac{1}{2}N\Big(1 - i\cot\frac{\pi k}{N}\Big),$$

$$k \in 1 : N-1.$$

Formula (2.3.3) is ascertained.

Let us calculate squares of norms of the signal x and its spectrum X. We have

$$\|x\|^2 = \sum_{j=1}^{N-1} j^2 = \frac{(N-1)N(2N-1)}{6}$$

(see the problem 1.15 from Chap. 1),

$$\|X\|^2 = \frac{1}{4}N^2(N-1)^2 + \frac{1}{4}N^2 \sum_{k=1}^{N-1} \frac{1}{\sin^2\frac{\pi k}{N}}.$$

On the ground of (2.3.2) we get

$$\frac{(N-1)(2N-1)}{6} = \frac{1}{4}(N-1)^2 + \frac{1}{4}\sum_{k=1}^{N-1}\frac{1}{\sin^2\frac{\pi k}{N}}.$$

After uncomplicated transformations we come to a remarkable formula

$$\sum_{k=1}^{N-1}\frac{1}{\sin^2\frac{\pi k}{N}} = \frac{N^2-1}{3}.$$

2.4 Sampling Theorem

2.4.1 As a *sample* we refer to a value $x(j)$ of a signal x for some fixed argument j. The theorem below shows that, under a certain assumption, a signal x can be completely restored from its samples on a grid coarser than $\mathbb{Z}$.

Let $N = mn$, where $n \geq 2$ and $m = 2\mu - 1$. We denote

$$h_m(j) = \frac{1}{m}\sum_{k=0}^{m-1}\omega_N^{kj}.$$

Theorem 2.4.1 (Sampling Theorem) *If the spectrum X of a signal x equals to zero on the set of indices $\mu : N-\mu$ then*

$$x(j) = \sum_{l=0}^{m-1} x(ln)\, h(j - ln), \qquad j \in \mathbb{Z}. \tag{2.4.1}$$

Proof By virtue of a DFT inversion formula and the theorem's hypothesis we have

$$\begin{aligned} x(j) &= \frac{1}{N}\Big(\sum_{k=0}^{\mu-1} X(k)\,\omega_N^{kj} + \sum_{k=N-\mu+1}^{N-1} X\big(-(N-k)\big)\,\omega_N^{-(N-k)j}\Big) \\ &= \frac{1}{N}\sum_{k=-\mu+1}^{\mu-1} X(k)\,\omega_N^{kj}. \end{aligned} \tag{2.4.2}$$

We fix an integer j and put $y(k) = \omega_N^{kj}$, $k \in -\mu+1 : \mu-1$. By extending y on $\mathbb{Z}$ periodically with a period of m we obtain a signal y that belongs to $\mathbb{C}_m$. Let us calculate its DFT. According to Lemma 2.1.2,

$$\begin{aligned} Y(l) &= \sum_{k=0}^{m-1} y(k-\mu+1)\,\omega_m^{-(k-\mu+1)l} = \sum_{k'=-\mu+1}^{\mu-1} \omega_N^{k'j}\,\omega_m^{-k'l} \\ &= \sum_{k=-\mu+1}^{\mu-1} \omega_N^{k(j-ln)} = m\, h_m(j-ln). \end{aligned}$$

The inversion formula yields

$$y(k) = \frac{1}{m}\sum_{l=0}^{m-1} Y(l)\,\omega_m^{lk} = \sum_{l=0}^{m-1} h_m(j-ln)\,\omega_m^{lk}, \quad k \in \mathbb{Z}.$$

Recalling a definition of the signal y we gain

$$\omega_N^{kj} = \sum_{l=0}^{m-1} h_m(j-ln)\,\omega_m^{lk}, \quad k \in -\mu+1 : \mu-1.$$

It remains to substitute this expression into (2.4.2). We come to the formula

$$\begin{aligned} x(j) &= \frac{1}{N}\sum_{k=-\mu+1}^{\mu-1} X(k)\sum_{l=0}^{m-1} h_m(j-ln)\,\omega_m^{lk} \\ &= \sum_{l=0}^{m-1} h_m(j-ln)\left\{\frac{1}{N}\sum_{k=-\mu+1}^{\mu-1} X(k)\,\omega_N^{k(ln)}\right\} \\ &= \sum_{l=0}^{m-1} h_m(j-ln)\,x(ln). \end{aligned}$$

The theorem is proved. □

In case of an odd $m = 2\mu - 1$ the kernel $h_m(j)$ can be represented as follows:

$$h_m(j) = \begin{cases} 1 & \text{for } j = 0; \\ \dfrac{\sin(\pi j/n)}{m \sin(\pi j/N)} & \text{for } j \in 1 : N-1. \end{cases} \tag{2.4.3}$$

The equality $h_m(0) = 1$ is obvious. Let $j \in 1 : N - 1$. Then, as it was shown in par. 2.2.5 during analysis of the Example 2.2.1,

$$\sum_{k=-\mu+1}^{\mu-1} \omega_N^{kj} = \frac{\sin\big(\pi j(2\mu - 1)/N\big)}{\sin(\pi j/N)}. \tag{2.4.4}$$

It remains to take into account that $2\mu - 1 = m$ and $N = mn$.

2.4.2 The sampling theorem is related to the following interpolation problem: *construct a signal $x \in \mathbb{C}_N$ that satisfies to the conditions*

$$\begin{aligned} x(ln) &= z(l), \quad l \in 0 : m-1, \\ X(k) &= 0, \quad k \in \mu : N - \mu, \end{aligned} \tag{2.4.5}$$

where $z(l)$ are given numbers (generally, complex ones).

Theorem 2.4.2 *The unique solution of the Problem* (2.4.5) *is a signal*

$$x(j) = \sum_{l=0}^{m-1} z(l)\, h_m(j - ln). \tag{2.4.6}$$

Proof The conditions (2.4.5) are in fact a system of N linear equations with respect to N variables $x(0), x(1), \ldots, x(N-1)$. Let us consider a homogeneous system

$$\begin{aligned} x(ln) &= 0, \quad l \in 0 : m-1, \\ X(k) &= 0, \quad k \in \mu : N - \mu. \end{aligned}$$

According to the sampling theorem, it has only zero solution. Therefore the system (2.4.5) is uniquely resolvable for any $z(l)$.

Formula (2.4.6) follows from (2.4.1). □

2.4.3 The interpolation formula (2.4.6) can be generalized to the case of an even m. For $m = 2\mu$ we put

$$h_m(j) = \frac{1}{m}\Big[\cos(\pi j/n) + \sum_{k=-\mu+1}^{\mu-1} \omega_N^{kj}\Big].$$

Theorem 2.4.3 *A signal*

$$x(j) = \sum_{l=0}^{m-1} z(l)\, h_m(j - ln)$$

satisfies to interpolation conditions

$$x(ln) = z(l), \quad l \in 0 : m - 1.$$

Proof Let us show that $h_m(ln) = \delta_m(l)$. We have

$$h_m(ln) = \frac{1}{m}\Big[(-1)^l + \sum_{k=0}^{\mu-1} \omega_m^{kl} + \sum_{k=-\mu+1}^{-1} \omega_m^{(k+m)l}\Big].$$

We replace an index in the latter sum by putting $k' = k + m$. When k goes from $-\mu + 1$ to -1, the index k' goes from $-\mu + 1 + m = \mu + 1$ to $m - 1$, thus

$$\sum_{k=-\mu+1}^{-1} \omega_m^{(k+m)l} = \sum_{k'=\mu+1}^{m-1} \omega_m^{k'l}.$$

The summand $(-1)^l$ can be written in a form $(-1)^l = \omega_m^{\mu l}$. As a result we come to the required formula

$$h_m(ln) = \frac{1}{m} \sum_{k=0}^{m-1} \omega_m^{kl} = \delta_m(l).$$

On the basis of Lemma 2.1.1, for $l \in 0 : m - 1$ we gain

$$x(ln) = \sum_{l'=0}^{m-1} z(l')\, h_m\big((l - l')n\big) = \sum_{l'=0}^{m-1} z(l')\, \delta_m(l - l') = z(l).$$

The theorem is proved. □

In case of an even $m = 2\mu$ the kernel $h_m(j)$ can be represented as follows:

$$h_m(j) = \begin{cases} 1 & \text{for } j = 0; \\ \frac{1}{m} \sin(\frac{\pi j}{n}) \cot(\frac{\pi j}{N}) & \text{for } j \in 1 : N - 1. \end{cases} \tag{2.4.7}$$

The equality $h_m(0) = 1$ is evident. Let $j \in 1 : N - 1$. Then according to (2.4.4)

$$\sum_{k=-\mu+1}^{\mu-1} \omega_N^{kj} = \frac{\sin\big(\pi j(m-1)/N\big)}{\sin(\pi j/N)}$$
$$= \frac{\sin(\pi j/n)\,\cos(\pi j/N) - \cos(\pi j/n)\,\sin(\pi j/N)}{\sin(\pi j/N)}$$
$$= \sin(\pi j/n)\,\cot(\pi j/N) - \cos(\pi j/n).$$

The remaining follows from the definition of $h_m(j)$.

2.5 Cyclic Convolution

2.5.1 Let x and y be signals of $\mathbb{C}_N$. A signal $u = x * y$ with samples

$$u(j) = \sum_{k=0}^{N-1} x(k)\, y(j-k), \qquad j \in \mathbb{Z}$$

is referred to as a *cyclic convolution* of signals x and y.

Theorem 2.5.1 (Convolution Theorem) *Let $X = \mathcal{F}_N(x)$ and $Y = \mathcal{F}_N(y)$. Then*

$$\mathcal{F}_N(x * y) = XY \tag{2.5.1}$$

where the right side is a component-wise product of spectra.

Proof According to Lemma 2.1.2 we have

$$[\mathcal{F}_N(x * y)](k) = \sum_{j=0}^{N-1} \left(\sum_{l=0}^{N-1} x(l)\, y(j-l) \right) \omega_N^{-k(j-l)-kl}$$
$$= \sum_{l=0}^{N-1} x(l)\, \omega_N^{-kl} \sum_{j=0}^{N-1} y(j-l)\, \omega_N^{-k(j-l)}$$
$$= \sum_{l=0}^{N-1} x(l)\, \omega_N^{-kl} \sum_{j=0}^{N-1} y(j)\, \omega_N^{-kj} = X(k)\, Y(k),$$

which conforms to (2.5.1). The theorem is proved. □

Corollary 2.5.1 *Valid is the formula*

$$x * y = \mathcal{F}_N^{-1}(XY). \tag{2.5.2}$$

Theorem 2.5.2 *A cyclic convolution is commutative and associative.*

Proof The equality $x * y = y * x$ follows directly from (2.5.2). Let us verify the associativity. Take three signals x_1, x_2, x_3 and denote their spectra by X_1, X_2, X_3. Relying on (2.5.1) and (2.5.2) we obtain

$$\begin{aligned}(x_1 * x_2) * x_3 &= \mathcal{F}_N^{-1}\big(\mathcal{F}_N(x_1 * x_2)\, X_3\big) = \mathcal{F}_N^{-1}\big((X_1 X_2)\, X_3\big)\\ &= \mathcal{F}_N^{-1}\big(X_1\,(X_2 X_3)\big) = \mathcal{F}_N^{-1}\big(X_1\, \mathcal{F}_N(x_2 * x_3)\big) = x_1 * (x_2 * x_3).\end{aligned}$$

The theorem is proved. □

2.5.2 A linear complex space $\mathbb{C}_N$ where component-wise product of signals as an operation of multiplication is replaced with a cyclic convolution constitutes another commutative algebra with unity. Here unity element is δ_N because according to Lemma 2.1.1

$$[x * \delta_N](j) = \sum_{k=0}^{N-1} x(k)\, \delta_N(j-k) = x(j).$$

An inverse element y to a signal x is defined by a condition

$$x * y = \delta_N. \tag{2.5.3}$$

It exists if and only if each component of the spectrum X is nonzero. In that case $y = \mathcal{F}_N^{-1}(X^{-1})$, where $X^{-1}(k) = [X(k)]^{-1}$. Let us verify this.

Applying an operation $\mathcal{F}_N$ to both sides of (2.5.3) we get an equation $XY = \mathbb{I}$ with respect to Y. This equation is equivalent to (2.5.3). Such a method is called a *transition into a spectral domain*. The latter equation is resolvable if and only if each component of the spectrum X is nonzero. The solution is written explicitly in a form $Y = X^{-1}$. The inversion formula yields $y = \mathcal{F}_N^{-1}(X^{-1})$. This very signal is inverse to x.

2.5.3 A transform $\mathcal{L}\colon \mathbb{C}_N \to \mathbb{C}_N$ is called *linear* if

$$\mathcal{L}(c_1 x_1 + c_2 x_2) = c_1 \mathcal{L}(x_1) + c_2 \mathcal{L}(x_2)$$

for any x_1, x_2 from $\mathbb{C}_N$ and any c_1, c_2 from $\mathbb{C}$. The simplest example of a linear transform is a shift operator $\mathcal{P}$ that maps a signal x to a signal $x' = \mathcal{P}(x)$ with samples $x'(j) = x(j-1)$.

A transform $\mathcal{L}\colon \mathbb{C}_N \to \mathbb{C}_N$ is referred to as *stationary* if $\mathcal{L}\big(\mathcal{P}(x)\big) = \mathcal{P}\big(\mathcal{L}(x)\big)$ for all $x \in \mathbb{C}_N$. It follows from the definition that

$$\mathcal{L}\big(\mathcal{P}^k(x)\big) = \mathcal{P}^k\big(\mathcal{L}(x)\big), \quad k = 0, 1, \ldots$$

Here $\mathcal{P}^0$ is an identity operator.

Theorem 2.5.3 *A transform* $\mathcal{L}\colon \mathbb{C}_N \to \mathbb{C}_N$ *is both linear and stationary if and only if there exists a signal* h *such that*

$$\mathcal{L}(x) = x * h \quad \textit{for all } x \in \mathbb{C}_N. \tag{2.5.4}$$

Proof
Necessity Taking into account that $\mathcal{P}^k(x) = x(\cdot - k)$ we rewrite formula (2.1.1) in a form

$$x = \sum_{k=0}^{N-1} x(k)\, \mathcal{P}^k(\delta_N).$$

According to the hypothesis, an operator $\mathcal{L}$ is linear and stationary. Hence

$$\mathcal{L}(x) = \sum_{k=0}^{N-1} x(k)\, \mathcal{L}\big(\mathcal{P}^k(\delta_N)\big) = \sum_{k=0}^{N-1} x(k)\, \mathcal{P}^k\big(\mathcal{L}(\delta_N)\big).$$

Denoting $h = \mathcal{L}(\delta_N)$ we obtain

$$\mathcal{L}(x) = \sum_{k=0}^{N-1} x(k)\, \mathcal{P}^k(h) = x * h.$$

Sufficiency The linearity of a convolution operator is evident. We will verify the stationarity. By virtue of the commutativity of a cyclic convolution we have

$$\mathcal{L}(x) = h * x = \sum_{k=0}^{N-1} h(k)\, \mathcal{P}^k(x).$$

Now we write

$$\mathcal{P}\big(\mathcal{L}(x)\big) = \sum_{k=0}^{N-1} h(k)\, \mathcal{P}^{k+1}(x) = \mathcal{L}\big(\mathcal{P}(x)\big).$$

The theorem is proved. □

It is affirmed that a linear stationary operator $\mathcal{L}$ can be represented in a form (2.5.4) where $h = \mathcal{L}(\delta_N)$. Such an operator is also referred to as a *filter*, and the signal h is referred to as its *impulse response*.

2.5.4 As an example we consider an operation of taking a finite difference of the r-th order:

$$[\Delta^r(x)](j) = \Delta^r x(j) = \sum_{l=0}^{r} (-1)^{r-l} \binom{r}{l} x(j+l).$$

We will show that $\Delta^r(x) = x * h_r$, where

$$h_r(j) = \sum_{l=0}^{r} (-1)^{r-l} \binom{r}{l} \delta_N(j+l). \tag{2.5.5}$$

According to (2.1.1) we have

$$\begin{aligned}
\Delta^r x(j) &= \sum_{l=0}^{r} (-1)^{r-l} \binom{r}{l} \sum_{k=0}^{N-1} x(k)\, \delta_N(j+l-k) \\
&= \sum_{k=0}^{N-1} x(k) \sum_{l=0}^{r} (-1)^{r-l} \binom{r}{l} \delta_N\big((j-k)+l\big) \\
&= \sum_{k=0}^{N-1} x(k)\, h_r(j-k) = [x * h_r](j),
\end{aligned}$$

as it was to be ascertained.

Thus, the operator $\Delta^r : \mathbb{C}_N \to \mathbb{C}_N$ is a filter with an impulse response h_r of a form (2.5.5). It is obvious that $h_r = \Delta^r(\delta_N)$.

2.5.5 We take a filter $\mathcal{L}(x) = x * h$ and denote $H = \mathcal{F}_N(h)$. Recall that $u_k(j) = \omega_N^{kj}$.

Theorem 2.5.4 *Valid is the equality*

$$\mathcal{L}(u_k) = H(k)\, u_k, \qquad k \in 0 : N-1.$$

Proof We have

$$\begin{aligned}
[\mathcal{L}(u_k)](j) = [h * u_k](j) &= \sum_{l=0}^{N-1} h(l)\, u_k(j-l) \\
&= \sum_{l=0}^{N-1} h(l)\, \omega_N^{k(j-l)} = \omega_N^{kj} \sum_{l=0}^{N-1} h(l)\, \omega_N^{-kl} = H(k)\, u_k(j).
\end{aligned}$$

The theorem is proved. □

Theorem 2.5.4 states that exponential functions u_k form a complete set of eigenfunctions of any filter $\mathcal{L}(x) = x * h$; in addition, an eigenfunction u_k corresponds to an eigenvalue $H(k) = [\mathcal{F}_N(h)](k)$, $k \in 0 : N-1$.

The signal H is referred to as a *frequency response* of the filter $\mathcal{L}$.

2.6 Cyclic Correlation

2.6.1 Let x and y be signals of $\mathbb{C}_N$. A signal R_{xy} with samples

$$R_{xy}(j) = \sum_{k=0}^{N-1} x(k)\,\overline{y}(k-j), \qquad j \in \mathbb{Z},$$

is referred to as a *cross-correlation* of signals x and y.

Put $y_1(j) = \overline{y}(-j)$. Then

$$R_{xy} = x * y_1. \tag{2.6.1}$$

Theorem 2.6.1 (Correlation Theorem) *The following formula is valid:*

$$\mathcal{F}_N(R_{xy}) = X\overline{Y}, \tag{2.6.2}$$

where $X = \mathcal{F}_N(x)$ *and* $Y = \mathcal{F}_N(y)$.

Proof By virtue of (2.6.1) and (2.5.1) we may write $\mathcal{F}_N(R_{xy}) = XY_1$, where $Y_1 = \mathcal{F}_N(y_1)$. It is remaining to verify that $Y_1 = \overline{Y}$. According to (2.1.4) we have

$$\begin{aligned} Y_1(k) &= \sum_{j=0}^{N-1} y_1(j)\,\omega_N^{-kj} = \sum_{j=0}^{N-1} \overline{y}(-j)\,\omega_N^{k(-j)} \\ &= \sum_{j=0}^{N-1} \overline{y}(j)\,\omega_N^{kj} = \overline{Y}(k). \end{aligned}$$

The theorem is proved. □

A signal R_{xx} is referred to as an *auto-correlation* of a signal x. According to (2.6.2)

$$\mathcal{F}_N(R_{xx}) = X\overline{X} = |X|^2. \tag{2.6.3}$$

We note that $|R_{xx}(j)| \le R_{xx}(0)$ for $j \in 1 : N-1$. Indeed, by virtue of Cauchy–Bunyakovskii inequality (2.1.6) and Lemma 2.1.2 we have

$$\begin{aligned} |R_{xx}(j)| &= \left| \sum_{k=0}^{N-1} x(k)\,\overline{x}(k-j) \right| \\ &\le \left(\sum_{k=0}^{N-1} |x(k)|^2 \right)^{1/2} \left(\sum_{k=0}^{N-1} |x(k-j)|^2 \right)^{1/2} \\ &= \sum_{k=0}^{N-1} |x(k)|^2 = R_{xx}(0). \end{aligned}$$

2.6.2 The orthonormal basis $\{\delta_N(\cdot - k)\}_{k=0}^{N-1}$ in a space $\mathbb{C}_N$ consists of shifts of the unit pulse. Are there any other signals whose shifts form orthonormal bases? This question can be answered positively.

Lemma 2.6.1 *Shifts* $\{x(\cdot - k)\}_{k=0}^{N-1}$ *of a signal* x *form an orthonormal basis in a space* $\mathbb{C}_N$ *if and only if* $R_{xx} = \delta_N$.

Proof Since

$$R_{xx}(l) = \sum_{j=0}^{N-1} x(j)\,\overline{x}(j-l) = \langle x,\, x(\cdot - l)\rangle,$$

the condition $R_{xx} = \delta_N$ is equivalent to the following:

$$\langle x,\, x(\cdot - l)\rangle = \delta_N(l), \qquad l \in 0 : N-1. \tag{2.6.4}$$

At the same time, for $k,\, k' \in 0 : N-1$ we have

$$\begin{aligned}\langle x(\cdot - k),\, x(\cdot - k')\rangle &= \sum_{j=0}^{N-1} x(j-k)\,\overline{x}\big((j-k)-(k'-k)\big)\\ &= \sum_{j'=0}^{N-1} x(j')\,\overline{x}\big(j' - \langle k'-k\rangle_N\big) = \big\langle x,\, x(\cdot - \langle k'-k\rangle_N)\big\rangle.\end{aligned}$$

Orthonormality condition $\langle x(\cdot - k),\, x(\cdot - k')\rangle = \delta_N(k'-k)$ takes a form

$$\big\langle x,\, x(\cdot - \langle k'-k\rangle_N)\big\rangle = \delta_N\big(\langle k'-k\rangle_N\big), \quad k,\, k' \in 0 : N-1. \tag{2.6.5}$$

Equivalence of the relations (2.6.4) and (2.6.5) guarantees the validity of the lemma's statement. □

Theorem 2.6.2 *Shifts* $\{x(\cdot - k)\}_{k=0}^{N-1}$ *of a signal* x *form an orthonormal basis in a space* $\mathbb{C}_N$ *if and only if* $|X(k)| = 1$ *for* $k \in 0 : N-1$.

Proof

Necessity If $\{x(\cdot - k)\}_{k=0}^{N-1}$ is an orthonormal basis then, by virtue of Lemma 2.6.1, $R_{xx} = \delta_N$. Hence $\mathcal{F}_N(R_{xx}) = \mathbb{I}$. On the strength of (2.6.3) we have $|X|^2 = \mathbb{I}$, thus $|X(k)| = 1$ for $k \in 0 : N-1$.

Sufficiency Let $|X| = \mathbb{I}$. Then, according to (2.6.3), $\mathcal{F}_N(R_{xx}) = \mathbb{I}$. It is possible only when $R_{xx} = \delta_N$. It is remaining to refer to Lemma 2.6.1. The theorem is proved. □

2.6.3 Let us take N complex numbers $Y(k)$, $k \in 0 : N-1$, whose moduli are equal to unity. With the aid of the inverse Fourier transform we construct a signal $y = \mathcal{F}_N^{-1}(Y)$. By virtue of Theorem 2.6.2 its shifts $\{y(\cdot - k)\}_{k=0}^{N-1}$ form an orthonormal basis in a space $\mathbb{C}_N$. Let us expand a signal x over this basis

$$x = \sum_{k=0}^{N-1} c(k)\, y(\cdot - k) \tag{2.6.6}$$

and calculate the coefficients $c(k)$. In order to do that we multiply both sides of (2.6.6) scalarly by $y(\cdot - l)$, $l \in 0 : N-1$. We gain $\langle x,\ y(\cdot - l)\rangle = c(l)$ or

$$c(l) = \sum_{j=0}^{N-1} x(j)\, \overline{y}(j-l) = R_{xy}(l).$$

We come to the formula

$$x = \sum_{k=0}^{N-1} R_{xy}(k)\, y(\cdot - k).$$

2.7 Optimal Interpolation

2.7.1 Let $N = mn$, where $n \geq 2$, and r be a natural number. We consider an extremal problem

$$\begin{gathered} f(x) := \|\Delta^r(x)\|^2 \to \min, \\ x(ln) = z(l), \ \ l \in 0 : m-1; \quad x \in \mathbb{C}_N. \end{gathered} \tag{2.7.1}$$

Here we need to construct the possibly smoothest signal that takes given values $z(l)$ in the nodes ln. The smoothness is characterized by the squared norm of the finite difference of the r-th order. Most commonly $r = 2$.

Let us perform a change of variables

$$X(k) = \sum_{j=0}^{N-1} x(j)\, \omega_N^{-kj}, \quad k \in 0 : N-1,$$

and rewrite the Problem (2.7.1) in new variables $X(k)$. We start with a goal function. As it was mentioned in par. 2.5.4, the equality $\Delta^r(x) = x * h_r$ holds, where h_r is determined by formula (2.5.5): $h_r = \Delta^r(\delta_N)$. Using the Parseval equality (2.3.2) and the Convolution Theorem 2.5.1 we obtain

$$\begin{aligned} \|\Delta^r(x)\|^2 &= \|x * h_r\|^2 = N^{-1}\|\mathcal{F}_N(x * h_r)\|^2 = N^{-1}\|X H_r\|^2 \\ &= N^{-1} \sum_{k=0}^{N-1} |X(k)\, H_r(k)|^2. \end{aligned}$$

Here

$$\begin{aligned}
H_r(k) &= \sum_{j=0}^{N-1} h_r(j)\,\omega_N^{-kj} \\
&= \sum_{l=0}^{r}(-1)^{r-l}\binom{r}{l}\sum_{j=0}^{N-1}\delta_N(j+l)\,\omega_N^{-k(j+l)+kl} \\
&= \sum_{l=0}^{r}(-1)^{r-l}\binom{r}{l}\,\omega_N^{kl}\sum_{j=0}^{N-1}\delta_N(j)\,\omega_N^{-kj} \\
&= \sum_{l=0}^{r}(-1)^{r-l}\binom{r}{l}\,\omega_N^{kl} = (\omega_N^k - 1)^r.
\end{aligned}$$

Denote

$$\alpha_k := |\omega_N^k - 1|^2 = \Big(\cos\frac{2\pi k}{N} - 1\Big)^2 + \sin^2\frac{2\pi k}{N} = 2\Big(1 - \cos\frac{2\pi k}{N}\Big) = 4\sin^2\frac{\pi k}{N}.$$

Then $|H_r(k)|^2 = \alpha_k^r$ and

$$\|\Delta^r(x)\|^2 = \frac{1}{N}\sum_{k=0}^{N-1}\alpha_k^r\,|X(k)|^2. \tag{2.7.2}$$

Now we turn to the constraints. We have

$$\begin{aligned}
x(ln) &= \frac{1}{N}\sum_{k=0}^{N-1} X(k)\,\omega_N^{kln} = \frac{1}{N}\sum_{p=0}^{m-1}\sum_{q=0}^{n-1} X(qm+p)\,\omega_m^{(qm+p)l} \\
&= \frac{1}{m}\sum_{p=0}^{m-1}\Bigg[\frac{1}{n}\sum_{q=0}^{n-1} X(p+qm)\Bigg]\omega_m^{pl}.
\end{aligned}$$

The constraints of the Problem (2.7.1) take a form

$$\frac{1}{m}\sum_{p=0}^{m-1}\Bigg[\frac{1}{n}\sum_{q=0}^{n-1} X(p+qm)\Bigg]\omega_m^{pl} = z(l), \quad l \in 0:m-1.$$

The latter formula is an expansion of a signal $z \in \mathbb{C}_m$ over the exponential basis. It is equivalent to

$$\frac{1}{n}\sum_{q=0}^{n-1} X(p+qm) = Z(p), \quad p \in 0:m-1, \tag{2.7.3}$$

where $Z = \mathcal{F}_m(z)$. On the basis of (2.7.2) and (2.7.3) we come to an equivalent setting of the Problem (2.7.1):

$$\begin{gathered}\frac{1}{N}\sum_{k=0}^{N-1}\alpha_k^r\,|X(k)|^2 \to \min,\\ \frac{1}{n}\sum_{q=0}^{n-1}X(p+qm) = Z(p),\quad p \in 0:m-1.\end{gathered} \tag{2.7.4}$$

2.7.2 The Problem (2.7.4) falls into m independent subproblems corresponding to different $p \in 0 : m-1$:

$$\begin{gathered}\frac{1}{N}\sum_{q=0}^{n-1}\alpha_{p+qm}^r\,|X(p+qm)|^2 \to \min,\\ \sum_{q=0}^{n-1}X(p+qm) = nZ(p).\end{gathered} \tag{2.7.5}$$

Since $\alpha_0 = 0$, we get the following problem for $p = 0$:

$$\begin{gathered}\frac{1}{N}\sum_{q=1}^{n-1}\alpha_{qm}^r\,|X(qm)|^2 \to \min,\\ \sum_{q=0}^{n-1}X(qm) = nZ(0).\end{gathered}$$

Its solution is evident:

$$X_*(0)=nZ(0),\quad X_*(m)=X_*(2m)=\cdots=X_*\big((n-1)m\big)=0. \tag{2.7.6}$$

The minimal value of the goal function equals to zero.

Let $p \in 1 : m-1$. In this case each coefficient α_{p+qm}, $q \in 0 : n-1$, is positive. According to Cauchy–Bunyakovskii inequality (2.1.6) we have

$$\begin{aligned}|nZ(p)|^2 &= \left|\sum_{q=0}^{n-1}\big(\alpha_{p+qm}^{r/2}\,X(p+qm)\big)\alpha_{p+qm}^{-r/2}\right|^2\\ &\le \left(\sum_{q=0}^{n-1}\alpha_{p+qm}^r\,|X(p+qm)|^2\right)\left(\sum_{q=0}^{n-1}\alpha_{p+qm}^{-r}\right).\end{aligned} \tag{2.7.7}$$

Denoting $\lambda_p = n\Big(\sum_{q=0}^{n-1} \alpha_{p+qm}^{-r}\Big)^{-1}$ we obtain

$$\frac{1}{N}\sum_{q=0}^{n-1} \alpha_{p+qm}^{r}\,|X(p+qm)|^2 \geq \frac{1}{m}\,\lambda_p\,|Z(p)|^2.$$

Inequality (2.7.7) turns into an equality if and only if

$$\alpha_{p+qm}^{r/2}\,X(p+qm) = c_p\,\alpha_{p+qm}^{-r/2},$$

or $X(p+qm) = c_p\,\alpha_{p+qm}^{-r}$, with some $c_p \in \mathbb{C}$ for all $q \in 0:n-1$. The variables $X(p+qm)$ must satisfy to the constraints of the Problem (2.7.5), so it is necessary that

$$c_p \sum_{q=0}^{n-1} \alpha_{p+qm}^{-r} = n Z(p).$$

Hence $c_p = \lambda_p\,Z(p)$.

It is affirmed that for every $p \in 1:m-1$ the unique solution of the Problem (2.7.5) is the sequence

$$X_*(p+qm) = \lambda_p\,Z(p)\,\alpha_{p+qm}^{-r}, \quad q \in 0:n-1. \tag{2.7.8}$$

Moreover, the minimal value of the goal function equals to $m^{-1}\lambda_p\,|Z(p)|^2$. Let us note that λ_p is a harmonic mean of the numbers

$$\alpha_p^r,\ \alpha_{p+m}^r,\ \ldots,\ \alpha_{p+(n-1)m}^r.$$

Formulae (2.7.6) and (2.7.8) define X_* on the whole main period $0:N-1$. The inversion formula yields the unique solution of the Problem (2.7.1):

$$x_*(j) = \frac{1}{N}\sum_{k=0}^{N-1} X_*(k)\,\omega_N^{kj}, \quad j \in \mathbb{Z}. \tag{2.7.9}$$

The minimal value of the goal function of the Problem (2.7.1) is a total of the minimal values of the goal functions of the Problems (2.7.5) for $p = 0,\ 1,\ \ldots,\ m-1$, so that

$$f(x_*) = \frac{1}{m}\sum_{p=1}^{m-1} \lambda_p\,|Z(p)|^2.$$

2.7.3 Let us modify formula (2.7.9) to the form more convenient for calculations. Represent indices $k,\ j \in 0 : N-1$ in a way $k = p + qm,\ j = s + ln$, where $p,\ l \in 0 : m-1$ and $q,\ s \in 0 : n-1$. In accordance with (2.7.6) and (2.7.8) we write

$$\begin{aligned} x_*(s+ln) &= \frac{1}{N}\sum_{p=0}^{m-1}\sum_{q=0}^{n-1} X_*(p+qm)\,\omega_N^{(p+qm)(s+ln)} \\ &= \frac{1}{m}\sum_{p=0}^{m-1}\left[\left(\frac{1}{n}\sum_{q=0}^{n-1} X_*(p+qm)\,\omega_n^{qs}\right)\omega_N^{ps}\right]\omega_m^{pl} \\ &= \frac{1}{m}Z(0) + \frac{1}{m}\sum_{p=1}^{m-1}\left[\lambda_p\, Z(p)\left(\frac{1}{n}\sum_{q=0}^{n-1}\alpha_{p+qm}^{-r}\,\omega_n^{qs}\right)\omega_N^{ps}\right]\omega_m^{pl}. \end{aligned}$$

We come to the following scheme of solving the Problem (2.7.1):

(1) we form two arrays of constants that depend only on m, n and r: one-dimensional

$$\lambda_p = n\left(\sum_{q=0}^{n-1}\alpha_{p+qm}^{-r}\right)^{-1}, \quad p \in 1 : m-1,$$

and (column-wise) two-dimensional

$$D[s,\ p] = \left(\frac{1}{n}\sum_{q=0}^{n-1}\alpha_{p+qm}^{-r}\,\omega_n^{qs}\right)\omega_N^{ps},$$

$$s \in 1 : n-1, \quad p \in 1 : m-1;$$

(2) we calculate $Z = \mathcal{F}_m(z)$ and $\widetilde{Z}(p) = \lambda_p\, Z(p)$ for $p \in 1 : m-1$;
(3) we introduce a two-dimensional array B with the columns

$$B[s, 0] = Z(0), \quad s \in 1 : n-1,$$

$$B[s,\ p] = \widetilde{Z}(p)\, D[s,\ p], \ \ s \in 1 : n-1, \ \ p \in 1 : m-1;$$

(4) applying the inverse DFT of order m to all $n-1$ rows of the matrix B we obtain a solution of the Problem (2.7.1):

$$x_*(ln) = z(l), \quad l \in 0 : m-1,$$

$$x_*(s+ln) = \frac{1}{m}\sum_{p=0}^{m-1} B[s,p]\,\omega_m^{pl},$$

$$l \in 0:m-1, \quad s \in 1:n-1.$$

2.8 Optimal Signal–Filter Pairs

2.8.1 We will proceed with a more detailed analysis of linear stationary operators (a. k. a. filters).

A filter $\mathcal{L}$ with an impulse response h is called *matched* with a signal x if

$$\mathcal{L}(x) := x * h = R_{xx}. \tag{2.8.1}$$

A matched filter exists. For instance, one may consider $h(j) = \overline{x}(-j),\ j \in \mathbb{Z}$. In this case

$$[x * h](j) = \sum_{k=0}^{N-1} x(k)\,\overline{x}(k-j) = R_{xx}(j).$$

Let us clarify the question of the uniqueness of a matched filter.

Theorem 2.8.1 *Let $x \in \mathbb{C}_N$ be a signal with all spectral components being nonzero. Then the impulse response h of a matched with the signal x filter is determined uniquely.*

Proof We take a signal h satisfying to the condition (2.8.1). Denote by X and H the spectra of the signals x and h, respectively. The convolution theorem and formula (2.6.3) yield $XH = \mathcal{F}_N(R_{xx}) = X\overline{X}$. As far as $X(k) \neq 0$ for all $k \in \mathbb{Z}$, it holds $H = \overline{X}$. Thus, by the inversion formula of DFT,

$$h(j) = \frac{1}{N}\sum_{k=0}^{N-1} \overline{X}(k)\,\omega_N^{kj}, \quad j \in \mathbb{Z}.$$

Now we have

$$\overline{h}(-j) = \frac{1}{N}\sum_{k=0}^{N-1} X(k)\,\omega_N^{kj} = x(j),$$

whence it follows that $h(j) = \overline{x}(-j),\ j \in \mathbb{Z}$. The theorem is proved. □

Provided the spectrum X of a signal x has zero components, a matched filter is not unique. One may put

$$H(k) = \begin{cases} \overline{X}(k), & \text{if } X(k) \neq 0, \\ c_k, & \text{if } X(k) = 0, \end{cases} \tag{2.8.2}$$

where c_k are arbitrary complex numbers. The inversion formula

$$h(j) = \frac{1}{N} \sum_{k=0}^{N-1} H(k)\, \omega_N^{kj}, \quad j \in \mathbb{Z},$$

gives the analytical representation of impulse responses of all filters matched with the signal x. Indeed, signals H of a form (2.8.2), and solely such signals, satisfy to a condition $X(H - \overline{X}) = \mathbb{O}$ which is equivalent to $XH = \mathcal{F}_N(R_{xx})$. Applying the operator $\mathcal{F}_N^{-1}$ to both sides of the latter equality we gain $x * h = R_{xx}$.

2.8.2 We denote by $\widetilde{R}_{xx} = \big(R_{xx}(0)\big)^{-1} R_{xx}$ a normalized auto-correlation of a nonzero signal x. If it holds

$$\widetilde{R}_{xx} = \delta_N, \tag{2.8.3}$$

the corresponding signal x is called *delta-correlated*. As long as $\widetilde{R}_{xx}(0) = 1$, the equality $\widetilde{R}_{xx}(0) = \delta_N(0)$ holds automatically. Thus, the condition (2.8.3) is equivalent to $R_{xx}(j) = 0$ for $j = 1, \ldots, N-1$.

It is not difficult to describe the whole set of all delta-correlated signals. In order to do that we transfer equality (2.8.3) into a spectral domain. Taking into account that $\mathcal{F}_N(R_{xx}) = |X|^2$ and $\mathcal{F}_N(\delta_N) = \mathbb{I}$, we gain $\big(R_{xx}(0)\big)^{-1}|X|^2 = \mathbb{I}$. It means that $|X(k)| = \sqrt{R_{xx}(0)}$ for all $k \in \mathbb{Z}$. We come to the following conclusion.

Theorem 2.8.2 *A nonzero signal x is delta-correlated if and only if it can be represented as*

$$x(j) = \frac{1}{N} \sum_{k=0}^{N-1} c_k\, \omega_N^{kj}, \quad j \in \mathbb{Z},$$

where c_k are nonzero complex coefficients whose moduli are pairwise equal.

Only the sufficiency needs proof. Let $|c(k)| \equiv \sqrt{A} > 0$. Since $c(k) = X(k)$, it holds $|X(k)| \equiv \sqrt{A}$. The Parseval equality (2.3.2) yields

$$R_{xx}(0) = \sum_{j=0}^{N-1} |x(j)|^2 = N^{-1} \sum_{k=0}^{N-1} |X(k)|^2 = A, \tag{2.8.4}$$

so that $|X(k)| \equiv \sqrt{R_{xx}(0)}$. The latter identity is equivalent to (2.8.3).

2.8.3 A value

$$E(x) = \sum_{j=0}^{N-1} |x(j)|^2$$

is referred to as the *energy* of a signal x. According to (2.8.4) it holds $E(x) = R_{xx}(0)$ and

$$E(x) = N^{-1} E(X), \tag{2.8.5}$$

where $X = \mathcal{F}_N(x)$.

Given a nonzero signal x, we associate with it a *side-lobe blanking filter* (SLB filter) whose impulse response h is determined by a condition

$$x * h = E(x)\, \delta_N. \tag{2.8.6}$$

A signal and its SLB filter form a *signal–filter pair.*

Transition of the equality (2.8.6) into a spectral domain yields $XH = E(x)\, 1\!\!1$. We see that SLB filter exists if and only if each component of the spectrum X of a signal x is nonzero. Moreover $H = E(x)\, X^{-1}$ and

$$h(j) = N^{-1} E(x) \sum_{k=0}^{N-1} [X(k)]^{-1}\, \omega_N^{kj}, \quad j \in \mathbb{Z}.$$

2.8.4 Let us consider a set of signals with a given energy. We will examine an extremal problem of selecting from this set a signal whose SLB filter has an impulse response with the smallest energy. This problem can be formalized as follows:

(P) *Minimize* $[E(x)]^{-1} E(h)$ *under constraints*

$$x * h = E(x)\, \delta_N; \;\; E(x) = A; \;\; x,\, h \in \mathbb{C}_N.$$

Here A is a fixed positive number. We rewrite the problem (P) in a more compact form:

$$\begin{gathered} \gamma := A^{-1} E(h) \to \min, \\ x * h = A\, \delta_N; \;\; E(x) = A; \;\; x,\, h \in \mathbb{C}_N. \end{gathered} \tag{2.8.7}$$

A solution (x_*, h_*) of this problem is referred to as an *optimal signal–filter pair.*

Theorem 2.8.3 *The minimal value of* γ *in the Problem* (2.8.7) *equals to unity. It is achieved on any delta-correlated signal* x_* *with* $E(x_*) = A$*. Moreover, the optimal SLB filter* h_* *is a matched filter.*

Proof Let us transfer the Problem (2.8.7) into a spectral domain. According to (2.8.5) we gain

$$\gamma := (AN)^{-1}E(H) \to \min,$$
$$XH = A\,\mathbb{I},\ \ (AN)^{-1}E(X) = 1.$$

The spectrum H can be excluded. Taking into account that $H = AX^{-1}$ we write

$$\gamma := \frac{A}{N}\sum_{k=0}^{N-1} |X(k)|^{-2} \to \min,$$
$$\frac{1}{AN}\sum_{k=0}^{N-1} |X(k)|^{2} = 1.$$

Denote $a_k = |X(k)|^2$. The classical inequality between a harmonic mean and an arithmetic mean yields

$$\frac{1}{A}\,\frac{N}{\sum_{k=0}^{N-1} a_k^{-1}} \le \frac{1}{A}\,\frac{\sum_{k=0}^{N-1} a_k}{N} = 1. \tag{2.8.8}$$

The left side of the inequality (2.8.8) consists of the value γ^{-1}, which is not greater than unity. So γ is not less than unity. The equality to unity is achieved if and only if all the values a_k are equal to each other. Since $\sum_{k=0}^{N-1} a_k = AN$ it means that $a_k \equiv A$, therefore $|X(k)| \equiv \sqrt{A}$. Taking into account Theorem 2.8.2 we come to the following conclusion: there holds the inequality $\gamma \ge 1$; the equality $\gamma = 1$ is achieved on all delta-correlated signals x_* with $E(x_*) = A$ and solely on them.

As to the optimal SLB filters for the indicated signals x_*, these are necessarily the matched filters. Indeed, as long as x_* is delta-correlated, we have

$$R_{x_*x_*} = R_{x_*x_*}(0)\,\delta_N = E(x_*)\,\delta_N.$$

On the other hand, the constraints of the problem (P) yield $x_* * h_* = E(x_*)\,\delta_N$. Therefore $x_* * h_* = R_{x_*x_*}$, which in accordance with (2.8.1) means that h_* is an impulse response of a matched with x_* filter. The theorem is proved. □

2.9 Ensembles of Signals

2.9.1 A finite set of signals x from $\mathbb{C}_N$ with the same energy will be referred to as an *ensemble of signals* and will be denoted as Q. For the definiteness sake we will assume that

$$E(x) = A \ \text{for all}\ x \in Q. \tag{2.9.1}$$

We introduce two characteristics of an ensemble of signals:

$$R_a = \max_{x \in Q} \max_{j \in 1:N-1} |R_{xx}(j)|,$$

$$R_c = \max_{\substack{x,y \in Q \\ x \neq y}} \max_{j \in 0:N-1} |R_{xy}(j)|.$$

Note that if $R_a = 0$ then every signal in Q is delta-correlated (see par. 2.8.2).

Theorem 2.9.1 *Let Q be an ensemble consisting of m signals, and let the condition* (2.9.1) *hold. Then*

$$N\left(\frac{R_c}{A}\right)^2 + \frac{N-1}{m-1}\left(\frac{R_a}{A}\right)^2 \geq 1. \tag{2.9.2}$$

The essential role in the proof of the above theorem is played by the following assertion.

Lemma 2.9.1 *For arbitrary signals x and y from $\mathbb{C}_N$ there holds an equality*

$$\sum_{j=0}^{N-1} |R_{xy}(j)|^2 = \sum_{j=0}^{N-1} R_{xx}(j)\, \overline{R_{yy}(j)}. \tag{2.9.3}$$

Proof On the basis of (2.3.2) and (2.6.2) we write

$$\begin{aligned}\sum_{j=0}^{N-1} |R_{xy}(j)|^2 &= \frac{1}{N}\sum_{k=0}^{N-1} \left|[\mathcal{F}_N(R_{xy})](k)\right|^2 \\ &= \frac{1}{N}\sum_{k=0}^{N-1} |X(k)\,\overline{Y}(k)|^2 = \frac{1}{N}\sum_{k=0}^{N-1} |X(k)|^2\, |Y(k)|^2.\end{aligned}$$

In addition to that, according to (2.3.1) and (2.6.3) we have

$$\begin{aligned}\sum_{j=0}^{N-1} R_{xx}(j)\, \overline{R_{yy}(j)} &= \frac{1}{N}\sum_{k=0}^{N-1} [\mathcal{F}_N(R_{xx})](k)\, \overline{[\mathcal{F}_N(R_{yy})](k)} \\ &= \frac{1}{N}\sum_{k=0}^{N-1} |X(k)|^2\, |Y(k)|^2.\end{aligned}$$

The right sides of the presented relations are equal, therefore the left sides are equal as well. The lemma is proved. □

Proof of the theorem The equality (2.9.3) yields

$$\sum_{j=0}^{N-1}\left|\sum_{x\in Q}R_{xx}(j)\right|^2=\sum_{j=0}^{N-1}\left(\sum_{x\in Q}R_{xx}(j)\right)\left(\sum_{y\in Q}\overline{R_{yy}(j)}\right)$$
$$=\sum_{x\in Q}\sum_{y\in Q}\sum_{j=0}^{N-1}R_{xx}(j)\,\overline{R_{yy}(j)}$$
$$=\sum_{x\in Q}\sum_{y\in Q}\sum_{j=0}^{N-1}|R_{xy}(j)|^2$$
$$=\sum_{x\in Q}\sum_{\substack{y\in Q\\ y\neq x}}\sum_{j=0}^{N-1}|R_{xy}(j)|^2+\sum_{x\in Q}\sum_{j=0}^{N-1}|R_{xx}(j)|^2. \quad (2.9.4)$$

Let us estimate the left and the right sides of this equality. Since $R_{xx}(0)=E(x)=A$ for all $x\in Q$, we have

$$\sum_{j=0}^{N-1}\left|\sum_{x\in Q}R_{xx}(j)\right|^2\geq\left|\sum_{x\in Q}R_{xx}(0)\right|^2=m^2\,A^2.$$

Further, by a definition of R_c and R_a,

$$\sum_{x\in Q}\sum_{\substack{y\in Q\\ y\neq x}}\sum_{j=0}^{N-1}|R_{xy}(j)|^2\leq R_c^2\,N\,(m-1)\,m,$$

$$\sum_{x\in Q}\sum_{j=0}^{N-1}|R_{xx}(j)|^2=\sum_{x\in Q}\sum_{j=1}^{N-1}|R_{xx}(j)|^2+\sum_{x\in Q}|R_{xx}(0)|^2$$
$$\leq R_a^2\,(N-1)\,m+m\,A^2.$$

Combining the derived estimates we come to inequality

$$m\,(m-1)\,A^2\leq m\,(m-1)\,N\,R_c^2+m\,(N-1)\,R_a^2.$$

Hence (2.9.2) follows evidently. The theorem is proved. □

The inequality (2.9.2) is referred to as *Sidel'nikov–Sarwate inequality*. It shows, in particular, that the values R_a and R_c cannot be arbitrary small simultaneously. Below we will consider two extreme cases when one of these values equals to zero and the other one gets the smallest possible magnitude.

2.9.2 Two signals $x,\ y$ from $\mathbb{C}_N$ are called *non-correlated* if $R_{xy}(k) \equiv 0$. Since

$$\overline{R_{yx}}(k) = \sum_{j=0}^{N-1} \overline{y}(j)\, x(j-k) = \sum_{j=0}^{N-1} x(j)\, \overline{y}(j+k) = R_{xy}(-k),$$

the identity $R_{xy}(k) \equiv 0$ holds together with $R_{yx}(k) \equiv 0$.

Let us rewrite the conditions $R_{xy} = \mathbb{O}$ and $R_{yx} = \mathbb{O}$ in the equivalent form

$$\langle x,\ y(\cdot - k)\rangle = 0, \quad \langle y,\ x(\cdot - k)\rangle = 0, \quad k \in \mathbb{Z}.$$

We come to the following conclusion: the signals x and y are non-correlated when the signal x is orthogonal to all shifts of the signal y and the signal y is orthogonal to all shifts of the signal x.

Let Q_c be an ensemble consisting of pairwise non-correlated signals. In this case $R_c = 0$. Furthermore, for such ensembles the equality (2.9.4) gets the form

$$\sum_{j=0}^{N-1} \Big| \sum_{x \in Q_c} R_{xx}(j) \Big|^2 = \sum_{x \in Q_c} \sum_{j=0}^{N-1} |R_{xx}(j)|^2. \tag{2.9.5}$$

Since non-correlated signals are at least orthogonal, the amount of signals in Q_c does not exceed N. We will show that in a space $\mathbb{C}_N$ there exist ensembles containing exactly N pairwise non-correlated signals.

Put $Q_c = \{u_p\}_{p=0}^{N-1}$, where $u_p(j) = \omega_N^{pj}$. This is an ensemble of signals with $A = N$ because $E(u_p) = N$ for all $p \in 0 : N-1$. Further,

$$R_{u_p u_{p'}}(k) = \sum_{j=0}^{N-1} u_p(j)\, \overline{u_{p'}}(j-k) = \sum_{j=0}^{N-1} \omega_N^{j(p-p')+p'k} = N\, \omega_N^{p'k}\, \delta_N(p-p').$$

It is evident that $R_{u_p u_{p'}}(k) \equiv 0$ for $p \neq p',\ p,\ p' \in 0 : N-1$, i.e. the ensemble Q_c consists of N pairwise non-correlated signals. Note as well that $R_{u_p u_p}(k) = N\, u_p(k)$. In particular, $|R_{u_p u_p}(k)| \equiv N$ for all $p \in 0 : N-1$. The latter identity has a general nature. Namely, the following theorem is true.

Theorem 2.9.2 *If an ensemble Q_c consists of N pairwise non-correlated signals, and $E(x) = A$ for all $x \in Q_c$, then*

$$|R_{xx}(k)| \equiv A \text{ for all } x \in Q_c. \tag{2.9.6}$$

Proof According to (2.9.5) we have

$$\sum_{x \in Q_c} \sum_{k=0}^{N-1} |R_{xx}(k)|^2 \geq \Big| \sum_{x \in Q_c} R_{xx}(0) \Big|^2 = N^2 A^2. \tag{2.9.7}$$

As it was mentioned in par. 2.6.1, valid are the relations $|R_{xx}(k)| \le R_{xx}(0) = A$. Assume that for some $x \in Q_c$ and $k \in 1 : N-1$ there holds $|R_{xx}(k)| < A$. Then

$$\sum_{x \in Q_c} \sum_{k=0}^{N-1} |R_{xx}(k)|^2 < N^2 A^2.$$

This contradicts with (2.9.7). The theorem is proved. □

Corollary 2.9.1 *For any ensemble consisting of N pairwise non-correlated signals, the inequality* (2.9.2) *is fulfilled as an equality.*

Indeed, we need to take into account that in this case $m = N$, $R_c = 0$, and, by virtue of Theorem 2.9.2, $R_a = A$.

2.9.3 Now we turn to ensembles Q_a consisting of delta-correlated signals. For such ensembles there holds $R_a = 0$, so that the inequality (2.9.2) gets a form $R_c \ge A/\sqrt{N}$. This estimation does not depend on the amount of signals in an ensemble. It, in particular, turns into an equality if

$$|R_{xy}(j)| \equiv A/\sqrt{N} \ \text{ for all } \ x, y \in Q_a, \ x \ne y. \tag{2.9.8}$$

We will present an example of an ensemble that satisfies the condition (2.9.8).

Consider a two-parametric collection of signals of a form

$$a_{kp}(j) = \omega_N^{k(j^2+pj)}, \ \ k, p \in 0 : N-1, \ \ \gcd(k, N) = 1. \tag{2.9.9}$$

Lemma 2.9.2 *Provided that N is odd, the signals a_{kp} are delta-correlated.*

Proof For the sake of simplicity we denote $x = a_{kp}$. We have

$$\begin{aligned} R_{xx}(j) &= \sum_{l=0}^{N-1} x(l)\,\overline{x}(l-j) = \sum_{l=0}^{N-1} x(l+j)\,\overline{x}(l) \\ &= \sum_{l=0}^{N-1} \omega_N^{k(l^2+2lj+j^2+pl+pj)-k(l^2+pl)} \\ &= \omega_N^{k(j^2+pj)} \sum_{l=0}^{N-1} \omega_N^{2klj} = N\,x(j)\,\delta_N(2kj). \end{aligned} \tag{2.9.10}$$

We will show that in our case

$$\delta_N(2kj) = \delta_N(j), \quad j \in \mathbb{Z}. \tag{2.9.11}$$

Rewrite (2.9.11) in an equivalent form

$$\delta_N\big(\langle 2kj\rangle_N\big) = \delta_N(j), \quad j \in 0 : N-1. \tag{2.9.12}$$

The number N is relatively prime with k (by the proviso) and is relatively prime with 2 (due to oddity), therefore N is relatively prime with the product $2k$. Since $\gcd(2k, N) = 1$, the mapping $j \to \langle 2kj\rangle_N$ is a permutation of a set $0 : N-1$ that maps zero to zero. This fact, along with the definition of the unit pulse δ_N, guarantees the validity of the equality (2.9.12) and, as a consequence, of (2.9.11).

On the basis of (2.9.10) and (2.9.11) we gain

$$R_{xx}(j) = Nx(j)\delta_N(j) = Nx(0)\delta_N(j) = N\delta_N(j).$$

The lemma is proved. □

Let N be an odd number. We take two signals of a form (2.9.9):

$$x(j) = \omega_N^{k(j^2+pj)}, \qquad y(j) = \omega_N^{s(j^2+pj)}.$$

To be definite, we assume that $k > s$.

Theorem 2.9.3 *Provided that* $\gcd(k-s,\ N) = 1$, *the following identity holds:*

$$|R_{xy}(j)| \equiv \sqrt{N}. \tag{2.9.13}$$

Proof We have

$$\begin{aligned}|R_{xy}(j)|^2 &= \sum_{l=0}^{N-1} x(l+j)\,\overline{y}(l) \sum_{q=0}^{N-1} \overline{x}(q+j)\,y(q)\\ &= \sum_{l=0}^{N-1}\sum_{q=0}^{N-1} \omega_N^{k(l^2+2lj+j^2+pl+pj)-s(l^2+pl)}\,\omega_N^{-k(q^2+2qj+j^2+pq+pj)+s(q^2+pq)}.\end{aligned}$$

The power is reduced to a form

$$k(l-q)(l+q+2j+p) - s(l-q)(l+q+p) = (k-s)(l-q)(l+q+p) + 2k(l-q)j.$$

Taking into account Lemma 2.1.2 we gain

$$\begin{aligned}|R_{xy}(j)|^2 &= \sum_{q=0}^{N-1}\sum_{l=0}^{N-1} \omega_N^{(k-s)(l-q)(l+q+p)+2k(l-q)j}\\ &= \sum_{q=0}^{N-1}\sum_{l=0}^{N-1} \omega_N^{(k-s)l(l+2q+p)+2klj}\\ &= \sum_{l=0}^{N-1} \omega_N^{(k-s)l(l+p)+2klj} \sum_{q=0}^{N-1} \omega_N^{2(k-s)lq}\\ &= N\sum_{l=0}^{N-1} \omega_N^{(k-s)l(l+p)+2klj}\, \delta_N\big(\langle 2(k-s)l\rangle_N\big).\end{aligned} \tag{2.9.14}$$

The number N is relatively prime with $k-s$ (by the hypothesis) and is relatively prime with 2 (due to oddity), therefore gcd $(2(k-s),\ N) = 1$. In this case the mapping $l \to \langle 2(k-s)l\rangle_N$ is a permutation of a set $0 : N-1$ that maps zero to zero. Using this fact and the definition of the unit pulse δ_N we conclude that the sum in the right side of (2.9.14) contains only one nonzero term that corresponds to $l = 0$. We come to the identity $|R_{xy}(j)|^2 \equiv N$, which is equivalent to (2.9.13). The theorem is proved. □

Signals a_{kp} of a form (2.9.9) have the same energy $E(a_{kp}) = N$. According to Lemma 2.9.2, provided that N is odd, any set of these signals constitutes an ensemble Q_a with $A = N$ and $R_a = 0$. Assume that signals a_{kp} from Q_a satisfy to two additional conditions:

- they have the same p;
- for any pair of signals a_{kp}, a_{sp} from Q_a with $k > s$, the difference $k-s$ is relatively prime with N.

Then, by virtue of Theorem 2.9.3, the following identity is valid:

$$|R_{xy}(j)| \equiv \sqrt{N} \ \text{ for all } \ x, y \in Q_a,\ x \neq y.$$

It coincides with (2.9.8) since in this case $A/\sqrt{N} = \sqrt{N}$.

Note that when N is prime, the following signals satisfy to all conditions formulated above: $a_{1,p},\ a_{2,p},\ \ldots,\ a_{N-1,p}$. The amount of these signals is $N-1$.

2.10 Uncertainty Principle

2.10.1 We refer to the following set as a *support* of a signal $x \in \mathbb{C}_N$:

$$\operatorname{supp} x = \{j \in 0 : N-1 \mid x(j) \neq 0\}.$$

We denote by $|\operatorname{supp} x|$ the number of indices contained in a support. Along with the support of a signal x we will consider the support of its spectrum X.

Theorem 2.10.1 (Uncertainty Principle) *Given any nonzero signal* $x \in \mathbb{C}_N$, *the following inequality holds:*

$$|\mathrm{supp\,x}| \times |\mathrm{supp\,X}| \geq \mathrm{N}. \tag{2.10.1}$$

The point of the inequality (2.10.1) is that the support of a nonzero signal and the support of its spectrum cannot both be small.

2.10.2 We precede the proof of Theorem 2.10.1 by an auxiliary statement.

Lemma 2.10.1 *Let* $m := |\mathrm{supp\,x}| > 0$. *Then for any* $q \in 0 : N-1$ *the sequence*

$$X(q+1),\ X(q+2),\ \ldots,\ X(q+m)$$

contains at least one nonzero element.

Proof Let $\operatorname{supp} x = \{j_1, \ldots, j_m\}$. We fix $q \in 0 : N-1$ and write

$$X(q+l) = \sum_{k=1}^{m} x(j_k)\, \omega_N^{-(q+l)j_k} = \sum_{k=1}^{m} z_k^{q+l}\, x(j_k), \quad l \in 1 : m, \tag{2.10.2}$$

where $z_k = \omega_N^{-j_k}$. It is clear that z_k are pairwise different points on a unit circle of a complex plane. We denote $a = \big(x(j_1), \ldots, x(j_m)\big)^{\top}, b = \big(X(q+1), \ldots, X(q+m)\big)^{\top}$, $Z = \{z_k^{q+l}\}_{l,k=1}^{m}$, and rewrite the equality (2.10.2) in a form $b = Za$. It is sufficient to show that the matrix Z is invertible. In this case the condition $a \neq \mathbb{O}$ will imply $b \neq \mathbb{O}$.

We have

$$Z = \begin{bmatrix} z_1^{q+1} & \cdots & z_m^{q+1} \\ \cdots & \cdots & \cdots \\ z_1^{q+m} & \cdots & z_m^{q+m} \end{bmatrix}.$$

The determinant Δ of this matrix can be transformed to a form

$$\Delta = \prod_{k=1}^{m} z_k^{q+1} \begin{vmatrix} 1 & \cdots & 1 \\ z_1 & \cdots & z_m \\ \cdots & \cdots & \cdots \\ z_1^{m-1} & \cdots & z_m^{m-1} \end{vmatrix}$$

where in the right side we can see a nonzero Vandermonde determinant. Therefore, $\Delta \neq 0$. This guarantees invertibility of the matrix Z.

The lemma is proved. □

2.10.3 Now we turn to proving Theorem 2.10.1. Let $\operatorname{supp} X = \{k_1, \ldots, k_n\}$, where $0 \le k_1 < k_2 < \cdots < k_n < N$. We fix $s \in 1 : n-1$. According to the lemma, the sequence $X(k_s+1), \ldots, X(k_s+m)$ contains a nonzero element. But the first nonzero element after $X(k_s)$ is $X(k_{s+1})$. Therefore,

$$k_{s+1} \le k_s + m, \quad s \in 1 : n-1. \tag{2.10.3}$$

Further, the sequence $X(k_n+1), \ldots, X(k_n+m)$ also contains a nonzero element, and by virtue of N-periodicity of a spectrum the first nonzero element after $X(k_n)$ is $X(k_1+N)$. Therefore,

$$k_1 + N \le k_n + m. \tag{2.10.4}$$

On the ground of (2.10.3) and (2.10.4) we gain

$$k_1 + N \le k_n + m \le k_{n-1} + 2m \le \cdots \le k_1 + nm. \tag{2.10.5}$$

Hence it follows that $mn \ge N$.

The theorem is proved. □

2.10.4 The signal $x = \delta_N$ turns the inequality (2.10.1) into an equality. In fact, we can describe the whole set of signals that turn the inequality (2.10.1) into an equality.

Theorem 2.10.2 *Let $x \in \mathbb{C}_N$ be a signal with properties*

$$m = |\operatorname{supp} \mathrm{x}|, \qquad \mathrm{n} = |\operatorname{supp} \mathrm{X}|,$$

and let the equality $mn = N$ holds for this signal. Then necessarily

$$x(j) = c\, \omega_N^{qj}\, \delta_n(j-p), \tag{2.10.6}$$

where $q \in 0 : m-1$, $p \in 0 : n-1$, and $c \in \mathbb{C}$, $c \neq 0$.

Proof Just like in the previous paragraph, we denote

$$\operatorname{supp} \mathcal{F}_N(x) = \{k_1, \ldots, k_n\}.$$

According to (2.10.5) and the equality $mn = N$ we have

$$k_n + m \le k_{n-1} + 2m \le \cdots \le k_1 + nm = k_1 + N \le k_n + m.$$

Hence

$$k_s = k_1 + (s-1)m, \quad s \in 1:n. \tag{2.10.7}$$

By a definition, $k_n = k_1 + (n-1)m \le N-1$, thus $k_1 \le m-1$. Denoting $q = k_1$, we rewrite (2.10.7) in a form $k_s = q + (s-1)m$.

Let us express the signal x through its spectrum:

$$x(j) = \frac{1}{N}\sum_{s=1}^{n} X(k_s)\,\omega_N^{k_s j} = \frac{1}{N}\sum_{s=0}^{n-1} X(q+sm)\,\omega_N^{(q+sm)j}$$
$$= \omega_N^{qj}\,\frac{1}{n}\sum_{s=0}^{n-1}\Big(\frac{1}{m}\,X(q+sm)\Big)\omega_n^{sj}.$$

Denote $H(s) = \frac{1}{m}\,X(q+sm)$, $s \in 0:n-1$, and $h = \mathcal{F}_n^{-1}(H)$. Then

$$x(j) = \omega_N^{qj}\,h(j). \tag{2.10.8}$$

Since x is a nonzero signal, according to (2.10.8) we can find an index p such that $h(p) \ne 0$. By virtue of n-periodicity of the signal h we can presume that $p \in 0:n-1$. We will show that

$$h(j) = c\,\delta_n(j-p), \tag{2.10.9}$$

where $c = h(p)$. The conclusion of the theorem will follow from this equality and from (2.10.8).

Again, by virtue of n-periodicity of the signal h and (2.10.8) we have

$$x(p+sn) = \omega_N^{q(p+sn)}\,h(p) \ne 0, \quad s \in 0:m-1.$$

That is, we pointed out m indices from the main period where the samples of the signal x are not zero. By the theorem hypothesis $|\operatorname{supp} x| = m$, so on other indices from $0:N-1$ the signal x is zero. In particular, for $j \in 0:n-1$, $j \ne p$, there will be $0 = x(j) = \omega_N^{qj}\,h(j)$. We derived that $h(j) = 0$ for all $j \in 0:n-1$, $j \ne p$. It means that the signal h can be represented in a form (2.10.9).

The theorem is proved. □

2.10.5 To make the picture complete, let us find the spectrum of a signal x of a form (2.10.6). We write

$$X(k) = c\sum_{j=0}^{N-1} \omega_N^{qj}\,\delta_n(j-p)\,\omega_N^{-kj} = c\sum_{s=0}^{m-1} \omega_N^{(q-k)(p+sn)}$$
$$= c\,\omega_N^{(q-k)p}\sum_{s=0}^{m-1} \omega_m^{(q-k)s} = m\,c\,\omega_N^{(q-k)p}\,\delta_m(k-q).$$

Therefore, the signal x of a form (2.10.6) is not equal to zero on the indices $j = p + sn$, $s \in 0 : m - 1$, while its spectrum X is not equal to zero on the indices $k = q + tm$, $t \in 0 : n - 1$.

Exercises

2.1 Prove that a signal $x \in \mathbb{C}_N$ is even if and only if the value $x(0)$ is real and $x(N - j) = \overline{x}(j)$ holds for $j \in 1 : N - 1$.

2.2 Prove that a signal $x \in \mathbb{C}_N$ is odd if and only if $\operatorname{Re} x(0) = 0$ and $x(N - j) = -\overline{x}(j)$ holds for $j \in 1 : N - 1$.

2.3 Prove that any signal can be uniquely represented as a sum of an even and an odd signal.

2.4 Prove that $\delta_{mn}(mj) = \delta_n(j)$ for all $j \in \mathbb{Z}$.

2.5 Prove that

$$\sum_{l=0}^{m-1} \delta_{mn}(j + ln) = \delta_n(j) \quad \text{for all } j \in \mathbb{Z}.$$

2.6 Prove that for $r \in 1 : N - 1$ there holds

$$\|\Delta^r(\delta_N)\|^2 = \sum_{s=0}^{r} \binom{r}{s}^2.$$

2.7 Let $N = mn$. Prove that for any signal $x \in \mathbb{C}_N$ there holds

$$\sum_{l=0}^{m-1} x(s + ln) = \sum_{j=0}^{N-1} x(j)\, \delta_n(s - j) \quad \text{for all } s \in \mathbb{Z}.$$

Numbers k and N in the Exercises 2.8 and 2.9 are natural relative primes.

2.8 Prove that $\delta_{kN}(j) = \delta_k(j)\, \delta_N(j)$ holds for all $j \in \mathbb{Z}$.

2.9 Prove that

$$\sum_{j=0}^{N-1} \delta_N(kj + l) = 1 \quad \text{for all } l \in \mathbb{Z}.$$

2.10 Prove that a signal $x \in \mathbb{C}_N$ is odd if and only if its spectrum X is pure imaginary.

2.11 Let a and b be two real signals from $\mathbb{C}_N$. We aggregate a complex signal $x = a + ib$. Prove that the spectra A, B, and X of these signals satisfy to the following relations

$$A(k) = \tfrac{1}{2}\,[X(k) + \overline{X}(N-k)],$$

$$B(k) = -\tfrac{1}{2}\,i\,[X(k) - \overline{X}(N-k)]$$

for all $k \in \mathbb{Z}$.

2.12 Let N be an even number. We associate a real signal x with a complex signal x_a with a spectrum

$$X_a(k) = \begin{cases} X(k) \text{ for } k = 0 \text{ and } k = N/2, \\ 2X(k) \text{ for } k \in 1 : N/2 - 1, \\ 0 \text{ for } k \in N/2 + 1 : N - 1. \end{cases}$$

Prove that $\operatorname{Re} x_a = x$.

2.13 Formulate and solve the problem analogous to the previous one for an odd N.

2.14 Prove that for an even N and for $k \in 0 : N/2 - 1$ there hold

$$X(k) = \sum_{j=0}^{N/2-1} \big[x(2j) + \omega_N^{-k}\, x(2j+1)\big]\, \omega_{N/2}^{-kj},$$

$$X(N/2 + k) = \sum_{j=0}^{N/2-1} \big[x(2j) - \omega_N^{-k}\, x(2j+1)\big]\, \omega_{N/2}^{-kj}.$$

2.15 Prove that for an even N and for $k \in 0 : N/2 - 1$ there hold

$$X(2k) = \sum_{j=0}^{N/2-1} \big[x(j) + x(N/2 + j)\big]\, \omega_{N/2}^{-kj},$$

$$X(2k+1) = \sum_{j=0}^{N/2-1} \big[x(j) - x(N/2 + j)\big]\, \omega_N^{-j}\, \omega_{N/2}^{-kj}.$$

In the Exercise 2.16 through 2.19 it is required to calculate the Fourier spectrum of given signals.

2.16 $x(j) = \sin \dfrac{\pi j}{N}$, $j \in 0 : N - 1$.

2.17 $x(j) = (-1)^j$, $j \in 0 : N - 1$. Consider the cases of $N = 2n$ and $N = 2n + 1$ separately.

2.18 $x(j) = \begin{cases} j \text{ for } j \in 0:n, \\ j-N \text{ for } j \in n+1:N-1 \quad (N = 2n+1). \end{cases}$

2.19 $x(j) = \begin{cases} j \text{ for } j \in 0:n, \\ N-j \text{ for } j \in n+1:N-1 \quad (N = 2n). \end{cases}$

2.20 Let $x(j) = \omega_N^{j^2}$. Find the amplitude spectrum $|X|$ of the signal x.

The Exercise 2.21 through 2.31 contain some transforms of a signal $x \in \mathbb{C}_N$. It is required to establish the relation between spectra of the given signal and the transformed one.

2.21 $x_l(j) = x(j+l)$, where $l \in \mathbb{Z}$.

2.22 $x_l(j) = \cos\dfrac{2\pi lj}{N}\, x(j)$, where $l \in \mathbb{Z}$.

2.23 $y_p(j) = x\big(\langle pj\rangle_N\big)$, in assumption that p and N are natural relative primes.

2.24 $x_n(j) = \begin{cases} x(j) \text{ for } j \in 0:N-1, \\ 0 \text{ for } j \in N:nN-1 \quad (x_n \in \mathbb{C}_{nN}). \end{cases}$

2.25 $x_n(j) = x\big(\langle j\rangle_N\big)$ for $j \in 0:nN-1 \quad (x_n \in \mathbb{C}_{nN})$.

The transforms presented in the Exercises 2.24 and 2.25 are referred to as *prolongations* of a signal.

2.26 $x_n(j) = \begin{cases} x(j/n) \text{ if } \langle j\rangle_n = 0, \\ 0 \text{ for others } j \in \mathbb{Z} \quad (x_n \in \mathbb{C}_{nN}). \end{cases}$

2.27 $x_n(j) = x\big(\lfloor j/n\rfloor\big) \quad (x_n \in \mathbb{C}_{nN})$.

The transforms presented in the Exercises 2.26 and 2.27 are referred to as *stretches* of a signal.

2.28 $y_n(j) = x(jm)$ for $N = mn \quad (y_n \in \mathbb{C}_n)$.

This transform is referred to as *subsampling*.

2.29 $y_n(j) = \sum_{p=0}^{m-1} x(j+pn)$ for $N = mn \quad (y_n \in \mathbb{C}_n)$.

2.30 $y_n(j) = \sum_{p=0}^{m-1} x(p+jm)$ for $N = mn \quad (y_n \in \mathbb{C}_n)$.

2.31 $y(j) = \sum_{p=0}^{m-1} x\big(p + \lfloor j/m \rfloor m\big)$ for $N = mn \quad (y \in \mathbb{C}_N)$.

2.32 Given the complex numbers $c_0, c_1, \ldots, c_{N-1}, c_N$, we form two signals $x_0(j) = c_j$, $j \in 0 : N-1$, and $x_1(j) = c_{j+1}$, $j \in 0 : N-1$. What is the relation between the spectra of these signals?

2.33 A spectrum X of a signal x is associated with a signal y with samples

$$y(k) = \begin{cases} X(0) \text{ for } k = 0, \\ X(N-k) \text{ for } k \in 1 : N-1. \end{cases}$$

Prove that $x = N^{-1}\mathcal{F}_N(y)$.

2.34 Prove that for any $x \in \mathbb{C}_N$ valid is the equality

$$\mathcal{F}_N^4(x) = N^2 x.$$

2.35 Prove the formula $\mathcal{F}_N(xy) = N^{-1}(X * Y)$.

2.36 Assume that all samples of a signal $x \in \mathbb{C}_N$ are nonzero. Introduce a signal $y = x^{-1}$. Prove that the DFTs X and Y of signals x and y are bound with a relation $X * Y = N^2\delta_N$.

2.37 Prove that convolution of two even signals is even.

2.38 Prove that auto-correlation R_{xx} is an even function for any $x \in \mathbb{C}_N$.

2.39 Prove that

$$\sum_{j=0}^{N-1} R_{xx}(j) = |X(0)|^2.$$

2.40 Let $u = x * y$. Prove that $R_{uu} = R_{xx} * R_{yy}$.

2.41 Take two delta-correlated signals x and y and their convolution $u = x * y$. Prove that $E(u) = E(x)\,E(y)$.

2.42 Prove that convolution of two delta-correlated signals is delta-correlated.

2.43 A Frank signal v belongs to a space $\mathbb{C}_{N^2}$ and is defined by the formula $v(j_1 N + j_0) = \omega_N^{j_1 j_0}$, $j_1, j_0 \in 0 : N-1$. Prove that a Frank signal is delta-correlated.

2.44 Consider a Zadoff–Chu signal

$$a(j) = \begin{cases} \omega_{2N}^{j^2+2qj} \text{ for an even } N, \\ \omega_{2N}^{j(j+1)+2qj} \text{ for an odd } N, \end{cases}$$

where $q \in \mathbb{Z}$ is a parameter. Prove that a signal a belongs to a space $\mathbb{C}_N$ and is delta-correlated.

2.45 A signal $x \in \mathbb{C}_N$ is called *binary* if it takes the values $+1$ and -1 only. Prove that there are no delta-correlated signals among binary ones if $N \neq 4p^2$, p being a natural number.

2.46 The Exercise 2.26 introduced a signal $x_n \in \mathbb{C}_{nN}$ that was a stretch of a signal $x \in \mathbb{C}_N$. What is the relation between auto-correlations of these signals?

2.47 Take four signals $x,\ y,\ w$, and z, and form four new signals $u_1 = R_{xy}$, $v_1 = R_{wz}$, $u_2 = R_{xw}$, and $v_2 = R_{yz}$. Prove that $R_{u_1 v_1} = R_{u_2 v_2}$.

2.48 Let x and y be non-correlated signals. Prove that signals R_{xw} and R_{yz} are also non-correlated regardless of w and z.

2.49 We remind that the signals from a basis of shifts of a unit pulse are pairwise orthogonal. Prove that they are pairwise correlated.

2.50 Prove that a system of shifts $\{x(j-k)\}_{k=0}^{N-1}$ is linearly independent on $\mathbb{Z}$ if and only if all the components of the spectrum X are nonzero.

2.51 Systems of shifts $\{x(\cdot - k)\}_{k=0}^{N-1}$ and $\{y(\cdot - k)\}_{k=0}^{N-1}$ are called *biorthogonal* if there holds $\langle x(\cdot - k),\ y(\cdot - k')\rangle = \delta_N(k-k')$. Prove that the criterion of biorthogonality is satisfying to the condition $R_{xy} = \delta_N$.

2.52 Let a system of shifts $\{x(\cdot - k)\}_{k=0}^{N-1}$ be linearly independent on $\mathbb{Z}$. Prove that there exist the unique signal $y \in \mathbb{C}_N$ such that the systems of shifts $\{x(\cdot - k)\}_{k=0}^{N-1}$ and $\{y(\cdot - k)\}_{k=0}^{N-1}$ are biorthogonal.

2.53 Let $x \in \mathbb{C}_N$ be a nonzero signal. A value

$$p(x) = \frac{\max\limits_{j \in 0:N-1} |x(j)|^2}{N^{-1} \sum\limits_{j=0}^{N-1} |x(j)|^2}$$

is referred to as a *peak factor* of a signal x. Prove that $1 \le p(x) \le N$. Clarify the cases when the inequalities are fulfilled as equalities.

2.54 Let $g \in \mathbb{C}_N$. Find a signal $x \in \mathbb{C}_N$ satisfying to the equation

$$-\Delta^2 x(j-1) + c\,x(j) = g(j), \quad j \in \mathbb{Z},$$

where $c > 0$ is a parameter.

2.55 Consider a complex-valued 1-periodic function of a real argument $f(t)$. Find the coefficients of a trigonometric polynomial of a form

$$T(t) = \sum_{k=-n}^{n} a(k) \exp(2\pi i k t)$$

that satisfies to the interpolation conditions $T(t_j) = f(t_j)$, where $t_j = j/(2n+1)$, $j \in \mathbb{Z}$.

Comments

In this chapter, we introduce the basic concepts of the discrete harmonic analysis such as discrete Fourier transform, cyclic convolution, and cyclic correlation. The peculiarity of the presentation is that we consider a signal as an element of the functional space $\mathbb{C}_N$.

We systematically use the N-periodic unit pulse δ_N. The expansion (2.1.1) of an arbitrary signal over the shifts of the unit pulse corresponds to the expansion of a vector over the unitary vectors. Lemma 2.1.1 is elementary; however, it lets us easily prove Theorem 2.5.3 about general form of a linear stationary operator. Theorem 2.6.2 is a generalization of Lemma 2.1.1. It makes it clear when shifts of a signal form an orthonormal basis in the space $\mathbb{C}_N$.

A solution of the optimal interpolation problem is obtained in [3]. More sophisticated question of this solution's behavior when $r \to \infty$ is investigated in the same paper. A similar approach is used in [2] for solving the problem of discrete periodic data smoothing.

The problem of the optimal signal–filter pair was studied in [11]. In our book we revise all the notions needed for the problem's setting and give its simple solution. A generalization of these results is presented in the paper [38].

The sections on ensembles of signals and the uncertainty principle are written on the basis of the survey [45] and the paper [9], respectively. The point of the uncertainty principle is that the number of indices comprising the support of a signal and the number of indices comprising the support of its spectrum cannot be simultaneously small. The more localized is a signal in time domain, the more dispersed is its frequency spectrum.

Additional exercises are proposed to be solved by the reader. They are intended to help in mastering the discrete harmonic analysis techniques. These exercises introduce, in particular, such popular signal transforms as prolongation, stretching, and subsampling. Some exercises prepare the reader for the further theory development. These are, first of all, the Exercises 2.14 and 2.15. Special signals are considered. We attract the reader's attention to Frank signal (Exercise 2.43). Detailed studying and generalization of this signal is undertaken in the paper [26].

Chapter 3
Spline Subspaces

3.1 Periodic Bernoulli Functions

3.1.1 Let $N \geq 2$ and $r \geq 0$ be integer numbers. A signal

$$b_r(j) = \frac{1}{N} \sum_{k=1}^{N-1} (\omega_N^k - 1)^{-r} \, \omega_N^{kj}, \quad j \in \mathbb{Z}, \tag{3.1.1}$$

is referred to as a *discrete periodic Bernoulli function of order* r. According to a definition we have

$$\big[\mathcal{F}_N(b_r)\big](k) = \begin{cases} 0 \ \text{for} \ k = 0, \\ (\omega_N^k - 1)^{-r} \ \text{for} \ k \in 1 : N-1. \end{cases}$$

A condition $\big[\mathcal{F}_N(b_r)\big](0) = 0$ means that

$$\sum_{j=0}^{N-1} b_r(j) = 0. \tag{3.1.2}$$

For $r = 0$ we have

$$b_0(j) = \frac{1}{N} \sum_{k=1}^{N-1} \omega_N^{kj} = \delta_N(j) - \frac{1}{N}. \tag{3.1.3}$$

Theorem 3.1.1 *For all* $r \geq 0$ *and* $j \in \mathbb{Z}$ *valid are the equalities*

$$\Delta b_{r+1}(j) := b_{r+1}(j+1) - b_{r+1}(j) = b_r(j), \tag{3.1.4}$$

$$b_r(r-j) = (-1)^r \, b_r(j). \tag{3.1.5}$$

V. N. Malozemov and S. M. Masharsky, *Foundations of Discrete Harmonic Analysis*, Applied and Numerical Harmonic Analysis,
https://doi.org/10.1007/978-3-030-47048-7_3

Proof According to (3.1.1) we write

$$\Delta b_{r+1}(j) = \frac{1}{N}\sum_{k=1}^{N-1}(\omega_N^k - 1)^{-r-1}\,\omega_N^{kj}\,(\omega_N^k - 1)$$
$$= \frac{1}{N}\sum_{k=1}^{N-1}(\omega_N^k - 1)^{-r}\,\omega_N^{kj} = b_r(j).$$

Further,

$$b_r(r-j) = \frac{1}{N}\sum_{k=1}^{N-1}(\omega_N^k - 1)^{-r}\,\omega_N^{k(r-j)}$$
$$= \frac{1}{N}\sum_{k=1}^{N-1}(1-\omega_N^{-k})^{-r}\,\omega_N^{-kj} = \frac{1}{N}\sum_{k=1}^{N-1}(1-\omega_N^{N-k})^{-r}\,\omega_N^{(N-k)j}$$
$$= \frac{1}{N}\sum_{k=1}^{N-1}(1-\omega_N^{k})^{-r}\,\omega_N^{kj} = (-1)^r\,b_r(j).$$

The theorem is proved. □

Lemma 3.1.1 *For $k \in 1 : N-1$ valid is the formula*

$$\sum_{j=0}^{N-1} b_{2r}(j+r)\,\omega_N^{-kj} = (-1)^r\Big(2\sin\frac{\pi k}{N}\Big)^{-2r}. \tag{3.1.6}$$

Proof Note that

$$\omega_N^{-k}\,(\omega_N^k - 1)^2 = \omega_N^{-k}\,(\omega_N^{2k} - 2\,\omega_N^k + 1) = \omega_N^k - 2 + \omega_N^{-k}$$
$$= -2\Big(1-\cos\frac{2\pi k}{N}\Big) = -4\sin^2\frac{\pi k}{N}. \tag{3.1.7}$$

Taking into account Lemma 2.1.2 we gain

$$\sum_{j=0}^{N-1} b_{2r}(j+r)\,\omega_N^{-k(j+r)+kr} = \omega_N^{kr}\sum_{j=0}^{N-1} b_{2r}(j)\,\omega_N^{-kj}$$
$$= \omega_N^{kr}\,(\omega_N^k - 1)^{-2r} = \big[\omega_N^{-k}(\omega_N^k - 1)^2\big]^{-r}$$
$$= (-1)^r\Big(2\sin\frac{\pi k}{N}\Big)^{-2r}.$$

The lemma is proved. □

3.1.2 Shifts of a Bernoulli function can be used for expansion of arbitrary signals.

Theorem 3.1.2 *Any signal $x \in \mathbb{C}_N$ for each $r \geq 0$ can be represented as*

$$x(j) = c + \sum_{k=0}^{N-1} \Delta^r x(k)\, b_r(j-k), \quad j \in \mathbb{Z}, \tag{3.1.8}$$

where $c = N^{-1} \sum_{j=0}^{N-1} x(j)$.

Proof We denote

$$I_r(j) = \sum_{k=0}^{N-1} \Delta^r x(k)\, b_r(j-k).$$

According to (3.1.3), for $r = 0$ we have

$$I_0(j) = \sum_{k=0}^{N-1} x(k)\, [\delta_N(j-k) - N^{-1}] = x(j) - c, \tag{3.1.9}$$

which conforms to (3.1.8).

Let $r \geq 1$. At first we will show that for arbitrary signals x and y from $\mathbb{C}_N$ there holds a *summation by parts* formula

$$\sum_{k=0}^{N-1} y(k)\, \Delta x(k) = -\sum_{k=0}^{N-1} x(k)\, \Delta y(k-1). \tag{3.1.10}$$

Indeed, Lemma 2.1.2 yields

$$\begin{aligned}\sum_{k=0}^{N-1} y(k)\, \Delta x(k) &= \sum_{k=0}^{N-1} y(k)\, [x(k+1) - x(k)] \\ &= \sum_{k=0}^{N-1} [y(k-1) - y(k)]\, x(k) = -\sum_{k=0}^{N-1} x(k)\, \Delta y(k-1).\end{aligned}$$

Recall that $\Delta^r(x) = \Delta\big(\Delta^{r-1}(x)\big)$. Taking into account (3.1.10) and (3.1.4) we gain

$$\begin{aligned}I_r(j) &= -\sum_{k=0}^{N-1} \Delta^{r-1} x(k)\, [b_r(j-k) - b_r(j-k+1)] \\ &= \sum_{k=0}^{N-1} \Delta^{r-1} x(k)\, b_{r-1}(j-k) = I_{r-1}(j).\end{aligned}$$

Thus, $I_r(j) = I_{r-1}(j) = \cdots = I_1(j) = I_0(j)$. It is remaining to refer to formula (3.1.9). The theorem is proved. □

Let us substitute $x(j) = b_{r+s}(j)$ into (3.1.8). According to (3.1.2) and (3.1.4) we gain

$$b_{r+s}(j) = \sum_{k=0}^{N-1} b_s(k)\, b_r(j-k).$$

This formula emphasizes a convolution nature of Bernoulli functions.

3.2 Periodic B-splines

3.2.1 We suppose that $N = mn$ and $m \geq 2$. Let us introduce a signal

$$x_1(j) = \begin{cases} n - j & \text{for } j \in 0 : n-1, \\ 0 & \text{for } j \in n : N-n, \\ j - N + n & \text{for } j \in N-n+1 : N-1, \end{cases} \tag{3.2.1}$$

and calculate its discrete Fourier transform $X_1 = \mathcal{F}_N(x_1)$.

Lemma 3.2.1 *The following equality holds:*

$$X_1(k) = \begin{cases} n^2 & \textit{for } k = 0, \\ \left(\dfrac{\sin(\pi k/m)}{\sin(\pi k/N)}\right)^2 & \textit{for } k \in 1 : N-1. \end{cases} \tag{3.2.2}$$

Proof A definition of DFT yields

$$\begin{aligned} X_1(k) &= \sum_{j=0}^{n-1} (n-j)\, \omega_N^{-kj} + \sum_{j=N-n+1}^{N-1} \big(n - (N-j)\big)\, \omega_N^{k(N-j)} \\ &= n + \sum_{j=1}^{n-1} (n-j)\, \omega_N^{-kj} + \sum_{j'=1}^{n-1} (n-j')\, \omega_N^{kj'} \\ &= n + 2\,\mathrm{Re} \sum_{j=1}^{n-1} (n-j)\, \omega_N^{k(n-j)-kn} = n + 2\,\mathrm{Re}\Big\{ \omega_m^{-k} \sum_{j=1}^{n-1} j\, \omega_N^{kj} \Big\}. \end{aligned}$$

For $k = 0$ we get $X_1(0) = n + n(n-1) = n^2$. Further we assume that $k \in 1 : N-1$.

Let us use the formula (see Sect. 1.6)

$$\sum_{j=1}^{n-1} j z^j = \frac{z}{(1-z)^2} \left[(n-1)\, z^n - n z^{n-1} + 1\right], \quad z \neq 1. \tag{3.2.3}$$

Substituting $z = \omega_N^k$ into (3.2.3) we gain

$$\sum_{j=1}^{n-1} j\, \omega_N^{kj} = \frac{\omega_N^k}{(1-\omega_N^k)^2} \left[(n-1)\, \omega_m^k - n\, \omega_m^k\, \omega_N^{-k} + 1\right].$$

Equality (3.1.7) yields

$$\omega_m^{-k} \sum_{j=1}^{n-1} j\, \omega_N^{kj} = -\frac{1}{4 \sin^2 \frac{\pi k}{N}} \left[n - 1 - n\, \omega_N^{-k} + \omega_m^{-k}\right].$$

Hence it follows that

$$X_1(k) = n - \frac{1}{2 \sin^2 \frac{\pi k}{N}} \left[n\left(1 - \cos \frac{2\pi k}{N}\right) - \left(1 - \cos \frac{2\pi k}{m}\right)\right] = \left(\frac{\sin(\pi k/m)}{\sin(\pi k/N)}\right)^2.$$

The lemma is proved. □

The signal x_1 is characterized by two properties: by the equality $x_1(ln) = n\, \delta_m(l)$ and by linearity on each interval $ln : (l+1)n$, $l \in \mathbb{Z}$.

3.2.2 We put

$$Q_1 = x_1; \quad Q_r = Q_1 * Q_{r-1}, \quad r = 2, 3, \ldots \tag{3.2.4}$$

A signal Q_r is referred to as a *discrete periodic* B*-spline of order* r. According to (3.2.1) and (3.2.4) it takes only non-negative integer values. Figure 3.1 depicts a graph of a B-spline $Q_r(j)$ for $m = 8$, $n = 5$, and $r = 2$.

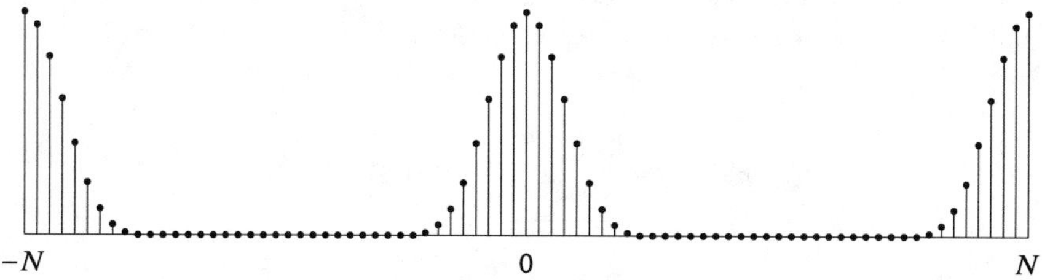

Fig. 3.1 Graph of a B-spline $Q_r(j)$ for $m = 8$, $n = 5$, and $r = 2$

Theorem 3.2.1 *For all natural r there holds*

$$Q_r(j) = \frac{1}{N} \sum_{k=0}^{N-1} X_1^r(k)\, \omega_N^{kj}, \quad j \in \mathbb{Z}. \tag{3.2.5}$$

Proof When $r = 1$, formula (3.2.5) coincides with the DFT inversion formula which reconstructs the signal x_1 from its spectrum X_1. We perform an induction step from r to $r+1$. From validity of (3.2.5) it follows that $\mathcal{F}_N(Q_r) = X_1^r$ holds. By virtue of the convolution theorem we write

$$\mathcal{F}_N(Q_{r+1}) = \mathcal{F}_N(Q_1 * Q_r) = X_1 X_1^r = X_1^{r+1}.$$

The DFT inversion formula yields

$$Q_{r+1}(j) = \frac{1}{N} \sum_{k=0}^{N-1} X_1^{r+1}(k)\, \omega_N^{kj}, \quad j \in \mathbb{Z}.$$

The theorem is proved. □

As it was noted earlier, B-spline $Q_r(j)$ takes only non-negative integer values. Hereto we should add that $Q_r(j)$ is an even signal. It follows from Theorems 3.1.2 and 2.2.3.

Formula (3.2.5) can be considered as a definition of B-spline of order r. It has meaning for $m = 1$ as well. In this case, equality (3.2.2) takes a form $X_1 = N^2\, \delta_N$ so that according to (3.2.5) we gain $Q_r(j) \equiv N^{2r-1}$.

We also note that for $m = N$ (and $n = 1$)

$$Q_r(j) = \frac{1}{N} \sum_{k=0}^{N-1} \omega_N^{kj} = \delta_N(j)$$

holds for all natural r.

3.2.3 Later on we will need the values $Q_r(pn)$ for $p \in 0 : m-1$. Let us calculate them. We will use the fact that every index $k \in 0 : N-1$ can be represented in a form $k = qm + l$, where $q \in 0 : n-1$ and $l \in 0 : m-1$. According to (3.2.5) we have

$$\begin{aligned} Q_r(pn) &= \frac{1}{N} \sum_{l=0}^{m-1} \sum_{q=0}^{n-1} X_1^r(qm+l)\, \omega_N^{pn(qm+l)} \\ &= \frac{1}{m} \sum_{l=0}^{m-1} \omega_m^{pl} \left\{ \frac{1}{n} \sum_{q=0}^{n-1} X_1^r(qm+l) \right\}. \end{aligned}$$

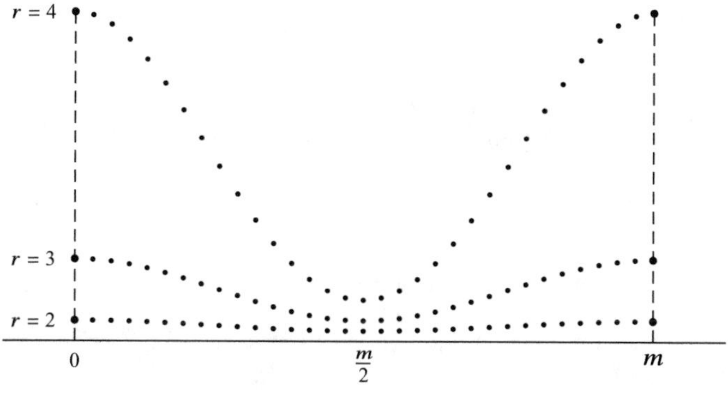

Fig. 3.2 Graphs of a signal $T_r(l)$ for $m = 512$, $n = 2$, and $r = 2, 3, 4$

Denoting

$$T_r(l) = \frac{1}{n} \sum_{q=0}^{n-1} X_1^r(qm + l) \tag{3.2.6}$$

we gain

$$Q_r(pn) = \frac{1}{m} \sum_{l=0}^{m-1} T_r(l)\, \omega_m^{pl}. \tag{3.2.7}$$

We note that a signal $T_r(l)$ is real, m-periodic, and even. Reality follows from the definition (3.2.6), and m-periodicity from Lemma 2.1.3. The formula (3.2.7) and Theorem 2.2.2 guarantee evenness of $T_r(l)$.

Figure 3.2 shows graphs of a signal $T_r(l)$ on the main period $0 : m - 1$ for $m = 512$, $n = 2$, and $r = 2, 3, 4$.

Let us transform formula (3.2.6). For $l \in 1 : m - 1$ we introduce a value

$$\Lambda_r(l) = \frac{1}{n} \sum_{q=0}^{n-1} \left(2 \sin \frac{\pi(qm + l)}{N}\right)^{-2r}.$$

Lemma 3.2.2 *Valid is the equality*

$$T_r(l) = \begin{cases} n^{2r-1} & \textit{for } l = 0, \\ \left(2 \sin \frac{\pi l}{m}\right)^{2r} \Lambda_r(l) & \textit{for } l \in 1 : m - 1. \end{cases} \tag{3.2.8}$$

Proof Equality (3.2.2) yields

$$X_1(m) = X_1(2m) = \cdots = X_1\big((n - 1)m\big) = 0,$$

therefore $T_r(0) = n^{-1}\, X_1^r(0) = n^{2r-1}$. For $l \in 1 : m-1$ we have

$$T_r(l) = \frac{1}{n}\sum_{q=0}^{n-1}\left(\frac{2\sin(\pi(qm+l)/m)}{2\sin(\pi(qm+l)/N)}\right)^{2r} = \left(2\sin\tfrac{\pi l}{m}\right)^{2r}\Lambda_r(l).$$

The lemma is proved. □

From (3.2.8) it follows, in particular, that $T_r(l) > 0$ for all $l \in \mathbb{Z}$.

3.2.4 We will determine the relation between discrete periodic B-splines and discrete periodic Bernoulli functions.

Theorem 3.2.2 *Valid is the formula*

$$Q_r(j) = \frac{1}{N}n^{2r} + \sum_{l=-r}^{r}(-1)^{r-l}\binom{2r}{r-l}\, b_{2r}(j+r-ln). \tag{3.2.9}$$

Proof According to (3.2.2) and (3.1.7) a value $X_1(k)$ for $k \in 1 : N-1$ can be represented in a form

$$X_1(k) = \frac{\omega_m^{-k}\,(\omega_m^k - 1)^2}{\omega_N^{-k}\,(\omega_N^k - 1)^2}.$$

Bearing this in mind we gain

$$\begin{aligned}
\frac{1}{N}\sum_{k=1}^{N-1} X_1^r(k)\,\omega_N^{kj} &= \frac{1}{N}\sum_{k=1}^{N-1}(\omega_m^k - 1)^{2r}(\omega_N^k - 1)^{-2r}\,\omega_N^{k(j+r-rn)} \\
&= \frac{1}{N}\sum_{k=1}^{N-1}(\omega_N^k - 1)^{-2r}\sum_{p=0}^{2r}(-1)^{2r-p}\binom{2r}{p}\omega_N^{k(j+r-(r-p)n)} \\
&= \frac{1}{N}\sum_{k=1}^{N-1}(\omega_N^k - 1)^{-2r}\sum_{l=-r}^{r}(-1)^{r-l}\binom{2r}{r-l}\omega_N^{k(j+r-ln)} \\
&= \sum_{l=-r}^{r}(-1)^{r-l}\binom{2r}{r-l}\left\{\frac{1}{N}\sum_{k=1}^{N-1}(\omega_N^k - 1)^{-2r}\,\omega_N^{k(j+r-ln)}\right\} \\
&= \sum_{l=-r}^{r}(-1)^{r-l}\binom{2r}{r-l}\, b_{2r}(j+r-ln).
\end{aligned}$$

Now the statement of the theorem follows from the formula (3.2.5). □

3.3 Discrete Periodic Splines

3.3.1 Let $N = mn$ and $N \geq 2$. A *discrete periodic spline* $S(j)$ *of order* r is defined as a linear combination of shifts of B-spline $Q_r(j)$ with complex coefficients:

$$S(j) = \sum_{p=0}^{m-1} c(p)\, Q_r(j - pn). \tag{3.3.1}$$

The set of splines of a form (3.3.1) is denoted by $\mathcal{S}_r^m$. Since $Q_r(j) = \delta_N(j)$ for $m = N$, Lemma 2.1.1 yields $\mathcal{S}_r^N = \mathbb{C}_N$ for all natural r. For $m = 1$ we have $Q_r(j) \equiv N^{2r-1}$, therefore $\mathcal{S}_r^1$ is a set of signals that are identically equal to a complex constant.

Lemma 3.3.1 *The basic signals* $Q_r(j - pn)$, $p \in 0{:}m-1$, *are linearly independent on* $\mathbb{Z}$.

Proof Let

$$S(j) := \sum_{p=0}^{m-1} c(p)\, Q_r(j - pn) = 0 \quad \forall j \in \mathbb{Z}$$

for some complex coefficients $c(p)$. We will show that all $c(p)$ are equal to zero. We have

$$0 = \sum_{j=0}^{N-1} S(j)\, \omega_N^{-kj} = \sum_{p=0}^{m-1} c(p) \sum_{j=0}^{N-1} Q_r(j - pn)\, \omega_N^{-k(j-pn)-kpn}$$
$$= \Big\{\sum_{p=0}^{m-1} c(p)\, \omega_m^{-kp}\Big\}\Big\{\sum_{j=0}^{N-1} Q_r(j)\, \omega_N^{-kj}\Big\}. \tag{3.3.2}$$

According to (3.2.5) there holds $\sum_{j=0}^{N-1} Q_r(j)\, \omega_N^{-kj} = X_1^r(k)$. We denote $C(k) = \sum_{p=0}^{m-1} c(p)\, \omega_m^{-kp}$. Then equality (3.3.2) can be rewritten as $C(k)X_1^r(k) = 0$. The formula (3.2.2) guarantees that $X_1^r(k) \neq 0$ for $k \in 0 : m-1$. Hence $C(k) = 0$ for the same indices k. The DFT inversion formula yields $c(p) = 0$ for $p \in 0 : m-1$. The lemma is proved. □

It is obvious that $\mathcal{S}_r^m$ is a linear complex space. It is a subspace of $\mathbb{C}_N$. On the basis of Lemma 3.3.1 one can ascertain that the dimension of $\mathcal{S}_r^m$ equals to m.

Lemma 3.3.2 *Valid is the identity*

$$\sum_{p=0}^{m-1} Q_r(j - pn) \equiv n^{2r-1}. \tag{3.3.3}$$

Proof According to (3.2.5) and (2.2.1) we gain

$$\sum_{p=0}^{m-1} Q_r(j-pn) = \frac{1}{N}\sum_{k=0}^{N-1} X_1^r(k)\sum_{p=0}^{m-1}\omega_N^{k(j-pn)}$$
$$= \frac{1}{n}\sum_{k=0}^{N-1} X_1^r(k)\,\omega_N^{kj}\left\{\frac{1}{m}\sum_{p=0}^{m-1}\omega_m^{-kp}\right\}$$
$$= \frac{1}{n}\sum_{k=0}^{N-1} X_1^r(k)\,\omega_N^{kj}\,\delta_m(k) = \frac{1}{n}\sum_{l=0}^{n-1} X_1^r(lm)\,\omega_n^{lj}.$$

It is remaining to take into account that $X_1(lm) = 0$ for $l \in 1 : n-1$, and $X_1(0) = n^2$. □

From (3.3.3) it follows, in particular, that a signal identically equal to a complex constant belongs to $\mathcal{S}_r^m$.

3.3.2 It is possible to give an equivalent definition of a discrete periodic spline with the aid of Bernoulli functions.

Theorem 3.3.1 *A signal S belongs to $\mathcal{S}_r^m$ if and only if it can be represented as*

$$S(j) = d + \sum_{l=0}^{m-1} d(l)\,b_{2r}(j+r-ln), \tag{3.3.4}$$

where $\sum_{l=0}^{m-1} d(l) = 0$.

Proof
Necessity According to (3.3.1) and (3.2.9) we have

$$S(j) = \sum_{p=0}^{m-1} c(p)\left\{\frac{1}{N}n^{2r} + \sum_{k=-r}^{r}(-1)^{r-k}\binom{2r}{r-k}b_{2r}\big(j+r-(k+p)n\big)\right\}$$
$$= \frac{n^{2r}}{N}\sum_{p=0}^{m-1} c(p) + \sum_{p=0}^{m-1}\sum_{k=-r}^{r} c(p)(-1)^{r-k}\binom{2r}{r-k}b_{2r}\big(j+r-\langle k+p\rangle_m n\big).$$

Collecting similar terms in the double sum we come to (3.3.4). Herein

$$\sum_{l=0}^{m-1} d(l) = \sum_{p=0}^{m-1} c(p)\sum_{k=-r}^{r}(-1)^{r-k}\binom{2r}{r-k} = 0.$$

Sufficiency As it was mentioned above, a signal $f(j) \equiv d$ belongs to $\mathcal{S}_r^m$. It is remaining to verify that the set $\mathcal{S}_r^m$ contains a signal

$$g(j) = \sum_{l=0}^{m-1} d(l)\, b_{2r}(j + r - ln) \tag{3.3.5}$$

which has $\sum_{l=0}^{m-1} d(l) = 0$. Let us calculate $G = \mathcal{F}_N(g)$. We have

$$\begin{aligned} G(k) &= \sum_{j=0}^{N-1} g(j)\, \omega_N^{-kj} = \sum_{l=0}^{m-1} d(l) \sum_{j=0}^{N-1} b_{2r}(j + r - ln)\, \omega_N^{-k(j-ln)-kln} \\ &= \Big\{ \sum_{l=0}^{m-1} d(l)\, \omega_m^{-kl} \Big\} \Big\{ \sum_{j=0}^{N-1} b_{2r}(j + r)\, \omega_N^{-kj} \Big\}. \end{aligned}$$

Denote $D(k) = \sum_{l=0}^{m-1} d(l)\, \omega_m^{-kl}$. The theorem's hypothesis yields $D(0) = 0$. Taking into account (3.1.6) we gain

$$G(k) = \begin{cases} 0 \ \text{ for } \ k = 0, \\ (-1)^r \big(2 \sin(\pi k/N)\big)^{-2r} D(k) \ \text{ for } \ k \in 1 : N - 1. \end{cases} \tag{3.3.6}$$

Note that due to m-periodicity there holds $D(m) = D(2m) = \cdots = D\big((n-1)m\big) = 0$, therefore

$$G(0) = G(m) = G(2m) = \cdots = G\big((n-1)m\big) = 0. \tag{3.3.7}$$

We introduce an m-periodic signal $A(k)$:

$$A(k) = \begin{cases} 0 \ \text{ for } \ k = 0, \\ (-1)^r \big(2 \sin(\pi k/m)\big)^{-2r} D(k) \ \text{ for } \ k \in 1 : m - 1. \end{cases}$$

We will show that

$$G(k) = A(k)\, X_1^r(k), \quad k \in 0 : N - 1. \tag{3.3.8}$$

For $k = 0,\ m,\ 2m,\ \ldots,\ (n-1)m$ this formula is true by virtue of (3.3.7) and the equalities $A(0) = 0$ and $X_1(m) = X_1(2m) = \cdots = X_1\big((n-1)m\big) = 0$. For other $k \in 1 : N - 1$, according to (3.3.6) and (3.2.2), we gain

$$G(k) = (-1)^r \big(2 \sin(\pi k/m)\big)^{-2r} X_1^r(k)\, D(k) = A(k)\, X_1^r(k).$$

Formula (3.3.8) allows conversion of a signal g to a form (3.3.1). Indeed, we put

$$a(p) = \frac{1}{m}\sum_{k=0}^{m-1} A(k)\,\omega_m^{kp}.$$

Then (3.2.5) yields

$$\begin{aligned} g(j) &= \frac{1}{N}\sum_{k=0}^{N-1} G(k)\,\omega_N^{kj} = \frac{1}{N}\sum_{k=0}^{N-1} A(k)\,X_1^r(k)\,\omega_N^{kj} \\ &= \frac{1}{N}\sum_{k=0}^{N-1}\Big(\sum_{p=0}^{m-1} a(p)\,\omega_m^{-kp}\Big)X_1^r(k)\,\omega_N^{kj} \\ &= \sum_{p=0}^{m-1} a(p)\Big\{\frac{1}{N}\sum_{k=0}^{N-1} X_1^r(k)\,\omega_N^{k(j-pn)}\Big\} \\ &= \sum_{p=0}^{m-1} a(p)\,Q_r(j-pn). \end{aligned} \tag{3.3.9}$$

The theorem is proved. □

Remark 3.3.1 The proof contains a scheme of transition from the expansion (3.3.5) of a signal g to the expansion (3.3.9). The scheme looks this way:

$$\{d(l)\} \to \{D(k)\} \to \{A(k)\} \to \{a(p)\}.$$

3.3.3 Let us present an important (for what follows) property of discrete periodic splines.

Theorem 3.3.2 *For an arbitrary spline $S \in \mathcal{S}_r^m$ and an arbitrary signal $x \in \mathbb{C}_N$ there holds*

$$\sum_{j=0}^{N-1} \Delta^r S(j)\,\Delta^r x(j) = (-1)^r \sum_{l=0}^{m-1} d(l)\,x(ln), \tag{3.3.10}$$

where $d(l)$ are the coefficients from the representation (3.3.4) *of the spline S.*

Proof We denote by $I_r(x)$ the expression in the left side of equality (3.3.10). According to (3.3.4) and (3.1.4) we have

$$I_r(x) = \sum_{j=0}^{N-1}\Big\{\sum_{l=0}^{m-1} d(l)\,b_r\big(r-(ln-j)\big)\Big\}\Delta^r x(j).$$

Using formulae (3.1.5) and (3.1.8) we gain

$$I_r(x) = (-1)^r \sum_{l=0}^{m-1} d(l) \Big\{ \sum_{j=0}^{N-1} \Delta^r x(j)\, b_r(ln - j) \Big\}$$
$$= (-1)^r \sum_{l=0}^{m-1} d(l)\, \big\{ x(ln) - c \big\},$$

where $c = N^{-1} \sum_{j=0}^{N-1} x(j)$. Taking into account the equality $\sum_{l=0}^{m-1} d(l) = 0$ we come to (3.3.10). The theorem is proved. □

3.4 Spline Interpolation

3.4.1 We consider the following interpolation problem on the set $\mathcal{S}_r^m$ of discrete periodic splines of order r:

$$S(ln) = z(l), \quad l \in 0 : m-1, \tag{3.4.1}$$

where $z(l)$ are arbitrary complex numbers. A detailed notation of the problem (3.4.1) by virtue of (3.3.1) looks this way:

$$\sum_{p=0}^{m-1} c(p)\, Q_r\big((l-p)n\big) = z(l), \quad l \in 0 : m-1. \tag{3.4.2}$$

Thus, a problem of discrete spline interpolation is reduced to solving a system of linear equations (3.4.2) with respect to spline's coefficients $c(p)$.

We introduce a signal $h(p) = Q_r(pn)$. By means of this signal we can rewrite the system (3.4.2) as follows:

$$\sum_{p=0}^{m-1} c(p)\, h(l-p) = z(l), \quad l \in 0 : m-1,$$

or in more compact form $c * h = z$. Going into a spectral domain we obtain the equivalent system of equations

$$C(k)\, H(k) = Z(k), \quad k \in 0 : m-1, \tag{3.4.3}$$

where $C = \mathcal{F}_m(c)$, $Z = \mathcal{F}_m(z)$, and

$$H(k) = \sum_{p=0}^{m-1} h(p)\, \omega_m^{-kp} = \sum_{p=0}^{m-1} Q_r(pn)\, \omega_m^{-kp}.$$

According to (3.2.7) we have $H(k) = T_r(k)$, where, as it was mentioned after the proof of Lemma 3.2.2, all values $T_r(k)$ are positive. The system (3.4.3) has a unique solution $C(k) = Z(k)/T_r(k)$, $k \in 0 : m-1$. The DFT inversion formula yields

$$c(p) = \frac{1}{m} \sum_{k=0}^{m-1} \big[Z(k)/\,T_r(k)\big]\, \omega_m^{kp}, \quad p \in 0 : m-1. \tag{3.4.4}$$

Let us summarize.

Theorem 3.4.1 *The interpolation problem* (3.4.1) *has a unique solution. Coefficients of the interpolation spline* S_* *are determined by formula* (3.4.4).

3.4.2 We will show that a discrete interpolation spline S_* has an extremal property. Incidentally we will clarify the role of the parameter r.

Consider an extremal problem

$$\begin{gathered} f(x) := \sum_{j=0}^{N-1} |\Delta^r x(j)|^2 \to \min, \\ x(ln) = z(l), \quad l \in 0 : m-1; \quad x \in \mathbb{C}_N. \end{gathered} \tag{3.4.5}$$

Theorem 3.4.2 *A unique solution of the problem* (3.4.5) *is a discrete interpolation spline* S_*.

Proof Let x be an arbitrary signal satisfying to the constraints of the problem (3.4.5). We put $\eta = x - S_*$. It is evident that $\eta(ln) = 0$ holds for $l \in 0 : m-1$. On the strength of linearity of a finite difference of order r we have

$$\begin{aligned} f(x) = f(S_* + \eta) &= \sum_{j=0}^{N-1} \big|\Delta^r S_*(j) + \Delta^r \eta(j)\big|^2 \\ &= f(S_*) + f(\eta) + 2\,\mathrm{Re} \sum_{j=0}^{N-1} \Delta^r S_*(j)\, \Delta^r \overline{\eta}(j). \end{aligned}$$

Theorem 3.3.2 yields

$$\sum_{j=0}^{N-1} \Delta^r S_*(j)\, \Delta^r \overline{\eta}(j) = (-1)^r \sum_{l=0}^{m-1} d_*(l)\, \overline{\eta}(ln) = 0.$$

Therefore, $f(x) = f(S_*) + f(\eta)$. Hence follows the inequality $f(x) \geq f(S_*)$ that guarantees optimality of S_*.

Let us verify the uniqueness of a solution of the problem (3.4.5). Assume that $f(x) = f(S_*)$. Then $f(\eta) = 0$. This is possible only when $\Delta^r \eta(j) = 0$ for all $j \in \mathbb{Z}$. Theorem 3.1.2 yields $\eta(j) \equiv \text{const}$. But $\eta(ln) = 0$ holds for $l \in 0 : m-1$, so $\eta(j) \equiv 0$. We gain $x = S_*$. The theorem is proved. □

3.5 Smoothing of Discrete Periodic Data

3.5.1 Let $N = mn$. We consider a problem of smoothing of discrete periodic data in the following setting:

$$f(x) := \sum_{j=0}^{N-1} |\Delta^r x(j)|^2 \to \min,$$
$$g(x) := \sum_{l=0}^{m-1} |x(ln) - z(l)|^2 \le \varepsilon, \quad x \in \mathbb{C}_N, \tag{3.5.1}$$

where $\varepsilon > 0$ is a fixed value (a parameter). Thus, it is required to find a signal $x_* \in \mathbb{C}_N$ that provides given accuracy of approximation $g(x_*) \le \varepsilon$ of data $z(l)$ on a coarse grid $\{ln\}_{l=0}^{m-1}$ and has the minimal squared norm of the finite difference of the r-th order. The latter condition characterizes "smoothness" of a desired signal.

Note that for $\varepsilon = 0$ the problem (3.5.1) is equivalent to the problem (3.4.5).

3.5.2 As a preliminary we will solve an auxiliary problem

$$q(c) := \sum_{l=0}^{m-1} |c - z(l)|^2 \to \min, \tag{3.5.2}$$

where the minimum is taken among all $c \in \mathbb{C}$. We have

$$q(c+h) = \sum_{l=0}^{m-1} \big|\big(c - z(l)\big) + h\big|^2$$
$$= q(c) + m\,|h|^2 + 2\,\mathrm{Re} \sum_{l=0}^{m-1} \big(c - z(l)\big)\,\overline{h}.$$

It is obvious that a unique minimum point c_* of the function $q(c)$ is determined from a condition

$$\sum_{l=0}^{m-1} \big(c - z(l)\big) = 0,$$

so that

$$c_* = \frac{1}{m} \sum_{l=0}^{m-1} z(l). \tag{3.5.3}$$

Herein

$$\varepsilon_* := q(c_*) = -\sum_{l=0}^{m-1} \big(c_* - z(l)\big)\,\overline{z}(l) = \sum_{l=0}^{m-1} |z(l)|^2 - m\,|c_*|^2. \tag{3.5.4}$$

The number ε_* is a critical value of the parameter ε. When $\varepsilon \geq \varepsilon_*$, a solution of the problem (3.5.1) is the signal $x_*(j) \equiv c_*$ because in this case there hold $g(x_*) = q(c_*) = \varepsilon_* \leq \varepsilon$ and $f(x_*) = 0$. Later on we assume that $0 < \varepsilon < \varepsilon_*$. In particular, $\varepsilon_* > 0$. This guarantees that $m \geq 2$ and $z(l) \not\equiv \text{const}$.

3.5.3 We fix a parameter $\alpha > 0$, introduce a function

$$F_\alpha(x) = \alpha f(x) + g(x),$$

and consider yet another auxiliary problem

$$F_\alpha(x) \to \min, \tag{3.5.5}$$

where the minimum is taken among all $x \in \mathbb{C}_N$. We take an arbitrary spline $S \in \mathcal{S}_r^m$ and write down an expansion

$$\begin{aligned} F_\alpha(S+H) &= \alpha f(S+H) + g(S+H) \\ &= \alpha\Big[f(S) + f(H) + 2\operatorname{Re}\sum_{j=0}^{N-1} \Delta^r S(j)\, \Delta^r \overline{H}(j)\Big] \\ &\quad + g(S) + \sum_{l=0}^{m-1} |H(ln)|^2 + 2\operatorname{Re}\sum_{l=0}^{m-1} [S(ln) - z(l)]\, \overline{H}(ln). \end{aligned}$$

According to Theorem 3.3.2 we have

$$\begin{aligned} F_\alpha(S+H) = F_\alpha(S) + \alpha f(H) + \sum_{l=0}^{m-1} |H(ln)|^2 \\ + 2\operatorname{Re}\sum_{l=0}^{m-1} [(-1)^r \alpha\, d(l) + S(ln) - z(l)]\, \overline{H}(ln). \end{aligned}$$

Here $d(l)$ are the coefficients of the expansion (3.3.4) of the spline S over the shifts of the Bernoulli function.

Suppose that there exists a spline $S_\alpha \in \mathcal{S}_r^m$ satisfying to the conditions

$$\begin{gathered} (-1)^r \alpha\, d(l) + S(ln) = z(l), \quad l \in 0 : m-1, \\ \sum_{l=0}^{m-1} d(l) = 0. \end{gathered} \tag{3.5.6}$$

Then for any $H \in \mathbb{C}_N$ there holds an equality

$$F_\alpha(S_\alpha + H) = F_\alpha(S_\alpha) + \alpha f(H) + \sum_{l=0}^{m-1} |H(ln)|^2.$$

It, in particular, yields $F_\alpha(S_\alpha + H) \geq F_\alpha(S_\alpha)$, so S_α is a solution of the problem (3.5.5). Moreover, this solution is unique. Indeed, assuming that $F_\alpha(S_\alpha + H) = F_\alpha(S_\alpha)$ we gain

$$\sum_{j=0}^{N-1} |\Delta^r H(j)|^2 = 0 \quad \text{and} \quad \sum_{l=0}^{m-1} |H(ln)|^2 = 0.$$

The former equality holds only when $H(j) \equiv \text{const}$ (see Theorem 3.1.2). According to the latter one we have $H(j) \equiv 0$.

It is remaining to verify that the system (3.5.6) has a unique solution in the class of splines S of a form (3.3.4). We take a solution d_0, $d_0(0)$, $d_0(1)$, ..., $d_0(m-1)$ of the homogeneous system

$$\begin{gathered} (-1)^r \alpha\, d(l) + S(ln) = 0, \quad l \in 0 : m-1, \\ \sum_{l=0}^{m-1} d(l) = 0. \end{gathered} \tag{3.5.7}$$

We denote the corresponding spline by S_0. According to Theorem 3.3.2 and (3.5.7) we have

$$\sum_{j=0}^{N-1} |\Delta^r S_0(j)|^2 = (-1)^r \sum_{l=0}^{m-1} d_0(l)\, \overline{S_0}(ln) = -\alpha \sum_{l=0}^{m-1} |d_0(l)|^2.$$

By virtue of positiveness of α this equality can be true only when all $d_0(l)$ are equal to zero. But in this case there holds $S_0(j) \equiv d_0$. At the same time, $S_0(ln) = (-1)^{r+1}\alpha\, d_0(l) = 0$ holds for $l \in 0 : m-1$, so that $d_0 = 0$. Thus it is proved that the homogeneous system (3.5.7) has only zero solution. As a consequence we gain that the system (3.5.6) has a unique solution for all $z(l)$, $l \in 0 : m-1$.

Let us summarize.

Theorem 3.5.1 *The auxiliary problem* (3.5.5) *has a unique solution S_α. This is a discrete periodic spline of a form* (3.3.4) *whose coefficients are determined from the system of linear equations* (3.5.6).

3.5.4 We will show that the system (3.5.6) can be solved explicitly. In order to do this we transit into a spectral domain:

$$\begin{aligned} (-1)^r \alpha \sum_{l=0}^{m-1} d(l)\, \omega_m^{-kl} &+ d \sum_{l=0}^{m-1} \omega_m^{-kl} \\ &+ \sum_{l=0}^{m-1} \sum_{p=0}^{m-1} d(p)\, b_{2r}\big((l-p)n + r\big)\, \omega_m^{-kl} \\ &= \sum_{l=0}^{m-1} z(l)\, \omega_m^{-kl}. \end{aligned}$$

We denote $D = \mathcal{F}_m(d)$ and $Z = \mathcal{F}_m(z)$. Taking into account that $kl = k(l-p) + kp$ we gain

$$(-1)^r \alpha D(k) + m\, d\, \delta_m(k) + \sum_{p=0}^{m-1} d(p)\, \omega_m^{-kp} \sum_{l=0}^{m-1} b_{2r}(ln+r)\, \omega_m^{-kl} = Z(k).$$

Note that $D(0) = \sum_{l=0}^{m-1} d(l) = 0$. Putting

$$B_r(k) = \sum_{l=0}^{m-1} b_{2r}(ln+r)\, \omega_m^{-kl}$$

we come to a system of linear equations with respect to d, $D(1), \ldots, D(m-1)$:

$$[(-1)^r \alpha + B_r(k)]\, D(k) + m\, d\, \delta_m(k) = Z(k), \qquad (3.5.8)$$

$$k \in 0 : m-1.$$

For $k = 0$ we have

$$d = \frac{1}{m} Z(0) = \frac{1}{m} \sum_{l=0}^{m-1} z(l) = c_*.$$

For $k \in 1 : m-1$ the equation (3.5.8) takes a form

$$[(-1)^r \alpha + B_r(k)]\, D(k) = Z(k). \qquad (3.5.9)$$

Lemma 3.5.1 *For $k \in 1 : m-1$ there holds*

$$B_r(k) = (-1)^r \Lambda_r(k), \qquad (3.5.10)$$

where, just like in par. 3.2.3,

$$\Lambda_r(k) = \frac{1}{n} \sum_{q=0}^{n-1} \left(2 \sin \frac{\pi(qm+k)}{N}\right)^{-2r}.$$

Proof We have

$$B_r(k) = \frac{1}{N} \sum_{l=0}^{m-1} \sum_{j=1}^{N-1} (\omega_N^j - 1)^{-2r}\, \omega_N^{(ln+r)j}\, \omega_m^{-kl}$$

$$= \frac{1}{n} \sum_{j=1}^{N-1} (\omega_N^j - 1)^{-2r} \, \omega_N^{rj} \left\{ \frac{1}{m} \sum_{l=0}^{m-1} \omega_m^{l(j-k)} \right\}$$

$$= \frac{1}{n} \sum_{j=1}^{N-1} (\omega_N^j - 1)^{-2r} \, \omega_N^{rj} \, \delta_m(j-k).$$

As far as $j \in 1 : N-1$ and $k \in 1 : m-1$, it holds

$$-m + 2 \leq j - k \leq N - 2.$$

The unit pulse δ_m in the latter sum is nonzero only when $j - k = qm$, $q \in 0 : n-1$. Taking into account this consideration and equality (3.1.7) we gain

$$B_r(k) = \frac{1}{n} \sum_{q=0}^{n-1} (\omega_N^{qm+k} - 1)^{-2r} \, \omega_N^{r(qm+k)}$$

$$= \frac{1}{n} \sum_{q=0}^{n-1} \left[\omega_N^{-(qm+k)} (\omega_N^{qm+k} - 1)^2 \right]^{-r}$$

$$= (-1)^r \frac{1}{n} \sum_{q=0}^{n-1} \left(2 \sin \frac{\pi (qm+k)}{N} \right)^{-2r} = (-1)^r \Lambda_r(k).$$

The lemma is proved. □

On the basis of (3.5.9) and (3.5.10) we write

$$D(k) = \begin{cases} 0 \ \text{for} \ k = 0, \\ \dfrac{(-1)^r Z(k)}{\alpha + \Lambda_r(k)} \ \text{for} \ k \in 1 : m-1. \end{cases}$$

The DFT inversion formula yields

$$d(l) = \frac{(-1)^r}{m} \sum_{k=1}^{m-1} \frac{Z(k) \, \omega_m^{kl}}{\alpha + \Lambda_r(k)}, \quad l \in 0 : m-1. \tag{3.5.11}$$

The explicit solution of the system (3.5.6) is found.

3.5.5 A solution of the auxiliary problem (3.5.5) is obtained in a form (3.3.4). Let us convert it to a form (3.3.1).

Theorem 3.5.2 *A smoothing spline $S_\alpha(j)$ can be represented as*

$$S_\alpha(j) = \sum_{p=0}^{m-1} c_\alpha(p)\, Q_r(j - pn), \tag{3.5.12}$$

where

$$c_\alpha(p) = \frac{1}{m} \sum_{k=0}^{m-1} \frac{Z(k)\,\omega_m^{kp}}{T_r(k) + \alpha\bigl(2\sin(\pi k/m)\bigr)^{2r}}. \tag{3.5.13}$$

Proof We have

$$S_\alpha(j) = d + \sum_{l=0}^{m-1} d(l)\, b_{2r}(j + r - ln), \tag{3.5.14}$$

where the coefficients $d(l)$ are calculated by formula (3.5.11) and

$$d = \frac{1}{m}\, Z(0) = \frac{1}{m} \sum_{p=0}^{m-1} \frac{Z(0)}{T_r(0)}\, Q_r(j - pn).$$

The latter equality is true by virtue of (3.2.8) and (3.3.3). Further, the remark to Theorem 3.3.1 yields

$$\sum_{l=0}^{m-1} d(l)\, b_{2r}(j + r - ln) = \sum_{p=0}^{m-1} a(p)\, Q_r(j - pn).$$

Here

$$\begin{aligned} a(p) &= \frac{1}{m} \sum_{k=1}^{m-1} \bigl[(-1)^r \bigl(2\sin(\pi k/m)\bigr)^{-2r} D(k)\bigr]\, \omega_m^{kp} \\ &= \frac{1}{m} \sum_{k=1}^{m-1} \frac{Z(k)\,\omega_m^{kp}}{\bigl(2\sin(\pi k/m)\bigr)^{2r}\bigl(\alpha + \Lambda_r(k)\bigr)} \\ &= \frac{1}{m} \sum_{k=1}^{m-1} \frac{Z(k)\,\omega_m^{kp}}{T_r(k) + \alpha\bigl(2\sin(\pi k/m)\bigr)^{2r}}. \end{aligned}$$

We used formula (3.2.8) again. Substituting derived expressions into (3.5.14) we come to (3.5.12). The theorem is proved. □

Note that when $\alpha = 0$ formula (3.5.13) for the coefficients of a smoothing spline coincides with the formula (3.4.4) for the coefficients of an interpolation spline.

3.5.6 We introduce a function $\varphi(\alpha) = g(S_\alpha)$. According to (3.5.1), (3.5.6) and the Parseval equality we have

$$\varphi(\alpha) = \alpha^2 \sum_{l=0}^{m-1} |d(l)|^2 = \frac{\alpha^2}{m} \sum_{k=0}^{m-1} |D(k)|^2$$
$$= \frac{\alpha^2}{m} \sum_{k=1}^{m-1} \frac{|Z(k)|^2}{\big(\alpha + \Lambda_r(l)\big)^2} = \frac{1}{m} \sum_{k=1}^{m-1} \frac{|Z(k)|^2}{\big(1 + \Lambda_r(k)/\alpha\big)^2}.$$

We remind that $z(l) \not\equiv \text{const}$, therefore at least one of the components $Z(1), \ldots, Z(m-1)$ of the discrete Fourier transform is nonzero.

The function $\varphi(\alpha)$ strictly increases on the semiaxis $(0, +\infty)$, whereby $\lim_{\alpha \to +0} \varphi(\alpha) = 0$. Let us determine the limit of $\varphi(\alpha)$ for $\alpha \to +\infty$. Taking into account (3.5.3) and (3.5.4) we gain

$$\lim_{\alpha \to +\infty} \varphi(\alpha) = \frac{1}{m} \sum_{k=1}^{m-1} |Z(k)|^2 = \frac{1}{m} \sum_{k=0}^{m-1} |Z(k)|^2 - \frac{1}{m} |Z(0)|^2$$
$$= \sum_{l=0}^{m-1} |z(l)|^2 - \frac{1}{m} \Big| \sum_{l=0}^{m-1} z(l) \Big|^2 = \sum_{l=0}^{m-1} |z(l)|^2 - m\, |c_*|^2 = \varepsilon_*,$$

where ε_* is the critical value of the parameter ε. Hence it follows, in particular, that the equation $\varphi(\alpha) = \varepsilon$ with $0 < \varepsilon < \varepsilon_*$ has a unique positive root α_*.

Theorem 3.5.3 *The discrete periodic spline S_{α_*} is a unique solution of the problem* (3.5.1).

Proof We take an arbitrary signal x satisfying to the constraints of the problem (3.5.1) and assume that there holds $f(x) \le f(S_{\alpha_*})$. Then

$$F_{\alpha_*}(x) = \alpha_* f(x) + g(x) \le \alpha_* f(S_{\alpha_*}) + \varepsilon$$
$$= \alpha_* f(S_{\alpha_*}) + \varphi(\alpha_*) = \alpha_* f(S_{\alpha_*}) + g(S_{\alpha_*}) = F_{\alpha_*}(S_{\alpha_*}).$$

Taking into account that S_{α_*} is a unique minimum point of F_{α_*} on $\mathbb{C}_N$ we conclude that $x(j) \equiv S_{\alpha_*}(j)$. It means that in case of $x(j) \not\equiv S_{\alpha_*}(j)$ there holds $f(x) > f(S_{\alpha_*})$. The theorem is proved. □

3.6 Tangent Hyperbolas Method

3.6.1 It is ascertained in par. 3.5.6 that a unique solution of the smoothing problem (3.5.1) for $0 < \varepsilon < \varepsilon_*$ is a discrete periodic spline $S_\alpha(j)$ of a form (3.5.12) with $\alpha = \alpha_*$, where α_* is a unique positive root of the equation $\varphi(\alpha) = \varepsilon$. Here we will consider a question of calculation of α_*.

We introduce a function

$$\psi(\beta) = \varphi\Big(\frac{1}{\beta}\Big) = \frac{1}{m}\sum_{k=1}^{m-1} \frac{|Z(k)|^2}{\big(1 + \Lambda_r(k)\,\beta\big)^2}.$$

We take an interval $(-\tau, +\infty)$, where $\tau = \min_{k\in 1:m-1}[\Lambda_r(k)]^{-1}$. On this interval there hold inequalities $\psi'(\beta) < 0$ and $\psi''(\beta) > 0$, therefore the function $\psi(\beta)$ is strictly decreasing and strictly convex on $(-\tau, +\infty)$. In addition to that we have $\psi(0) = \varepsilon_*$ and $\lim\limits_{\beta\to+\infty} \psi(\beta) = 0$. If β_* is a positive root of the equation $\psi(\beta) = \varepsilon$ then $\alpha_* = 1/\beta_*$. Thus, instead of $\varphi(\alpha) = \varepsilon$ we can solve the equation $\psi(\beta) = \varepsilon$.

Let us consider the equivalent equation $[\psi(\beta)]^{-1/2} = \varepsilon^{-1/2}$. We will solve it by the Newton method with an initial approximation $\beta_0 = 0$. Working formula of the method looks this way:

$$\begin{aligned}\beta_{k+1} &= \beta_k - \frac{\psi^{-1/2}(\beta_k) - \varepsilon^{-1/2}}{(-1/2)\,\psi^{-3/2}(\beta_k)\,\psi'(\beta_k)} \\ &= \beta_k + \frac{2\,\psi(\beta_k)}{\psi'(\beta_k)}\Big[1 - \Big(\frac{\psi(\beta_k)}{\varepsilon}\Big)^{1/2}\Big], \quad k = 0, 1, \ldots \end{aligned} \tag{3.6.1}$$

Let us find out what this method corresponds to when it is applied to the equation $\psi(\beta) = \varepsilon$.

3.6.2 We will need the following properties of the function $[\psi(\beta)]^{-1/2}$.

Lemma 3.6.1 *The function $[\psi(\beta)]^{-1/2}$ strictly increases and is concave on the interval $(-\tau, +\infty)$.*

Proof We have

$$\big[\psi^{-1/2}(\beta)\big]' = -\tfrac{1}{2}\,\psi^{-3/2}(\beta)\,\psi'(\beta),$$

$$\begin{aligned}\big[\psi^{-1/2}(\beta)\big]'' &= -\tfrac{1}{2}\big[-\tfrac{3}{2}\,\psi^{-5/2}(\beta)\,\big(\psi'(\beta)\big)^2 + \psi^{-3/2}(\beta)\,\psi''(\beta)\big] \\ &= \tfrac{1}{4}\,\psi^{-5/2}(\beta)\,\big[3\,\big(\psi'(\beta)\big)^2 - 2\,\psi(\beta)\,\psi''(\beta)\big].\end{aligned}$$

It is obvious that $[\psi^{-1/2}(\beta)]' > 0$. The inequality $[\psi^{-1/2}(\beta)]'' \le 0$ is equivalent to the following:

$$3\,\big(\psi'(\beta)\big)^2 \le 2\,\psi(\beta)\,\psi''(\beta). \tag{3.6.2}$$

To verify (3.6.2) we introduce notations

$$\eta_k = m^{-1}\,\big(|Z(k)|\,/\,\Lambda_r(k)\big)^2, \qquad \theta_k = [\Lambda_r(k)]^{-1}.$$

Then

$$\psi(\beta) = \sum_{k=1}^{m-1} \frac{\eta_k}{(\beta+\theta_k)^2}, \qquad \psi'(\beta) = -2\sum_{k=1}^{m-1} \frac{\eta_k}{(\beta+\theta_k)^3},$$

$$\psi''(\beta) = 6\sum_{k=1}^{m-1} \frac{\eta_k}{(\beta+\theta_k)^4}.$$

The inequality (3.6.2) takes a form

$$\left(\sum_{k=1}^{m-1} \frac{\sqrt{\eta_k}}{\beta+\theta_k}\,\frac{\sqrt{\eta_k}}{(\beta+\theta_k)^2}\right)^2 \leq \left(\sum_{k=1}^{m-1} \frac{\eta_k}{(\beta+\theta_k)^2}\right)\left(\sum_{k=1}^{m-1} \frac{\eta_k}{(\beta+\theta_k)^4}\right). \tag{3.6.3}$$

The latter is true on the strength of Cauchy–Bunyakovskii inequality. The lemma is proved. □

Remark 3.6.1 By virtue of Lemma 2.1.5 the inequality (3.6.3) is fulfilled as an equality if and only if $\beta + \theta_k \equiv \text{const}$. If not all values $\theta_k = [\Lambda_r(k)]^{-1}$ for $k \in 1 : m-1$ are equal to each other then the inequality (3.6.3) holds strictly. In this case the function $[\psi(\beta)]^{-1/2}$ is strictly concave.

3.6.3 According to concavity of the function $[\psi(\beta)]^{-1/2}$, for $\beta \geq 0$ we have

$$\psi^{-1/2}(\beta) - \psi^{-1/2}(\beta_k) \leq -\tfrac{1}{2}\,\psi^{-3/2}(\beta_k)\,\psi'(\beta_k)\,(\beta - \beta_k),$$

so

$$0 < \psi^{-1/2}(\beta) \leq \psi^{-1/2}(\beta_k)\left[1 - \frac{\psi'(\beta_k)}{2\,\psi(\beta_k)}\,(\beta - \beta_k)\right].$$

Raising to the power of -2 we gain

$$\psi(\beta) \geq \psi(\beta_k)\left[1 - \frac{\psi'(\beta_k)}{2\,\psi(\beta_k)}\,(\beta - \beta_k)\right]^{-2}. \tag{3.6.4}$$

We denote the function in the right side of the inequality (3.6.4) by $\zeta_k(\beta)$. A graph of this function is a hyperbola. By virtue of (3.6.4) this hyperbola lies under the graph of the function $\psi(\beta)$. Since $\zeta_k(\beta_k) = \psi(\beta_k)$ and $\zeta_k'(\beta_k) = \psi'(\beta_k)$, the mentioned graphs are tangent to each other when $\beta = \beta_k$ (see Fig. 3.3). Moreover, the root β_{k+1} of the equation $\zeta_k(\beta) = \varepsilon$ is calculated with the aid of the formula (3.6.1).

According to what has been said it is reasonable to refer to the iterative method (3.6.1) for solving the equation $\psi(\beta) = \varepsilon$ as a *tangent hyperbolas method*.

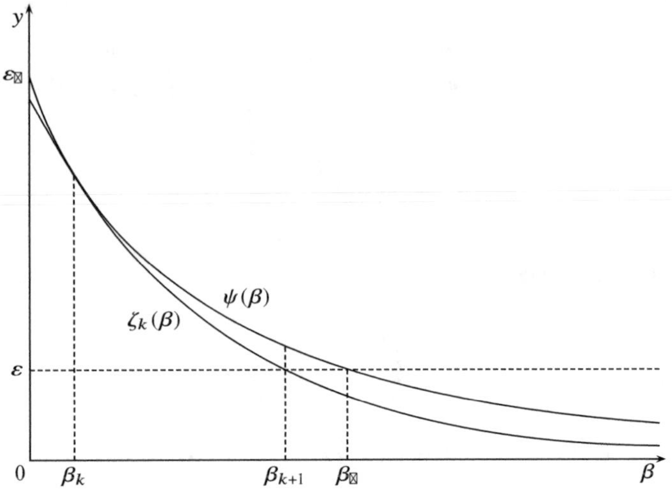

Fig. 3.3 Tangent hyperbolas method

3.7 Calculation of Discrete Spline's Values

3.7.1 We start with a discrete periodic spline of the first order

$$S_1(j) = \sum_{p=0}^{m-1} c(p)\, Q_1(j - pn). \tag{3.7.1}$$

We assume that the coefficients $c(p)$ are continued with a period m on all integer indices p. In particular, $c(m) = c(0)$.

Lemma 3.7.1 *The values $S_1(0),\ S_1(1),\ \dots,\ S_1(N)$ are calculated consecutively by the scheme*

$$S_1(0) = n\, c(0);$$

$$S_1(ln + k + 1) = S_1(ln + k) + \Delta c(l), \tag{3.7.2}$$

$$k \in 0 : n - 1, \quad l \in 0 : m - 1.$$

Proof Let $j = ln + k,\ k \in 0 : n - 1,\ l \in 0 : m - 1$. Formula (3.7.1) yields

$$S_1(j) = \sum_{p=0}^{m-1} c(p)\, Q_1(k - (p - l)n) = \sum_{p=0}^{m-1} c(p + l)\, Q_1(k - pn)$$
$$= \sum_{p=0}^{m-1} c(p + l)\, Q_1(k + (m - p)n).$$

For $p \in 2 : m - 1$ we have

$$n \le k + (m - p)n \le n - 1 + (m - 2)n = N - n - 1.$$

Hence (see (3.2.4) and (3.2.1)) $Q_1(k + (m - p)n) = 0$ holds for the given p. We gain

$$S_1(j) = c(l)\, Q_1(k) + c(l + 1)\, Q_1(k + N - n).$$

Recall that $Q_1(k) = n - k$ for $k \in 0 : n - 1$. Furthermore, $Q_1(j) = j - N + n$ for $j \in N - n : N - 1$. Since $N - n \le k + N - n \le N - 1$ for $k \in 0 : n - 1$, it holds $Q_1(k + N - n) = k$. Thus, for $k \in 0 : n - 1$ and $l \in 0 : m - 1$ we have

$$S_1(ln + k) = c(l)(n - k) + c(l + 1)k = nc(l) + k\Delta c(l). \tag{3.7.3}$$

In particular, $S_1(ln) = nc(l)$ holds for $l \in 0 : m - 1$. By virtue of periodicity the latter equality is true for $l = m$ as well.

Note that formula (3.7.3) is true for $k = n$. In this case it takes a form $S_1\big((l + 1)n\big) = nc(l + 1)$, $l \in 0 : m - 1$. Replacing k by $k + 1$ in (3.7.3) we write

$$S_1(ln + k + 1) = nc(l) + (k + 1)\Delta c(l), \quad k \in 0 : n - 1. \tag{3.7.4}$$

On the basis of (3.7.3) and (3.7.4) we come to (3.7.2). The lemma is proved. □

Below we present a program that implements calculations along the scheme (3.7.2).

Program Code

```
s1(0) := n * c(0);  j := 0;
for l := 0 to m − 1 do
begin h := c(l + 1) − c(l);
  for k := 1 to n do
  begin j := j + 1;
        s1(j) := s1(j − 1) + h end
end
```

We see that calculation of values $S_1(j)$ for $j = 0, 1, \ldots, N$ requires one multiplication by n and $(n + 1)m$ additions.

3.7.2 Now we turn to a general case of a discrete periodic spline of order r. With the aid of cyclic convolution we introduce a sequence of signals

$$S_\nu = Q_1 * S_{\nu-1}, \quad \nu = 2, 3, \ldots \tag{3.7.5}$$

Here S_1 is a spline of a form (3.7.1).

Theorem 3.7.1 *There holds an equality*

$$S_r(j) = \sum_{p=0}^{m-1} c(p)\, Q_r(j - pn), \quad j \in \mathbb{Z}. \tag{3.7.6}$$

Proof When $r = 1$, formula (3.7.6) coincides with (3.7.1). We perform an induction step from $r - 1$ to r. According to the inductive hypothesis, (3.7.5), and (3.2.4), we have

$$\begin{aligned} S_r(j) &= \sum_{l=0}^{N-1} Q_1(l)\, S_{r-1}(j - l) \\ &= \sum_{p=0}^{m-1} c(p) \sum_{l=0}^{N-1} Q_1(l)\, Q_{r-1}(j - l - pn) = \sum_{p=0}^{m-1} c(p)\, Q_r(j - pn). \end{aligned}$$

The theorem is proved. □

3.7.3 Theorem 3.7.1 shows that calculation of values of $S_r(j)$ is reduced to calculation of values of $S_1(j)$ and consecutive convolution with B-spline $Q_1(j)$. We will consider a question of calculation of a convolution with Q_1 separately. Let

$$y(j) = \sum_{k=0}^{N-1} x(k)\, Q_1(j - k).$$

Theorem 3.7.2 *Valid is the equality*

$$y(j) = nx(j) + \sum_{k=1}^{n-1} (n - k)[x(j + k) + x(j - k)], \tag{3.7.7}$$

$$j \in 0 : N - 1.$$

Proof Since Q_1 is even, we have

$$\begin{aligned} y(j) &= \sum_{k=0}^{N-1} x(k)\, Q_1(k - j) = \sum_{k=0}^{N-1} x(k + j)\, Q_1(k) \\ &= \sum_{k=0}^{n-1} (n - k)x(k + j) + \sum_{k=N-n+1}^{N-1} \big(n - (N - k)\big)x\big(j - (N - k)\big) \\ &= nx(j) + \sum_{k=1}^{n-1} (n - k)[x(j + k) + x(j - k)]. \end{aligned}$$

The theorem is proved. □

We fix $j \in 0 : N-1$ and introduce notations

$$d_0 = x(j); \quad d_k = x(j+k) + x(j-k),\ k \in 1 : n-1; \quad t_k = n-k.$$

Formula (3.7.7) can be rewritten in a form

$$y(j) = \sum_{k=0}^{n-1} d_k\, t_k.$$

We construct a sequence of numbers $\{h_k\}$ by a rule

$$h_k = d_k + h_{k-1},\ k = 0, 1, \ldots, n-1; \quad h_{-1} = 0. \tag{3.7.8}$$

Taking into account that $t_k - t_{k+1} = 1$ we gain

$$\begin{aligned} y(j) &= \sum_{k=0}^{n-1} (h_k - h_{k-1})\, t_k = \sum_{k=0}^{n-1} h_k\, t_k - \sum_{k=-1}^{n-2} h_k\, t_{k+1} \\ &= h_{n-1}\, t_{n-1} + \sum_{k=0}^{n-2} h_k = \sum_{k=0}^{n-1} h_k. \end{aligned}$$

Thus,

$$y(j) = \sum_{k=0}^{n-1} h_k, \tag{3.7.9}$$

where h_k are calculated with the recurrent formula (3.7.8).

Below we present a program of calculation of values $y(j)$ for $j \in 0 : N-1$ which is based on the representation (3.7.9).

Program Code

```
for j := 0 to n − 1 do
begin h := x(j); s := h;
  for k := 1 to n − 1 do
  begin h := h + (x(j + k) + x(j − k));
     s := s + h end;
  y(j) := s
end
```

The program uses only additions. The number of additions is $3(n-1)N$.

The values $x(j)$ for j from $(-n+1)$ to $N+n-2$ must be given explicitly. By virtue of periodicity one should put

$$x(j) = x(N+j) \quad \text{for } j \in -n+1 : -1;$$
$$x(j) = x(j-N) \quad \text{for } j \in N : N+n-2.$$

3.8 Orthogonal Basis in a Space of Splines

3.8.1 We consider a discrete periodic spline

$$S(j) = \sum_{p=0}^{m-1} c(p)\, Q_r(j-pn) \tag{3.8.1}$$

and transform its coefficients by a rule $\xi = \mathcal{F}_m(c)$. Taking into account the DFT inversion formula we write

$$\begin{aligned} S(j) &= \frac{1}{m} \sum_{p=0}^{m-1} \left\{ \sum_{k=0}^{m-1} \xi(k)\, \omega_m^{kp} \right\} Q_r(j-pn) \\ &= \sum_{k=0}^{m-1} \xi(k) \left\{ \frac{1}{m} \sum_{p=0}^{m-1} \omega_m^{kp}\, Q_r(j-pn) \right\}. \end{aligned} \tag{3.8.2}$$

We introduce a notation

$$\mu_k(j) = \frac{1}{m} \sum_{p=0}^{m-1} \omega_m^{kp}\, Q_r(j-pn), \quad k \in 0 : m-1. \tag{3.8.3}$$

Formula (3.8.2) takes a form

$$S(j) = \sum_{k=0}^{m-1} \xi(k)\, \mu_k(j). \tag{3.8.4}$$

It is obvious that the signals μ_k belong to $\mathcal{S}_r^m$. According to (3.8.4) they form a basis in $\mathcal{S}_r^m$. We will show that this basis is orthogonal.

3.8.2 As a precursor, let us obtain an expansion of the signal μ_k over the exponential basis.

Lemma 3.8.1 *Valid is the equality*

$$\mu_k(j) = \frac{1}{N} \sum_{q=0}^{n-1} X_1^r(qm+k)\, \omega_N^{(qm+k)j}. \tag{3.8.5}$$

Proof According to (3.2.5) and (3.8.3) we have

$$\begin{aligned}
\mu_k(j) &= \frac{1}{mN} \sum_{p=0}^{m-1} \omega_m^{kp} \left\{ \sum_{l=0}^{N-1} X_1^r(l)\, \omega_N^{l(j-pn)} \right\} \\
&= \frac{1}{N} \sum_{l=0}^{N-1} X_1^r(l)\, \omega_N^{lj} \left\{ \frac{1}{m} \sum_{p=0}^{m-1} \omega_m^{(k-l)p} \right\} \\
&= \frac{1}{N} \sum_{l=0}^{N-1} X_1^r(l)\, \omega_N^{lj}\, \delta_m(k-l).
\end{aligned}$$

The index l can be represented as $l = qm + k'$, where $q \in 0 : n-1$ and $k' \in 0 : m-1$. Bearing this in mind we gain

$$\begin{aligned}
\mu_k(j) &= \frac{1}{N} \sum_{q=0}^{n-1} \sum_{k'=0}^{m-1} X_1^r(qm+k')\, \omega_N^{(qm+k')j}\, \delta_m(k-k') \\
&= \frac{1}{N} \sum_{q=0}^{n-1} X_1^r(qm+k)\, \omega_N^{(qm+k)j}.
\end{aligned}$$

The lemma is proved. □

Theorem 3.8.1 *The following relations hold:*

$$\begin{aligned}
\langle \mu_k, \mu_{k'} \rangle &= 0 \text{ for } k \neq k', \\
\|\mu_k\|^2 &= \tfrac{1}{m}\, T_{2r}(k), \quad k \in 0 : m-1.
\end{aligned} \tag{3.8.6}$$

Proof On the basis of (3.8.5) we write

$$\begin{aligned}
\langle \mu_k, \mu_{k'} \rangle &= \sum_{j=0}^{N-1} \mu_k(j)\, \overline{\mu}_{k'}(j) \\
&= \frac{1}{N} \sum_{q,q'=0}^{n-1} X_1^r(qm+k)\, X_1^r(q'm+k') \left\{ \frac{1}{N} \sum_{j=0}^{N-1} \omega_N^{(qm+k-q'm-k')j} \right\} \\
&= \frac{1}{N} \sum_{q,q'=0}^{n-1} X_1^r(qm+k)\, X_1^r(q'm+k')\, \delta_N\big((q-q')m + k - k'\big).
\end{aligned}$$

It is evident that $|k - k'| \le m - 1$ and $|(q - q')m + k - k'| \le N - 1$. If $k \ne k'$ then the argument of the unit pulse δ_N is nonzero for all $q, q' \in 0 : n - 1$. In this case there holds $\langle \mu_k, \mu_{k'} \rangle = 0$. When $k = k'$, according to (3.2.6) we gain

$$\|\mu_k\|^2 = \frac{1}{N} \sum_{q=0}^{n-1} X_1^{2r}(qm + k) = \frac{1}{m} T_{2r}(k).$$

The theorem is proved. □

3.8.3 It is ascertained that the splines $\{\mu_k\}_{k=0}^{m-1}$ form an orthogonal basis in a space $\mathcal{S}_r^m$. A transition from the expansion (3.8.1) to the expansion (3.8.4) is based on the coefficients transform $\xi = \mathcal{F}_m(c)$. An inverse transition is related to the inversion formula for DFT: $c = \mathcal{F}_m^{-1}(\xi)$.

Note that (3.8.5) and (3.2.2) yield

$$\mu_0(j) \equiv N^{-1} n^{2r}. \tag{3.8.7}$$

Lemma 3.8.2 *For $k \in 1 : m - 1$ valid is the equality*

$$\mu_{m-k}(j) = \overline{\mu}_k(j), \quad j \in \mathbb{Z}. \tag{3.8.8}$$

Proof According to (3.8.3) we have

$$\overline{\mu}_{m-k}(j) = \frac{1}{m} \sum_{p=0}^{m-1} \omega_m^{-(m-k)p}\, Q_r(j - pn) = \mu_k(j),$$

which is equivalent to (3.8.8). The lemma is proved. □

Lemma 3.8.3 *For all integer l there holds*

$$\mu_k(j + ln) = \omega_m^{kl}\, \mu_k(j), \quad j \in \mathbb{Z}. \tag{3.8.9}$$

Indeed,

$$\mu_k(j + ln) = \frac{1}{m} \sum_{p=0}^{m-1} \omega_m^{k(p-l)+kl}\, Q_r\big(j - (p - l)n\big) = \omega_m^{kl}\, \mu_k(j).$$

Theorem 3.8.2 *Let $k \in 1 : m - 1$ and p be a natural number. If a product kp is not divisible by m then there holds*

$$\sum_{j=0}^{N-1} \big[\mu_k(j)\big]^p = 0.$$

Proof On the basis of (3.8.9) we write

$$\sum_{j=0}^{N-1} \big[\mu_k(j)\big]^p = \sum_{l=0}^{m-1}\sum_{q=0}^{n-1} \big[\mu_k(q+ln)\big]^p$$
$$= \sum_{l=0}^{m-1} \omega_m^{klp} \sum_{q=0}^{n-1} \big[\mu_k(q)\big]^p = m\,\delta_m(kp) \sum_{q=0}^{n-1} \big[\mu_k(q)\big]^p.$$

Hence the required equality follows immediately. □

Let us consider special cases. For $p = 1$ we have

$$\sum_{j=0}^{N-1} \mu_k(j) = 0, \quad k \in 1 : m-1.$$

Let $p = 2$. For $k \in 1 : m-1$ the condition $\langle 2k\rangle_m \neq 0$ is reduced to a relation $2k \neq m$. Thus, for $k \in 1 : m-1$, $k \neq m/2$ valid is the equality

$$\sum_{j=0}^{N-1} \big[\mu_k(j)\big]^2 = 0.$$

We see that for the mentioned indices k there always exist some complex values among $\mu_k(j)$. Along with that, if m is even then all values $\mu_{m/2}(j)$ are real because there holds

$$\mu_{m/2}(j) = \frac{1}{m} \sum_{p=0}^{m-1} (-1)^p\, Q_r(j - pn). \tag{3.8.10}$$

3.8.4 Let us return to formula (3.8.3) and rewrite it as follows:

$$\mu_k(j) = \frac{1}{m} \sum_{p=0}^{m-1} Q_r(j - pn)\, \omega_m^{pk}.$$

This formula has a form of the DFT inversion formula, therefore

$$Q_r(j - pn) = \sum_{k=0}^{m-1} \mu_k(j)\, \omega_m^{-pk}, \quad p \in 0 : m-1.$$

In particular,

$$Q_r(j) = \sum_{k=0}^{m-1} \mu_k(j), \quad j \in \mathbb{Z}. \tag{3.8.11}$$

3.9 Bases of Shifts

3.9.1 As it was noted in the par. 3.3.1, shifts of a B-spline $\{Q_r(j-pn)\}_{p=0}^{m-1}$ form a basis in a space $\mathcal{S}_r^m$. Are there any other splines with the similar property? This question can be answered completely.

We take a spline $\varphi \in \mathcal{S}_r^m$ and expand it over the orthogonal basis:

$$\varphi(j) = \sum_{k=0}^{m-1} \xi(k)\, \mu_k(j). \tag{3.9.1}$$

Formula (3.8.9) yields

$$\varphi(j-pn) = \sum_{k=0}^{m-1} \xi(k)\, \omega_m^{-kp}\, \mu_k(j). \tag{3.9.2}$$

Hence the shifts $\varphi(j-pn)$ also belong to $\mathcal{S}_r^m$.

Theorem 3.9.1 *A system of shifts* $\{\varphi(j-pn)\}_{p=0}^{m-1}$ *forms a basis in* $\mathcal{S}_r^m$ *if and only if each coefficient* $\xi(k)$ *in the expansion* (3.9.1) *is nonzero.*

Proof We rewrite (3.9.2) in a form

$$\varphi(j-pn) = \sum_{k=0}^{m-1} \xi(k)\, \mu_k(j)\, \omega_m^{-kp}.$$

The DFT inversion formula yields

$$\xi(k)\, \mu_k(j) = \frac{1}{m} \sum_{p=0}^{m-1} \omega_m^{kp}\, \varphi(j-pn), \quad k \in 0:m-1. \tag{3.9.3}$$

If every $\xi(k)$ is nonzero then we can divide (3.9.3) by $\xi(k)$ and thus gain an expansion of all splines $\mu_k(j)$ over the system $\{\varphi(j-pn)\}_{p=0}^{m-1}$. Therefore this system is a basis in $\mathcal{S}_r^m$.

Conversely, let $\{\varphi(j-pn)\}_{p=0}^{m-1}$ be a basis in $\mathcal{S}_r^m$. If at least one coefficient $\xi(k)$ in the expansion (3.9.1) is equal to zero then according to (3.9.3) the system $\{\varphi(j-pn)\}_{p=0}^{m-1}$ is linearly dependent. But this contradicts with a definition of a basis. The theorem is proved. □

3.9.2 Two splines φ and ψ from $\mathcal{S}_r^m$ are called *dual* if for all $p, q \in 0:m-1$ there holds

$$\langle \varphi(\cdot - pn),\, \psi(\cdot - qn) \rangle = \delta_m(p-q). \tag{3.9.4}$$

Thus, duality of splines φ and ψ is characterized by the fact that the systems of their shifts $\{\varphi(j-pn)\}_{p=0}^{m-1}$ and $\{\psi(j-pn)\}_{p=0}^{m-1}$ are biorthogonal.

Along with (3.9.1) we write an expansion

$$\psi(j) = \sum_{k=0}^{m-1} \eta(k)\, \mu_k(j).$$

We note that (3.9.2) and (3.8.6) yield

$$\begin{aligned}\bigl\langle \varphi(\cdot - pn),\, \psi(\cdot - qn)\bigr\rangle &= \Bigl\langle \sum_{k=0}^{m-1} \xi(k)\, \omega_m^{-kp}\, \mu_k,\, \sum_{l=0}^{m-1} \eta(l)\, \omega_m^{-lq}\, \mu_l \Bigr\rangle \\ &= \sum_{k=0}^{m-1} \xi(k)\, \overline{\eta}(k)\, \omega_m^{k(q-p)}\, \|\mu_k\|^2 \\ &= \frac{1}{m} \sum_{k=0}^{m-1} \xi(k)\, \overline{\eta}(k)\, T_{2r}(k)\, \omega_m^{k(q-p)}. \qquad (3.9.5)\end{aligned}$$

Theorem 3.9.2 *Splines φ and ψ from S_r^m are dual if and only if their coefficients $\xi(k)$, $\eta(k)$ in the expansions over the orthogonal basis satisfy to the condition*

$$\xi(k)\, \overline{\eta}(k) = \bigl[T_{2r}(k)\bigr]^{-1}, \quad k \in 0 : m-1. \qquad (3.9.6)$$

Proof
Necessity We take (3.9.4) and put $p = 0$ there. According to (3.9.5) we gain

$$\frac{1}{m} \sum_{k=0}^{m-1} \xi(k)\, \overline{\eta}(k)\, T_{2r}(k)\, \omega_m^{kq} = \delta_m(q).$$

Therefore

$$\xi(k)\, \overline{\eta}(k)\, T_{2r}(k) = \sum_{q=0}^{m-1} \delta_m(q)\, \omega_m^{-kq} = 1, \quad k \in 0 : m-1,$$

which is equivalent to (3.9.6).

Sufficiency obviously follows from (3.9.5) and (3.9.6). The theorem is proved. □

3.9.3 Theorem 3.9.2 lets us introduce a *self-dual spline*. It is obtained when $\xi(k) = \eta(k)$, $k \in 0 : m-1$. In this case the condition (3.9.6) takes a form

$$|\xi(k)|^2 = \bigl[T_{2r}(k)\bigr]^{-1}, \quad k \in 0 : m-1.$$

The simplest self-dual spline is defined by the formula (see Fig. 3.4)

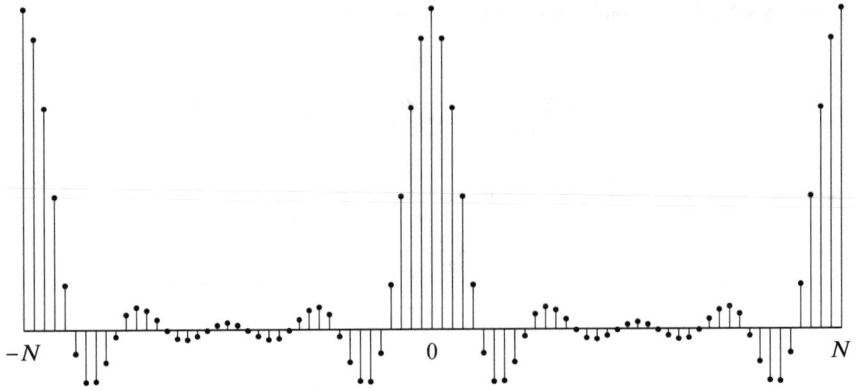

Fig. 3.4 Graph of a self-dual spline $\varphi_r(j)$ for $m = 8$, $n = 5$, and $r = 2$

$$\varphi_r(j) = \sum_{k=0}^{m-1} \frac{\mu_k(j)}{\sqrt{T_{2r}(k)}}, \qquad j \in \mathbb{Z}.$$

According to (3.9.5) we have $\langle \varphi_r(\cdot - pn),\, \varphi_r(\cdot - qn)\rangle = \delta_m(q-p)$. The latter means that the shifts $\{\varphi_r(j - pn)\}_{p=0}^{m-1}$ form an orthonormal system.

3.9.4 According to (3.8.11) each coefficient in the expansion of a discrete periodic B-spline $Q_r(j)$ over the orthogonal basis is equal to unity. By virtue of (3.9.6) a dual to $Q_r(j)$ spline $R_r(j)$ looks this way (see Fig. 3.5):

$$R_r(j) = \sum_{k=0}^{m-1} \frac{\mu_k(j)}{T_{2r}(k)}. \tag{3.9.7}$$

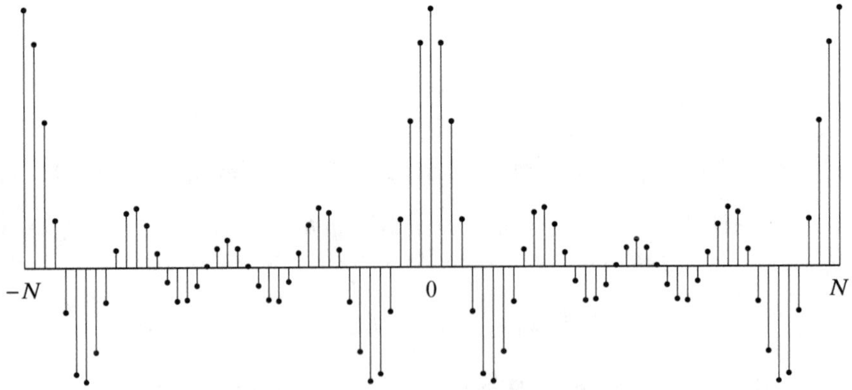

Fig. 3.5 Graph of a spline $R_r(j)$ dual to a B-spline $Q_r(j)$ for $m = 8$, $n = 5$, and $r = 2$

We will show how the dual splines $Q_r(j)$ and $R_r(j)$ help in solving a problem of spline processing of discrete periodic data with the least squares method.

Consider an extremal problem

$$F(S) := \sum_{j=0}^{N-1} |S(j) - z(j)|^2 \to \min, \tag{3.9.8}$$

where the minimum is taken among all $S \in \mathcal{S}_r^m$. Given an arbitrary $H \in \mathcal{S}_r^m$, we have

$$F(S+H) = \|(S-z)+H\|^2 = F(S) + \|H\|^2 + 2\,\mathrm{Re}\langle S-z,\ H\rangle.$$

If we manage to construct a spline $S_* \in \mathcal{S}_r^m$ such that the difference $S_* - z$ is orthogonal to any element H of $\mathcal{S}_r^m$, then S_* will be the unique solution of the problem (3.9.8).

Let us use a representation

$$S_*(j) = \sum_{q=0}^{m-1} d(q)\, R_r(j - qn). \tag{3.9.9}$$

A condition of orthogonality can be written as

$$\Big\langle \sum_{q=0}^{m-1} d(q)\, R_r(\cdot - qn) - z,\ Q_r(\cdot - pn)\Big\rangle = 0, \quad p \in 0 : m-1.$$

On the strength of duality of the splines Q_r and R_r we gain

$$d(p) = \big\langle z,\ Q_r(\cdot - pn)\big\rangle = \sum_{j=0}^{N-1} z(j)\, Q_r(j - pn), \tag{3.9.10}$$

$$p \in 0 : m-1.$$

Thus, a unique solution of the problem (3.9.8) is the spline (3.9.9) with the coefficients being calculated with formula (3.9.10).

The problem (3.9.8) can be interpreted as a problem of orthogonal projection of a signal z on a subspace $\mathcal{S}_r^m$.

3.9.5 We can transit from the expansion (3.9.9) of the spline $S_*(j)$ over the basis $\big\{R_r(j - qn)\big\}_{q=0}^{m-1}$ to the expansion over the basis $\big\{Q_r(j - pn)\big\}_{p=0}^{m-1}$. In order to do this we use formulae (3.9.7) and (3.9.2) and write down

$$S_*(j) = \sum_{q=0}^{m-1} d(q)\, R_r(j - qn) = \sum_{q=0}^{m-1} d(q) \sum_{k=0}^{m-1} \frac{\omega_m^{-kq}}{T_{2r}(k)}\, \mu_k(j)$$
$$= \sum_{k=0}^{m-1} \frac{\mu_k(j)}{T_{2r}(k)} \sum_{q=0}^{m-1} d(q)\, \omega_m^{-kq}.$$

We denote $D = \mathcal{F}_m(d)$, $\xi(k) = D(k)/T_{2r}(k)$, $c = \mathcal{F}_m^{-1}(\xi)$. Then

$$S_*(j) = \sum_{p=0}^{m-1} c(p)\, Q_r(j - pn).$$

3.10 Wavelet Subspaces

3.10.1 We will carry out further analysis in assumption that $m = 2^t$, where t is a natural number. We put $m_\nu = m/2^\nu$ and $n_\nu = 2^\nu n$. In this case $m_\nu n_\nu = N$ for all $\nu = 0, 1, \ldots, t$.

We denote orthogonal splines corresponding to parameters m_ν, n_ν by μ_k^ν. In particular, $\mu_k^0 = \mu_k$. We have by a definition that $\mu_k^\nu \in \mathcal{S}_r^{m_\nu}$. On the strength of (3.8.3) we may consider the splines μ_k^ν being defined for all integer k. It is clear that they are m_ν-periodic on k.

Theorem 3.10.1 *The following recurrent formula holds for* $\nu = 0, 1, \ldots, t-1$*:*

$$\mu_k^{\nu+1}(j) = c_\nu(k)\, \mu_k^\nu(j) + c_\nu(m_{\nu+1} + k)\, \mu_{m_{\nu+1}+k}^\nu(j), \tag{3.10.1}$$

where $c_\nu(l) = \big(2\cos(\pi l/m_\nu)\big)^{2r}$.

Proof We introduce a notation

$$y_\nu(l) = \begin{cases} n_\nu^{2r} & \text{for } l = 0, \\ \left(\dfrac{\sin(\pi l/m_\nu)}{\sin(\pi l/N)}\right)^{2r} & \text{for } l \in 1 : N-1. \end{cases}$$

Equality (3.8.5) yields

$$\mu_k^\nu(j) = \frac{1}{N} \sum_{q=0}^{n_\nu - 1} y_\nu(q m_\nu + k)\, \omega_N^{(q m_\nu + k) j}.$$

We note that

$$y_{\nu+1}(l) = c_\nu(l)\, y_\nu(l), \quad l \in 0 : N-1. \tag{3.10.2}$$

Bearing in mind m_ν-periodicity of the signal c_ν we gain

$$\begin{aligned}
\mu_k^{\nu+1}(j) &= \frac{1}{N}\sum_{q=0}^{2n_\nu-1} y_{\nu+1}(qm_{\nu+1}+k)\,\omega_N^{(qm_{\nu+1}+k)j} \\
&= \frac{1}{N}\sum_{q=0}^{n_\nu-1} y_{\nu+1}(2qm_{\nu+1}+k)\,\omega_N^{(2qm_{\nu+1}+k)j} \\
&+ \frac{1}{N}\sum_{q=0}^{n_\nu-1} y_{\nu+1}\big((2q+1)m_{\nu+1}+k\big)\,\omega_N^{((2q+1)m_{\nu+1}+k)j} \\
&= c_\nu(k)\,\frac{1}{N}\sum_{q=0}^{n_\nu-1} y_\nu(qm_\nu+k)\,\omega_N^{(qm_\nu+k)j} \\
&+ c_\nu(m_{\nu+1}+k)\,\frac{1}{N}\sum_{q=0}^{n_\nu-1} y_\nu(qm_\nu+m_{\nu+1}+k)\,\omega_N^{(qm_\nu+m_{\nu+1}+k)j} \\
&= c_\nu(k)\,\mu_k^\nu(j)+c_\nu(m_{\nu+1}+k)\,\mu_{m_{\nu+1}+k}^\nu(j).
\end{aligned}$$

The theorem is proved. □

Formula (3.10.1), in particular, yields an inclusion $\mathcal{S}_r^{m_{\nu+1}} \subset \mathcal{S}_r^{m_\nu}$.

3.10.2 Let us construct a nonzero spline $w_k^{\nu+1} \in \mathcal{S}_r^{m_\nu}$, $k \in 0 : m_{\nu+1}-1$, of a form

$$w_k^{\nu+1}(j) = a_\nu(k)\,\mu_k^\nu(j) + a_\nu(m_{\nu+1}+k)\,\mu_{m_{\nu+1}+k}^\nu(j) \tag{3.10.3}$$

being orthogonal to $\mu_k^{\nu+1}$. Since

$$\langle w_k^{\nu+1},\, \mu_k^{\nu+1}\rangle = a_\nu(k)\,c_\nu(k)\,\|\mu_k^\nu\|^2 + a_\nu(m_{\nu+1}+k)\,c_\nu(m_{\nu+1}+k)\,\|\mu_{m_{\nu+1}+k}^\nu\|^2,$$

the condition $\langle w_k^{\nu+1},\, \mu_k^{\nu+1}\rangle = 0$ can be written down with the aid of a determinant of order two:

$$\begin{vmatrix} a_\nu(k) & c_\nu(m_{\nu+1}+k)\,\|\mu_{m_{\nu+1}+k}^\nu\|^2 \\ a_\nu(m_{\nu+1}+k) & -c_\nu(k)\,\|\mu_k^\nu\|^2 \end{vmatrix} = 0. \tag{3.10.4}$$

The second column of the determinant is nonzero, therefore the equality (3.10.4) is possible only if there exists a number $\lambda_\nu(k)$ such that

$$a_\nu(k) = \lambda_\nu(k)\,c_\nu(m_{\nu+1}+k)\,\|\mu_{m_{\nu+1}+k}^\nu\|^2,$$

$$a_\nu(m_{\nu+1}+k) = -\lambda_\nu(k)\,c_\nu(k)\,\|\mu_k^\nu\|^2.$$

Putting

$$\lambda_\nu(m_{\nu+1}+k) = -\lambda_\nu(k), \quad k \in 0 : m_{\nu+1}-1, \tag{3.10.5}$$

we come to a single formula

$$a_\nu(k) = \lambda_\nu(k)\, c_\nu(m_{\nu+1}+k)\, \|\mu^\nu_{m_{\nu+1}+k}\|^2, \tag{3.10.6}$$

$$k \in 0 : m_\nu - 1.$$

Thus, a spline $w_k^{\nu+1}$ of a form (3.10.3) is orthogonal to $\mu_k^{\nu+1}$ if and only if coefficients $a_\nu(k)$ can be represented by (3.10.6), where the numbers $\lambda_\nu(k)$ are of the property (3.10.5). A condition $w_k^{\nu+1}(j) \not\equiv 0$ is equivalent to $\lambda_\nu(k) \neq 0$, $k \in 0 : m_{\nu+1}-1$.

According to (3.10.5) numbers $\rho_\nu(k) = \lambda_\nu(k)\, \omega_{m_\nu}^{-k}$ satisfy to the equality $\rho_\nu(m_{\nu+1}+k) = \rho_\nu(k)$, $k \in 0 : m_{\nu+1}-1$. It means that $\lambda_\nu(k)$ can be represented in a form $\lambda_\nu(k) = \rho_\nu(k)\, \omega_{m_\nu}^k$, where $\{\rho_\nu(k)\}$ is an arbitrary $m_{\nu+1}$-periodic sequence whose members are all nonzero.

We will consider the simplest case of $\rho_\nu(k) \equiv 1$. It corresponds to splines $w_k^{\nu+1}$ of a form (3.10.3) with the coefficients

$$a_\nu(k) = \omega_{m_\nu}^k\, c_\nu(m_{\nu+1}+k)\, \|\mu^\nu_{m_{\nu+1}+k}\|^2. \tag{3.10.7}$$

Formula (3.10.7) lets us consider the splines $w_k^{\nu+1}$ being defined for all integer k. In addition to that, according to (3.10.3), the sequence $\{w_k^{\nu+1}(j)\}$ is $m_{\nu+1}$-periodic on k.

We note that

$$w_0^{\nu+1}(j) = a_\nu(m_{\nu+1})\, \mu^\nu_{m_{\nu+1}}(j) = -2^{2r}\, \|\mu_0^\nu\|^2\, \mu^\nu_{m_{\nu+1}}(j).$$

Lemma 3.10.1 *For all integer numbers l there holds*

$$w_k^{\nu+1}(j + l n_{\nu+1}) = \omega_{m_{\nu+1}}^{kl}\, w_k^{\nu+1}(j). \tag{3.10.8}$$

Proof According to (3.10.3) and (3.8.9) we write

$$\begin{aligned} w_k^{\nu+1}(j + l n_{\nu+1}) &= a_\nu(k)\, \mu_k^\nu(j + 2l n_\nu) + a_\nu(m_{\nu+1}+k)\, \mu^\nu_{m_{\nu+1}+k}(j + 2l n_\nu) \\ &= a_\nu(k)\, \omega_{m_\nu}^{2lk}\, \mu_k^\nu(j) + a_\nu(m_{\nu+1}+k)\, \omega_{m_\nu}^{2l(m_{\nu+1}+k)}\, \mu^\nu_{m_{\nu+1}+k}(j) \\ &= \omega_{m_{\nu+1}}^{kl}\, w_k^{\nu+1}(j). \end{aligned}$$

The lemma is proved. □

Theorem 3.10.2 *The splines* $\{w_k^{\nu+1}\}_{k=0}^{m_{\nu+1}-1}$ *form an orthogonal system. Moreover,*

$$\begin{aligned}\langle w_k^{\nu+1},\ \mu_{k'}^{\nu+1}\rangle &= 0 \text{ for all } k, k' \in 0 : m_{\nu+1}-1,\\ \|w_k^{\nu+1}\| &= \|\mu_k^{\nu}\|\,\|\mu_{m_{\nu+1}+k}^{\nu}\|\,\|\mu_k^{\nu+1}\|.\end{aligned} \tag{3.10.9}$$

Proof The equalities $\langle w_k^{\nu+1},\ w_{k'}^{\nu+1}\rangle = 0$ and $\langle w_k^{\nu+1},\ \mu_{k'}^{\nu+1}\rangle = 0$ for $k \neq k',\ k, k' \in 0 : m_{\nu+1}-1$, follow from (3.10.3), (3.10.1), and orthogonality of the system $\{\mu_k^{\nu}\}_{k=0}^{m_\nu-1}$. The equality $\langle w_k^{\nu+1},\ \mu_k^{\nu+1}\rangle = 0$ is provided by the choice of a spline $w_k^{\nu+1}$. The norm of $w_k^{\nu+1}$ is calculated directly on the basis of formula (3.10.7). Indeed,

$$\begin{aligned}\|w_k^{\nu+1}\|^2 &= |a_\nu(k)|^2\,\|\mu_k^{\nu}\|^2 + |a_\nu(m_{\nu+1}+k)|^2\,\|\mu_{m_{\nu+1}+k}^{\nu}\|^2\\ &= \|\mu_k^{\nu}\|^2\,\|\mu_{m_{\nu+1}+k}^{\nu}\|^2 \Big(|c_\nu(m_{\nu+1}+k)|^2\,\|\mu_{m_{\nu+1}+k}^{\nu}\|^2 + |c_\nu(k)|^2\,\|\mu_k^{\nu}\|^2\Big)\\ &= \|\mu_k^{\nu}\|^2\,\|\mu_{m_{\nu+1}+k}^{\nu}\|^2\,\|\mu_k^{\nu+1}\|^2.\end{aligned}$$

It is remaining to extract the square root. The theorem is proved. □

3.10.3 We denote by $\mathcal{W}_r^{m_{\nu+1}}$ a linear hull spanned by the splines $w_k^{\nu+1}$, $k \in 0 : m_{\nu+1}-1$. As long as each $w_k^{\nu+1}$ belongs to $\mathcal{S}_r^{m_\nu}$, we have $\mathcal{W}_r^{m_{\nu+1}} \subset \mathcal{S}_r^{m_\nu}$. As it was noted earlier, the inclusion $\mathcal{S}_r^{m_{\nu+1}} \subset \mathcal{S}_r^{m_\nu}$ holds as well. According to Theorem 3.10.2 the splines

$$\mu_0^{\nu+1},\ \mu_1^{\nu+1},\ \ldots,\ \mu_{m_{\nu+1}-1}^{\nu+1},\ w_0^{\nu+1},\ w_1^{\nu+1},\ \ldots,\ w_{m_{\nu+1}-1}^{\nu+1}$$

form an orthogonal basis in $\mathcal{S}_r^{m_\nu}$. The space $\mathcal{S}_r^{m_\nu}$ itself can be considered as an orthogonal sum of the subspaces $\mathcal{S}_r^{m_{\nu+1}}$ and $\mathcal{W}_r^{m_{\nu+1}}$, i.e.

$$\mathcal{S}_r^{m_\nu} = \mathcal{S}_r^{m_{\nu+1}} \oplus \mathcal{W}_r^{m_{\nu+1}}. \tag{3.10.10}$$

Applying formula (3.10.10) consecutively for $\nu = 0,\ 1,\ \ldots,\ t-1$, we come to the expansion

$$\begin{aligned}\mathcal{S}_r^{m} &= \mathcal{S}_r^{m_0} = \mathcal{S}_r^{m_1} \oplus \mathcal{W}_r^{m_1} = \big(\mathcal{S}_r^{m_2} \oplus \mathcal{W}_r^{m_2}\big) \oplus \mathcal{W}_r^{m_1}\\ &= \mathcal{S}_r^{m_2} \oplus \big(\mathcal{W}_r^{m_2} \oplus \mathcal{W}_r^{m_1}\big) = \ldots\\ &= \mathcal{S}_r^{m_t} \oplus \mathcal{W}_r^{m_t} \oplus \mathcal{W}_r^{m_{t-1}} \oplus \cdots \oplus \mathcal{W}_r^{m_1}.\end{aligned}$$

Here $\mathcal{S}_r^{m_t} = \mathcal{S}_r^{1}$ is a one-dimensional space consisting of signals that are identically equal to a complex constant (see par. 3.3.1).

Let us formulate the obtained result as a theorem.

Theorem 3.10.3 *A space of discrete periodic splines* $\mathcal{S}_r^m$ *with* $m = 2^t$ *can be decomposed into orthogonal sum*

$$\mathcal{S}_r^{m} = \mathcal{S}_r^{1} \oplus \mathcal{W}_r^{m_t} \oplus \mathcal{W}_r^{m_{t-1}} \oplus \cdots \oplus \mathcal{W}_r^{m_1}.$$

According to this theorem any spline $S \in \mathcal{S}_r^m$ with $m = 2^t$ can be represented as

$$S(j) = \alpha + \sum_{\nu=1}^{t} \sum_{k=0}^{m_\nu - 1} \alpha_\nu(k)\, w_k^\nu(j),$$

where α and $\alpha_\nu(k)$ are complex coefficients.

The subspaces $\mathcal{W}_r^{m_\nu}$, $\nu \in 1 : t$, are called *wavelet* ones.

Formula (3.10.10) shows that $\mathcal{W}_r^{m_{\nu+1}}$ is the orthogonal complement of $\mathcal{S}_r^{m_{\nu+1}}$ to $\mathcal{S}_r^{m_\nu}$.

3.11 First Limit Theorem

3.11.1 We return to the problem of discrete spline interpolation (see the Sect. 3.4). We denote the only spline from the set $\mathcal{S}_r^m$ that satisfies to interpolation conditions

$$S(ln) = z(l), \quad l \in 0 : m - 1,$$

by $S_{r,n}(j)$. By this we emphasize dependency of the interpolating spline on the parameters r and n (with fixed $m \geq 2$). We are interested in behavior of the spline $S_{r,n}(j)$ whether $r \to \infty$ or $n \to \infty$.

In this section we consider the case $r \to \infty$.

3.11.2 Recall that

$$S_{r,n}(j) = \sum_{p=0}^{m-1} c(p)\, Q_r(j - pn), \tag{3.11.1}$$

whereby

$$[\mathcal{F}_m(c)](k) = Z(k)/T_r(k), \quad k \in 0 : m - 1. \tag{3.11.2}$$

Here $Z(k) = [\mathcal{F}_m(z)](k)$ and

$$T_r(k) = \begin{cases} n^{2r-1} & \text{for } k = 0, \\ \dfrac{1}{n} \displaystyle\sum_{s=0}^{n-1} \left(\frac{\sin(\pi k/m)}{\sin(\pi(sm + k)/N)} \right)^{2r} & \text{for } k \in 1 : m - 1. \end{cases} \tag{3.11.3}$$

Let us find the discrete Fourier transform of $S_{r,n}$.

Lemma 3.11.1 *The following formula holds for the spectrum* $X_r = \mathcal{F}_N(S_{r,n})$*:*

$$X_r = Z V_r, \tag{3.11.4}$$

where $Z = \mathcal{F}_m(z)$ *and for* $q \in 0 : n - 1$ *there holds*

$$V_r(qm+l) = \begin{cases} n\,\delta_n(q) & \textit{for } l = 0; \\ n\Bigl(\sum_{s=0}^{n-1}\Bigl(\frac{\sin(\pi(qm+l)/N)}{\sin(\pi(sm+l)/N)}\Bigr)^{2r}\Bigr)^{-1} & \textit{for } l \in 1:m-1. \end{cases} \tag{3.11.5}$$

Proof According to (3.11.1) we have

$$X_r(k) = \sum_{j=0}^{N-1}\Bigl(\sum_{p=0}^{m-1} c(p)\,Q_r(j-pn)\Bigr)\omega_N^{-k(j-pn)-kpn} =$$
$$= \sum_{p=0}^{m-1} c(p)\,\omega_m^{-kp}\sum_{j=0}^{N-1} Q_r(j)\,\omega_N^{-kj} = \bigl[\mathcal{F}_m(c)\bigr](k)\,\bigl[\mathcal{F}_N(Q_r)\bigr](k).$$

Taking into account (3.11.2) we gain

$$X_r(k) = Z(k)V_r(k),$$

where

$$V_r(k) = \bigl[\mathcal{F}_N(Q_r)\bigr](k)/T_r(k). \tag{3.11.6}$$

We will show that the signals V_r can be represented in a form (3.11.5).
As you know (see (3.2.5) and (3.2.2)),

$$[\mathcal{F}_N(Q_r)](k) = \begin{cases} n^{2r} & \text{for } k = 0, \\ \Bigl(\frac{\sin(\pi k/m)}{\sin(\pi k/N)}\Bigr)^{2r} & \text{for } k \in 1:N-1. \end{cases} \tag{3.11.7}$$

For another thing, $T_r(0) = n^{2r-1}$ and $T_r(k) > 0$ for all $k \in \mathbb{Z}$. Therefore,

$$V_r(0) = n \text{ and } V_r(m) = V_r(2m) = \cdots = V_r\bigl((n-1)m\bigr) = 0.$$

This fact can be written as $V_r(qm) = n\delta_n(q)$, $q \in 0:n-1$.

For $k = qm + l$, where $l \neq 0$, m-periodicity of the signal $T_r(k)$ and formulae (3.11.7) and (3.11.3) yield

$$V_r(qm+l) = \frac{[\mathcal{F}_N(Q_r)](qm+l)}{T_r(l)} = \frac{\bigl(\sin(\pi l/m)\bigr)^{2r}\bigl(T_r(l)\bigr)^{-1}}{\bigl(\sin(\pi(qm+l)/N)\bigr)^{2r}}$$
$$= n\Bigl(\sum_{s=0}^{n-1}\Bigl(\frac{\sin(\pi(qm+l)/N)}{\sin(\pi(sm+l)/N)}\Bigr)^{2r}\Bigr)^{-1}.$$

The lemma is proved. □

3.11.3 As it follows from (3.11.6) and (3.11.5), the signal V_r is N-periodic, real, and even. For all natural r the following equalities hold: $V_r(qm) = n\delta_n(q), q \in 0 : n-1$, and

$$\sum_{q=0}^{n-1} V_r(qm + l) = n, \quad l \in 1 : m-1. \tag{3.11.8}$$

Equality (3.11.8) also holds for $l = 0$.

We will show that, for each fixed $k \in \mathbb{Z}$, the sequence $\{V_r(k)\}_{r=1}^{\infty}$ has a limit when $r \to \infty$.

To do so, we introduce a spectrum V_* with the following components:

$$V_*(k) = \begin{cases} n & \text{for } k = 0, 1, \dots, \lfloor (m-1)/2 \rfloor, \\ 0 & \text{for } k = \lfloor m/2 \rfloor + 1, \dots, \lfloor N/2 \rfloor. \end{cases}$$

In case of an even m we additionally put $V_*(m/2) = n/2$. With the aid of the equality $V_*(N-k) = V_*(k)$ we spread V_* onto the whole main period $0 : N-1$. Figure 3.6 depicts the graphs of $V_*(k)$ for $m = 3, n = 3$, and $m = 4, n = 3$.

It is evident that $V_*(qm) = n\,\delta_n(q), q \in 0 : n-1$, because

$$\lfloor m/2 \rfloor + 1 \le m \quad \text{and} \quad (n-1)m \le N - \lfloor m/2 \rfloor - 1.$$

Therefore, for all natural r there holds

$$V_r(qm) = V_*(qm), \qquad q \in 0 : n-1. \tag{3.11.9}$$

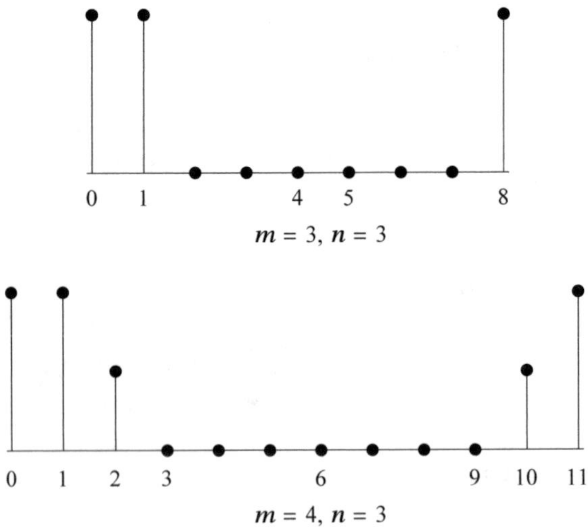

Fig. 3.6 Graphs of the spectrum $V_*(k)$ on the main period

Lemma 3.11.2 *Valid is the limit relation*

$$\lim_{r\to\infty} V_r(k) = V_*(k), \quad k \in \mathbb{Z}. \tag{3.11.10}$$

Proof Let $k = qm + l$, where $q \in 0 : n-1$ and $l \in 0 : m-1$. In case of $l = 0$ the conclusion of the lemma is a trivial consequence of the equality (3.11.9). Thereafter we assume that $l \in 1 : m-1$.

Denote $\alpha_k = \sin^2(k\pi/N)$, $k \in 0 : N-1$. We will need a few properties of these numbers.

1. The biggest value of α_k equals to $\alpha_{\lfloor N/2 \rfloor}$. In case of an odd N it is achieved on two indices $k = (N-1)/2$ and $k = (N+1)/2$; in case of an even N the only critical point is $k = N/2$.
2. For $s \in 1 : n-1$ and $l \in 1 : \lfloor (m-1)/2 \rfloor$ there holds $\alpha_{sm+l} > \alpha_l$. It follows from the inequalities $2l < m$ and

$$l < sm + l \le (n-1)m + l < nm - l$$

   and from the equality $\alpha_l = \alpha_{N-l}$ (see Fig. 3.7).
3. If m is even then $\alpha_{sm+m/2} > \alpha_{m/2}$ holds for $s \in 1 : n-2$. Indeed,

$$m/2 < sm + m/2 \le (n-2)m + m/2 < nm - m/2.$$

Let us verify validity of the limit relation (3.11.10) for $q = 0$ and $l \in 1 : \lfloor (m-1)/2 \rfloor$. According to (3.11.3) we have

$$V_r(l) = n\left(1 + \sum_{s=1}^{n-1}\left(\frac{\alpha_l}{\alpha_{sm+l}}\right)^r\right)^{-1}.$$

On the strength of the property (2) of the numbers α_k we gain

$$\lim_{r\to\infty} V_r(l) = n = V_*(l), \quad l \in 1 : \lfloor (m-1)/2 \rfloor.$$

Further, from (3.11.8) it follows that

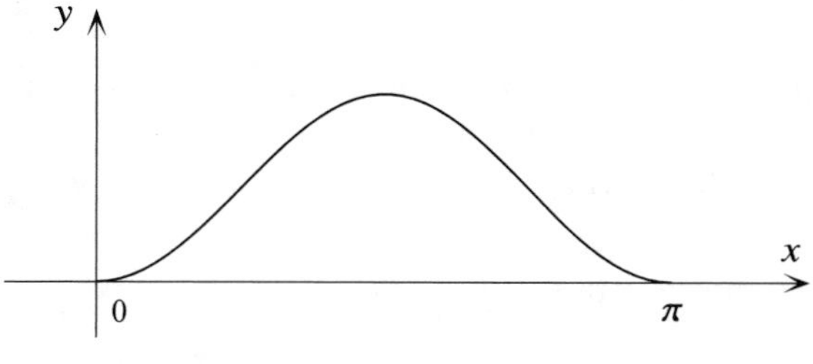

Fig. 3.7 Graph of the function $y = \sin^2 x$ on its period

$$\sum_{q=0}^{n-1} V_r(qm+l) = n, \quad l \in 1 : \lfloor (m-1)/2 \rfloor.$$

All the values $V_r(qm+l)$ are non-negative and $V_r(l) \to n$ when $r \to \infty$, so for $q \in 1 : n-1$ there holds

$$\lim_{r\to\infty} V_r(qm+l) = 0, \quad l \in 1 : \lfloor (m-1)/2 \rfloor.$$

Note that for $q \in 1 : n-1$ and $l \in 1 : \lfloor (m-1)/2 \rfloor$ the index $qm+l$ varies from $m+1$ to $(n-1)m + \lfloor (m-1)/2 \rfloor$. Bearing in mind the equality

$$\lfloor (m-1)/2 \rfloor + \lfloor m/2 \rfloor = m-1$$

we obtain

$$(n-1)m + \lfloor (m-1)/2 \rfloor = N - \lfloor m/2 \rfloor - 1.$$

Furthermore $m+1 > \lfloor m/2 \rfloor + 1$, so

$$\lfloor m/2 \rfloor + 1 < qm + l \le N - \lfloor m/2 \rfloor - 1.$$

By a definition, $V_*(qm+l) = 0$ for the given q and l. Hence

$$\lim_{r\to\infty} V_r(qm+l) = 0 = V_*(qm+l),$$

$$q \in 1 : n-1, \quad l \in 1 : \lfloor (m-1)/2 \rfloor.$$

If m is even, we should additionally consider the case of $l = m/2$ where $V_*(l) = n/2$. According to (3.11.5) we have

$$V_r(\tfrac{m}{2}) = n\left(2 + \sum_{s=1}^{n-2}\left(\frac{\alpha_{m/2}}{\alpha_{sm+m/2}}\right)^r\right)^{-1}.$$

By virtue of the property (3) of the numbers α_k we gain

$$\lim_{r\to\infty} V_r(\tfrac{m}{2}) = \tfrac{n}{2} = V_*(\tfrac{m}{2}).$$

It is also clear that, due to the fact that both signals V_r and V_* are even, there holds

$$\lim_{r\to\infty} V_r\big((n-1)m + \tfrac{m}{2}\big) = \lim_{r\to\infty} V_r(\tfrac{m}{2}) = V_*(\tfrac{m}{2}) = V_*\big((n-1)m + \tfrac{m}{2}\big).$$

According to (3.11.8)

$$\sum_{q=0}^{n-1} V_r(qm + \tfrac{m}{2}) = n.$$

Taking into account the relations $V_r(m/2) \to n/2$ and $V_r\big((n-1)m + m/2\big) \to n/2$ we conclude that

$$\lim_{r\to\infty} V_r(qm + \tfrac{m}{2}) = 0 = V_*(qm + \tfrac{m}{2}), \quad q \in 1 : n-2.$$

Thus, for all $q \in 0 : n-1$ the limit relation

$$\lim_{r\to\infty} V_r(qm + \tfrac{m}{2}) = V_*(qm + \tfrac{m}{2}) \tag{3.11.11}$$

holds.

When $m = 2$, the set $1 : m-1$ consists of a single index $l = 1 = m/2$. In this case the relation (3.11.11) proves the lemma. Thereafter we assume that $m \geq 3$. This guarantees that the set $\lfloor m/2 \rfloor + 1 : m-1$ is not empty.

According to (3.11.8),

$$\sum_{q=0}^{n-1} V_r(qm + l) = n, \quad l \in \lfloor m/2 \rfloor + 1 : m-1. \tag{3.11.12}$$

Furthermore,

$$\lim_{r\to\infty} V_r\big((n-1)m + l\big) = \lim_{r\to\infty} V_r(m-l) = n = V_*(m-l) = V_*\big((n-1)m + l\big).$$

We took into account that the following inequalities hold for the given l:

$$1 \leq m - l \leq \big\lfloor (m-1)/2 \big\rfloor.$$

Now (3.11.12) yields

$$\lim_{r\to\infty} V_r(qm + l) = 0 = V_*(qm + l),$$

$$q \in 0 : n-2, \quad l \in \lfloor m/2 \rfloor + 1 : m-1.$$

Therefore, the limit relation

$$\lim_{r\to\infty} V_r(qm + l) = V_*(qm + l), \quad l \in \lfloor m/2 \rfloor + 1 : m-1,$$

holds for all $q \in 0 : n-1$.

The lemma is proved. □

3.11.4 We introduce the notations

$$X_* = ZV_*, \quad t_{m,n} = \mathcal{F}_N^{-1}(X_*).$$

First Limit Theorem *Valid is the limit relation*

$$\lim_{r\to\infty} S_{r,n}(j) = t_{m,n}(j), \quad j \in \mathbb{Z}. \tag{3.11.13}$$

Proof Using the inversion formula and the fact that $|\omega_N^{kj}| = 1$ holds for all integer k and j we gain

$$\begin{aligned}|S_{r,n}(j) - t_{m,n}(j)| &\le \frac{1}{N}\sum_{k=0}^{N-1} |X_r(k) - X_*(k)| \times |\omega_N^{kj}| \\ &= \frac{1}{N}\sum_{k=0}^{N-1} |Z(k)| \times |V_r(k) - V_*(k)|.\end{aligned}$$

Now the conclusion of the theorem immediately follows from Lemma 3.11.2. □

3.11.5 Let us find out the nature of the limit signal $t_{m,n}$. Consider two cases depending upon whether m is even.

Case of $m = 2\mu - 1$. We have $\lfloor m/2 \rfloor + 1 = \mu$. By a definition, the even signal V_* takes the following values:

$$V_*(k) = \begin{cases} 0 & \text{for } k \in \mu : N-\mu, \\ n & \text{for the other } k \in 0 : N-1. \end{cases}$$

Therefore,

$$X_*(k) = \begin{cases} 0 & \text{for } k \in \mu : N-\mu, \\ nZ(k) & \text{for the other } k \in 0 : N-1. \end{cases}$$

We will use the sampling theorem (see Sect. 2.4 of Chap. 2). According to it there holds

$$t_{m,n}(j) = \sum_{l=0}^{N-1} t_{m,n}(ln)\, h_{m,n}(j - ln), \tag{3.11.14}$$

where

$$h_{m,n}(j) = \frac{1}{m}\sum_{k=-\mu+1}^{\mu-1} \omega_N^{kj}. \tag{3.11.15}$$

We will show that

$$t_{m,n}(ln) = z(l), \quad l \in 0 : m-1. \tag{3.11.16}$$

The inversion formula yields

$$t_{m,n}(ln) = \frac{1}{N}\sum_{k=0}^{N-1} X_*(k)\,\omega_N^{kln} = \frac{1}{m}\Big(\sum_{k=0}^{\mu-1} Z(k)\,\omega_m^{kl} + \sum_{k=N-\mu+1}^{N-1} Z(k)\,\omega_m^{kl}\Big).$$

Replacing the index $k' = k - N + m$ in the second sum and bearing in mind m-periodicity of the spectrum Z we gain

$$t_{m,n}(ln) = \frac{1}{m}\sum_{k=0}^{m-1} Z(k)\,\omega_m^{kl} = z(l).$$

Formula (3.11.16) is ascertained.

Now the expression (3.11.14) takes a form

$$t_{m,n}(j) = \sum_{l=0}^{N-1} z(l)\,h_{m,n}(j - ln). \tag{3.11.17}$$

On the basis of (3.11.17) and (3.11.16) we conclude that the limit signal $t_{m,n}(j)$ is an interpolating trigonometric polynomial defined on a set of integer numbers.

Case of $m = 2\mu$. We have $\lfloor m/2 \rfloor + 1 = \mu + 1$ and

$$V_*(k) = \begin{cases} 0 & \text{for } k \in \mu+1 : N-\mu-1, \\ \dfrac{n}{2} & \text{for } k = \mu \text{ and } k = N-\mu, \\ n & \text{for the other } k \in 0 : N-1. \end{cases}$$

We introduce a signal $h_{m,n} = \mathcal{F}_N^{-1}(V_*)$. By a definition of discrete Fourier transform, the following formula is true:

$$\begin{aligned} h_{m,n}(j) &= \frac{1}{N}\Big[\sum_{k=0}^{\mu-1} n\,\omega_N^{kj} + \frac{n}{2}\big(\omega_N^{\mu j} + \omega_N^{(N-\mu)j}\big) + \sum_{k=N-\mu+1}^{N-1} n\,\omega_N^{kj}\Big] \\ &= \frac{1}{m}\Big[\cos(\pi j/n) + \sum_{k=-\mu+1}^{\mu-1} \omega_N^{kj}\Big]. \end{aligned} \tag{3.11.18}$$

We will show that the representation (3.11.17) still holds for the limit signal $t_{m,n}$, however, unlike with the case of $m = 2\mu - 1$, the kernel $h_{m,n}$ has a form (3.11.18). This statement is equivalent to the following: for signal (3.11.17) there holds $\mathcal{F}_N(t_{m,n}) = X_*$. Let us verify this equality.

We write

$$[\mathcal{F}_N(t_{m,n})](k) = \sum_{j=0}^{N-1} \Big(\sum_{l=0}^{m-1} z(l)\, h_{m,n}(j - ln) \Big) \omega_N^{-k(j-ln)-kln}$$
$$= \sum_{l=0}^{m-1} z(l)\, \omega_m^{-kl} \sum_{j=0}^{N-1} h_{m,n}(j)\, \omega_N^{-kj} = Z(k)\, V_*(k) = X_*(k).$$

This is exactly what had to be ascertained.

Note that for $m = 2\mu$, by virtue of Theorem 2.4.3, the signal of a form (3.11.17) with the kernel (3.11.18) satisfies to the interpolation conditions

$$t_{m,n}(ln) = z(l), \quad l \in 0 : m-1.$$

Let us summarize.

Lemma 3.11.3 *The limit signal $t_{m,n}(j)$ is an interpolating trigonometric polynomial defined on the set of integer numbers. The kernel $h_{m,n}$ in the representation (3.11.17) of this signal has a form (3.11.15) when $m = 2\mu - 1$ and a form (3.11.18) when $m = 2\mu$.*

Section 2.4 of Chap. 2 denotes compact representations for the kernels $h_{m,n}$; namely, (2.4.3) for an odd m and (2.4.7) for an even m.

3.12 Second Limit Theorem

3.12.1 Now we turn to examination of limit behavior of the interpolating spline $S_{r,n}(j)$ in a case of $n \to \infty$ (and fixed r and m). For this end we present $S_{r,n}(j)$ in the following form:

$$S_{r,n}(j) = \sum_{p=0}^{m-1} c(p)\, \widetilde{Q}_{r,n}(j - pn), \tag{3.12.1}$$

where $\widetilde{Q}_{r,n}$ is a normalized B-spline defined by the formula

$$\widetilde{Q}_{r,n}(j) = \frac{1}{n^{2r-1}}\, Q_{r,n}(j), \quad j \in \mathbb{Z}.$$

By virtue of Lemma 3.3.2,

$$\sum_{p=0}^{m-1} \widetilde{Q}_{r,n}(j - pn) \equiv 1.$$

Taking into account non-negativity of $Q_{r,n}(j)$ we conclude that the values of $\widetilde{Q}_{r,n}(j)$ belong to the segment [0, 1].

Let us find out how the normalized B-spline $\widetilde{Q}_{r,n}(j)$ behaves when n is growing.

3.12.2 We introduce a continuous periodic B-spline of the first order. We define it as an m-periodic function of a real argument that is determined on the main period $[0, m]$ by the following formula:

$$B_1(x) = \begin{cases} 1 - x & \text{for } x \in [0, 1], \\ 0 & \text{for } x \in [1, m-1], \\ x - m + 1 & \text{for } x \in [m-1, m]. \end{cases}$$

Periodic B-splines of higher orders are defined with the aid of convolution:

$$B_\nu(x) = \int_0^m B_{\nu-1}(t)\, B_1(x - t)\, dt, \quad \nu = 2, 3, \ldots$$

It is clear that $B_\nu(x) \geq 0$ for any natural ν and real x.

Lemma 3.12.1 *The following identity holds for all natural* ν*:*

$$\sum_{l=0}^{m-1} B_\nu(x - l) \equiv 1. \tag{3.12.2}$$

Proof We will make use of an induction on ν.

Denote the left side of the identity (3.12.2) by $L_\nu(x)$. When $\nu = 1$, the function $L_1(x)$ is a continuous m-periodic polygonal path that is fully identified by its values at the nodes $x \in \{0, 1, \ldots, m\}$. By a definition,

$$B_1(k) = \begin{cases} 1 & \text{for } k = 0 \text{ and } k = m, \\ 0 & \text{for } k \in 1 : m-1, \end{cases}$$

and for $l \in 1 : m - 1$

$$B_1(k - l) = \begin{cases} 1 & \text{for } k = l, \\ 0 & \text{for the others } k \in 0 : m. \end{cases}$$

Hence it follows that $L_1(k) = 1$ for all $k \in 0 : m$. This guarantees validity of the identity (3.12.2) for $\nu = 1$.

We perform an induction step from ν to $\nu + 1$. We have

$$L_{\nu+1}(x) = \sum_{l=0}^{m-1} B_{\nu+1}(x - l) = \sum_{l=0}^{m-1} \int_0^m B_\nu(t)\, B_1(x - l - t)\, dt.$$

For any m-periodic function $f(t)$ that is integrable on the main period $[0, m]$, the following formula holds:

$$\int_0^m f(t-\xi)\,dt = \int_0^m f(t)\,dt \qquad \forall \xi \in \mathbb{R}.$$

Using this formula and the induction hypothesis we gain

$$L_{\nu+1}(x) = \sum_{l=0}^{m-1} \int_0^m B_\nu(t-l)\, B_1(x-t)\,dt = \int_0^m B_1(x-t)\Big(\sum_{l=0}^{m-1} B_\nu(t-l)\Big)dt$$
$$= \int_0^m B_1(x-t)\,dt = \int_0^m B_1(-t)\,dt = 1.$$

The lemma is proved. □

As it was noted, $B_\nu(x) \geq 0$ for all $x \in \mathbb{R}$. Lemma 3.12.1 yields $B_\nu(x) \leq 1$. Thus, values of a B-spline $B_\nu(x)$ belong to the segment $[0, 1]$ for all natural ν and all real x.

Lemma 3.12.2 *A periodic B-spline $B_\nu(x)$ is a $(2\nu-2)$ times continuously differentiable function of a real variable $\big(B_\nu \in C^{2\nu-2}(\mathbb{R})\big)$. On each segment $[k, k+1]$ for an integer k, the spline $B_\nu(x)$ coincides with some algebraic polynomial of order not higher than $2\nu-1$.*

Proof When $\nu = 1$, the stated properties of the B-spline $B_1(x)$ hold. Let us perform an induction step from ν to $\nu+1$.

We have

$$B_{\nu+1}(x) = \int_0^m B_\nu(t)\, B_1(x-t)\,dt = \int_{x-m/2}^{x+m/2} B_\nu(t)\, B_1(x-t)\,dt.$$

By a definition of a B-spline of the first order,

$$B_1(x-t) = \begin{cases} 1-(x-t), & \text{if } (x-t) \in [0,1], \\ 1+(x-t), & \text{if } (x-t) \in [-1,0], \\ 0, & \text{if } 1 \leq |x-t| \leq m/2. \end{cases}$$

Therefore,

$$B_{\nu+1}(x) = \int_{x-1}^{x} B_\nu(t)\,(1-x+t)\,dt + \int_x^{x+1} B_\nu(t)\,(1+x-t)\,dt.$$

Using the definition of a derivative we gain

$$B'_{\nu+1}(x) = -\int_{x-1}^{x} B_\nu(t)\,dt + \int_x^{x+1} B_\nu(t)\,dt. \tag{3.12.3}$$

Yet another differentiation yields

$$B''_{\nu+1}(x) = B_\nu(x-1) + B_\nu(x+1) - 2B_\nu(x). \tag{3.12.4}$$

By the induction hypothesis, $B_\nu \in C^{2\nu-2}(\mathbb{R})$. It follows from (3.12.3) and (3.12.4) that $B_{\nu+1} \in C^{2\nu}(\mathbb{R})$.

If $x \in [k, k+1]$, where k is an integer number, then $(x-1) \in [k-1, k]$ and $(x+1) \in [k+1, k+2]$. According to (3.12.4) and the induction hypothesis, $B''_{\nu+1}(x)$ on the segment $[k, k+1]$ coincides with some algebraic polynomial of order not higher than $2\nu - 1$. Hence $B_{\nu+1}(x)$ on this segment is a polynomial of order not higher than $2\nu + 1$.

The lemma is proved. □

Lemma 3.12.3 *For any natural* ν *and all real* x, y *there holds*

$$|B_\nu(x) - B_\nu(y)| \le |x - y|.$$

Proof For $\nu = 1$ the inequality is evident. Replacing the index $\nu + 1$ by ν in formula (3.12.3) we come to the fact that the derivative $B'_\nu(x)$ equals to a difference of two integrals taken over the segments of unit length, whereby the values of the integrand $B_{\nu-1}(t)$ for $\nu \ge 2$ fall into the segment $[0, 1]$. It is clear that $|B'_\nu(x)| \le 1$. This guarantees validity of the required inequality. □

3.12.3 Let us find out how a continuous periodic B-spline $B_\nu(x)$ and a discrete normalized B-spline $\widetilde{Q}_{r,n}(j)$ are related.

Lemma 3.12.4 *For any given order* ν *there exists a non-negative* A_ν *such that*

$$\left|\widetilde{Q}_{\nu,n}(j) - B_\nu\left(\frac{j}{n}\right)\right| \le \frac{A_\nu}{n} \quad \text{for all } j \in \mathbb{Z} \text{ and } n \ge 2. \tag{3.12.5}$$

Proof For $\nu = 1$ we have $\widetilde{Q}_{1,n}(j) = B_1(\frac{j}{n})$, so we can take $A_1 = 0$.

Let the inequality (3.12.5) hold for some $\nu \ge 1$. We will verify that it holds for the next ν too. We write

$$\begin{aligned}\widetilde{Q}_{\nu+1,n}(j) &= \frac{1}{n}\sum_{k=0}^{N-1} \widetilde{Q}_{\nu,n}(j-k)\,\widetilde{Q}_{1,n}(k) = \frac{1}{n}\sum_{k=0}^{N-1} \widetilde{Q}_{\nu,n}(j-k)\,B_1\left(\frac{k}{n}\right)\\ &= \frac{1}{n}\sum_{k=0}^{N-1}\int_k^{k+1} \widetilde{Q}_{\nu,n}(j-k)\,B_1\left(\frac{k}{n}\right)dt.\end{aligned}$$

Further,

$$\begin{aligned}B_{\nu+1}(x) &= \int_0^m B_\nu(t)\,B_1(x-t)\,dt = \int_0^m B_\nu(t+x)\,B_1(-t)\,dt\\ &= \int_0^m B_\nu\big(x-(m+t)\big)\,B_1(m-t)\,dt = \int_0^m B_\nu(x-t)\,B_1(t)\,dt,\end{aligned}$$

therefore

$$B_{\nu+1}\left(\frac{j}{n}\right) = \int_0^m B_\nu\left(\frac{j}{n} - t\right) B_1(t)\,dt = \frac{1}{n}\int_0^N B_\nu\left(\frac{j-t}{n}\right) B_1\left(\frac{t}{n}\right) dt$$
$$= \frac{1}{n}\sum_{k=0}^{N-1}\int_k^{k+1} B_\nu\left(\frac{j-t}{n}\right) B_1\left(\frac{t}{n}\right) dt.$$

We have

$$\left|\widetilde{Q}_{\nu+1,n}(j) - B_{\nu+1}\left(\frac{j}{n}\right)\right| \le \frac{1}{n}\sum_{k=0}^{N-1}\int_k^{k+1}\left|\widetilde{Q}_{\nu,n}(j-k)\,B_1\left(\frac{k}{n}\right) - B_\nu\left(\frac{j-t}{n}\right) B_1\left(\frac{t}{n}\right)\right| dt$$

$$= \frac{1}{n}\sum_{k=0}^{N-1}\int_k^{k+1}\left|\left(\widetilde{Q}_{\nu,n}(j-k) - B_\nu\left(\frac{j-t}{n}\right)\right) B_1\left(\frac{k}{n}\right) + B_\nu\left(\frac{j-t}{n}\right)\left(B_1\left(\frac{k}{n}\right) - B_1\left(\frac{t}{n}\right)\right)\right| dt.$$

We know that $B_1\left(\frac{k}{n}\right)$ and $B_\nu\left(\frac{j-t}{n}\right)$ do not exceed unity in modulus. Lemma 3.12.3 yields

$$\left|B_1\left(\frac{k}{n}\right) - B_1\left(\frac{t}{n}\right)\right| \le \left|\frac{k-t}{n}\right| \le \frac{1}{n} \quad \text{for } t \in [k, k+1].$$

Finally, by virtue of the induction hypothesis, for the same t we have

$$\left|\widetilde{Q}_{\nu,n}(j-k) - B_\nu\left(\frac{j-t}{n}\right)\right| \le \left|\widetilde{Q}_{\nu,n}(j-k) - B_\nu\left(\frac{j-k}{n}\right)\right| + \left|B_\nu\left(\frac{j-k}{n}\right) - B_\nu\left(\frac{j-t}{n}\right)\right|$$
$$\le \frac{A_\nu}{n} + \left|\frac{t-k}{n}\right| \le \frac{A_\nu + 1}{n}.$$

We come to an inequality

$$\left|\widetilde{Q}_{\nu+1,n}(j-k) - B_{\nu+1}(\frac{j}{n})\right| \le m\,\frac{A_\nu + 2}{n}.$$

To finish the proof of the lemma it is remaining to put $A_{\nu+1} = m(A_\nu + 2)$. □

3.12.4 Recall that the spline $S_{r,n}$ of a form (3.12.1) satisfies to interpolation conditions

$$S_{r,n}(ln) = z(l), \quad l \in 0 : m-1. \tag{3.12.6}$$

Hence, just like in Sect. 3.4, we can obtain an expression for $C_n = \mathcal{F}_m(c_n)$. Let us do it. Denote $Z = \mathcal{F}_m(z)$,

$$g_n(l) = Q_r(ln), \quad \widetilde{g}_n(l) = \widetilde{Q}_r(ln),$$

$$G_n = \mathcal{F}_m(g_n), \quad \widetilde{G}_n = \mathcal{F}_m(\widetilde{g}_n).$$

By a definition of a normalized B-spline, there hold

$$\widetilde{g}_n(l) = \frac{1}{n^{2r-1}}\, g_n(l), \quad \widetilde{G}_n = \frac{1}{n^{2r-1}}\, G_n.$$

Equation (3.12.6) in a spectral domain takes a form (see Sect. 3.4)

$$C_n \widetilde{G}_n = Z.$$

Therefore, $C_n = Z/\widetilde{G}_n$.

Lemma 3.12.5 *A sequence of spectra* $\{\widetilde{G}_n\}$ *converges to a spectrum* $\widetilde{G}$ *with components*

$$\widetilde{G}(k) = \sum_{l=0}^{m-1} B_r(l)\, \omega_m^{-kl}. \tag{3.12.7}$$

Furthermore, $\widetilde{G}(k) > 0$ *for all* $k \in \mathbb{Z}$.

Proof According to Lemma 3.12.4 we have

$$\lim_{n\to\infty} \widetilde{g}_n(l) = \lim_{n\to\infty} \widetilde{Q}_r(ln) = B_r(l).$$

As a consequence,

$$\lim_{n\to\infty} \widetilde{G}_n(k) = \lim_{n\to\infty} \sum_{l=0}^{m-1} \widetilde{g}_n(l)\, \omega_m^{-kl} = \sum_{l=0}^{m-1} B_r(l)\, \omega_m^{-kl} =: \widetilde{G}(k).$$

Let us verify that all components of the limit spectrum $\widetilde{G}$ are positive.

It follows from (3.12.7) and (3.12.2) that

$$\widetilde{G}(0) = \sum_{l=0}^{m-1} B_r(l) = \sum_{l=0}^{m-1} B_r(m-1-l) = 1.$$

For $k \in 1 : m-1$, on the basis of (3.2.7) and (3.2.8) we write

$$\widetilde{G}_n(k) = \Big(\sin\frac{\pi k}{m}\Big)^{2r} \sum_{q=0}^{n-1} \Big(n\, \sin\frac{\pi(qm+k)}{mn}\Big)^{-2r}.$$

Dropping all summands but the one corresponding to $q = 0$ we come to an inequality

$$\widetilde{G}_n(k) \geq \Big(\sin\frac{\pi k}{m}\Big)^{2r} \Big(n\, \sin\frac{\pi k}{mn}\Big)^{-2r}.$$

Since $\sin x < x$ for $x > 0$, we have

$$\widetilde{G}_n(k) \geq \left(\sin\frac{\pi k}{m}\right)^{2r}\left(\frac{\pi k}{m}\right)^{-2r}.$$

Passing to the limit as $n \to \infty$ we gain $\widetilde{G}(k) > 0$, $k \in 1 : m-1$.
The lemma is proved. □

3.12.5 According to Lemma 3.12.5, coefficients $c_n(p)$ of an interpolating spline $S_{r,n}(j)$ of a form (3.12.1) converge as $n \to \infty$. Indeed, the inversion formula yields

$$c_n(p) = \frac{1}{m}\sum_{k=0}^{m-1} C_n(k)\,\omega_m^{kp} = \frac{1}{m}\sum_{k=0}^{m-1}\frac{Z(k)}{\widetilde{G}_n(k)}\,\omega_m^{kp}.$$

Passing to the limit we gain

$$\lim_{n\to\infty} c_n(p) = \frac{1}{m}\sum_{k=0}^{m-1}\frac{Z(k)}{\widetilde{G}(k)}\,\omega_m^{kp} =: c_*(p), \tag{3.12.8}$$

where the denominator $\widetilde{G}(k)$ is defined by formula (3.12.7).
We introduce an m-periodic spline

$$S_r(x) = \sum_{p=0}^{m-1} c_*(p)\,B_r(x-p).$$

It follows from Lemma 3.12.2 that $S_r \in C^{2r-2}(\mathbb{R})$.

Second Limit Theorem *For all real x the following limit relation holds:*

$$\lim_{n\to\infty} S_{r,n}(\lfloor nx\rfloor) = S_r(x). \tag{3.12.9}$$

Proof Fix an arbitrary $x \in \mathbb{R}$. We have

$$\left|S_{r,n}(\lfloor nx\rfloor) - S_r(x)\right| \leq \left|S_r\Big(\frac{\lfloor nx\rfloor}{n}\Big) - S_r(x)\right| + \left|S_{r,n}(\lfloor nx\rfloor) - S_r\Big(\frac{\lfloor nx\rfloor}{n}\Big)\right|. \tag{3.12.10}$$

The first summand in the right side of (3.12.10) vanishes as $n \to \infty$ by virtue of continuity of the spline S_r and by inequalities

$$-\frac{1}{n} \leq \frac{\lfloor nx\rfloor}{n} - x \leq 0.$$

Let us make an estimate of the second summand. We write

$$\left|S_{r,n}(\lfloor nx\rfloor) - S_r\Big(\frac{\lfloor nx\rfloor}{n}\Big)\right| = \left|\sum_{p=0}^{m-1}\Big(c_n(p)\,\widetilde{Q}_{r,n}(\lfloor nx\rfloor - pn) - c_*(p)\,B_r\Big(\frac{\lfloor nx\rfloor}{n} - p\Big)\right|$$

$$\leq \sum_{p=0}^{m-1} \Big(\Big| c_n(p)\Big(\widetilde{Q}_{r,n}\big(\lfloor nx \rfloor - pn\big) - B_r\Big(\frac{\lfloor nx \rfloor}{n} - p\Big)\Big| + \Big|\big(c_n(p) - c_*(p)\big)\, B_r\Big(\frac{\lfloor nx \rfloor}{n} - p\Big)\Big| \Big).$$

Lemma 3.12.4 yields

$$\Big| \Big(\widetilde{Q}_{r,n}\big(\lfloor nx \rfloor - pn\big) - B_r\Big(\frac{\lfloor nx \rfloor}{n} - p\Big)\Big| \leq \frac{A_r}{n}.$$

By virtue of (3.12.8), $c_n(p) \to c_*(p)$ as $n \to \infty$ for all $p \in 0 : m-1$. Hereto we should add that all the sequences $\{c_n(p)\}$ are bounded and $B_r\big(\frac{\lfloor nx \rfloor}{n} - p\big)$ does not exceed unity in modulus. These facts guarantee that the second summand in the right side of (3.12.10) vanishes as $n \to \infty$. Therefore, the limit relation (3.12.9) holds.

The theorem is proved. □

It follows from (3.12.6) and (3.12.9) that

$$S_r(l) = z(l), \quad l \in 0 : m-1.$$

In other words, the limit spline S_r satisfies to the same interpolation conditions as all discrete splines $S_{r,n}$.

Exercises

3.1 Prove that discrete Bernoulli functions are real.

3.2 Prove that the discrete Bernoulli function of the first order $b_1(j)$ can be represented as

$$b_1(j) = \frac{1}{N}\left(\frac{N+1}{2} - j\right) \quad \text{for} \quad j \in 1 : N.$$

3.3 Prove that the discrete Bernoulli function of the second order $b_2(j)$ can be represented as

$$b_2(j) = -\frac{N^2-1}{12N} + \frac{(j-1)(N-j+1)}{2N}$$

for $j \in 1 : N$.

3.4 Prove that for $n = 2$ there holds

$$T_r(l) = 2^{2r-1}\Big[\Big(\cos\frac{\pi l}{N}\Big)^{2r} + \Big(\sin\frac{\pi l}{N}\Big)^{2r}\Big], \quad l \in 0 : m-1.$$

3.5 Prove that for $p, p' \in 0 : m-1$ there holds

$$\big\langle Q_r(\cdot - pn),\, Q_r(\cdot - p'n)\big\rangle = Q_{2r}\big((p - p')n\big).$$

3.6 Prove that

$$\Delta^{2r} Q_r(j) = \sum_{l=-r}^{r} (-1)^{r-l} \binom{2r}{r-l} \delta_N(j + r - ln).$$

3.7 Let $2r(n-1) \le N-2$. Prove that B-spline $Q_r(j)$ has the following properties:

- $Q_r(j) > 0$ for $j \in 0{:}r(n-1)$ and $j \in N - r(n-1){:}N-1$,
- $Q_r(j) = 0$ for $j \in r(n-1)+1 : N - r(n-1) - 1$,
- $Q_r\big(r(n-1)\big) = 1$.

(A statement of the exercise guarantees that the set of arguments where $Q_r(j) = 0$ holds is not empty.)

3.8 Let $N = mn$, $m \ge 2$. We put

$$Q_{1/2}(j) = \frac{1}{N}\left[n + \sum_{k=1}^{N-1} \frac{\sin(\pi k/m)}{\sin(\pi k/N)}\, \omega_N^{kj}\right].$$

Prove that if n is odd then there holds

$$Q_{1/2}(j) = \begin{cases} 1, & \text{for } j \in 0 : (n-1)/2 \\ & \text{and } j \in N - (n-1)/2 : N-1, \\ 0 & \text{for the others } j \in 0 : N-1. \end{cases}$$

3.9 Prove that the signal $\widetilde{Q}_{1/2}(j) = Q_{1/2}(j) - 1/m$ is pure imaginary if n is even.

3.10 Consider a spline $S(j)$ of a form (3.3.1). Let its coefficients $c(p)$ after m-periodic continuation form an even signal. Prove that the spline $S(j)$ is even as well.

3.11 We take a signal $x \in \mathbb{C}_N$ and consider the extremal problem

$$\big\|\Delta^r(x - S)\big\|^2 \to \min,$$

where the minimum is taken among all $S \in \mathcal{S}_r^m$. Prove that a unique (up to an additive constant) solution of this problem is the interpolation spline S_* satisfying to the conditions $S_*(ln) = x(ln)$, $l \in 0 : m-1$.

3.12 Prove that the smoothing spline S_α from par. 3.5.3 is real-valued if the initial data $z(l)$, $l \in 0 : m-1$, are real.

3.13 Prove that the orthogonal spline $\mu_k(j)$ is even with respect to j.

3.14 Formula (3.8.3) defines $\mu_k(j)$ for all $k \in \mathbb{Z}$. Prove that with j being fixed the m-periodic with respect to k sequence $\{\mu_k(j)\}$ is even.

3.15 Construct an orthogonal basis in $\mathcal{S}_r^m$ with $m=n=2$ and $r=1$.

3.16 Consider the expansion (3.8.4) of a spline S over the orthogonal basis. Let the m-periodically continued coefficients $\xi(k)$ form an even signal. Prove that in this case $S(j)$ takes only real values.

3.17 Under the conditions of the previous exercise, let the signal ξ composed of the coefficients of the expansion (3.8.4) be real and even. Prove that this guarantees reality and evenness of the spline $S(j)$.

3.18 Prove that a self-dual spline $\varphi_r(j)$ defined by formula (3.9.7) is real and even.

3.19 Prove that the spline $R_r(j)$ dual to B-spline $Q_r(j)$ is real and even.

3.20 Prove that

$$\sum_{q=0}^{m-1} R_r(j - qn) \equiv n^{-2r}.$$

3.21 Prove that the numbers

$$T_r^{\nu}(k) = \frac{1}{n_\nu} \sum_{q=0}^{n_\nu - 1} y_\nu(q m_\nu + k)$$

(see par. 3.10.1) satisfy to the recurrent relation

$$T_r^{\nu+1}(k) = \tfrac{1}{2}\big[c_\nu(k)\, T_r^{\nu}(k) + c_\nu(m_{\nu+1} + k)\, T_r^{\nu}(m_{\nu+1} + k)\big].$$

3.22 Prove that

$$c_\nu(l) = \sum_{p=-r}^{r} \binom{2r}{r-p} \omega_{m_\nu}^{-lp}.$$

3.23 Prove that B-splines

$$Q_r^{\nu}(j) = \frac{1}{N} \sum_{l=0}^{N-1} y_\nu(l)\, \omega_N^{lj}$$

satisfy to the recurrent relation

$$Q_r^{\nu+1}(j) = \sum_{p=-r}^{r} \binom{2r}{r-p} Q_r^{\nu}(j - p n_\nu).$$

3.24 Prove that an m_ν-periodic on k sequence $\{a_\nu(k)\}$ defined by formula (3.10.7) is even.

3.25 Prove that an $m_{\nu+1}$-periodic on k sequence $\{w_k^{\nu+1}(j)\}$ of a form (3.10.3) with the coefficients (3.10.7) is even.

3.26 Let $w_k^{\nu+1}$ be the splines introduced in par. 3.10.2 and $k \in 1 : m_{\nu+1} - 1,\ k \neq m_{\nu+1}/2$. Prove that there holds

$$\sum_{j=0}^{N-1} \left[w_k^{\nu+1}(j)\right]^2 = 0.$$

3.27 Prove that the spline $w_{m_{\nu+2}}^{\nu+1}(j)$ is real-valued.

3.28 Let a spline φ belong to a wavelet subspace $\mathcal{W}_r^{m_\nu}$. Prove that the system of shifts $\left\{\varphi(\cdot - l n_\nu)\right\}_{l=0}^{m_\nu-1}$ forms a basis in $\mathcal{W}_r^{m_\nu}$ if and only if each coefficient in the expansion of φ over the basis $\left\{w_k^\nu\right\}_{k=0}^{m_\nu-1}$ is nonzero.

3.29 Spline wavelets $\varphi,\ \psi \in \mathcal{W}_r^{m_\nu}$ are called dual if their shifts $\left\{\varphi(\cdot - l n_\nu)\right\}_{l=0}^{m_\nu-1}$ and $\left\{\psi(\cdot - l n_\nu)\right\}_{l=0}^{m_\nu-1}$ are biorthogonal. Prove that φ and ψ are dual if and only if their coefficients $\beta_\nu(k),\ \gamma_\nu(k)$ in the expansions over the basis $\left\{w_k^\nu\right\}_{k=0}^{m_\nu-1}$ satisfy to the condition

$$\beta_\nu(k)\,\overline{\gamma}_\nu(k) = \left[m_\nu\, \|w_k^\nu\|^2\right]^{-1}, \quad k = 0, 1, \ldots, m_\nu - 1.$$

3.30 Let $\nu \in 0 : t - 1$. By analogy with (3.8.11) we introduce a B-wavelet

$$P_r^{\nu+1}(j) = \sum_{k=0}^{m_{\nu+1}-1} w_k^{\nu+1}(j).$$

Prove that

$$P_r^{\nu+1}(j) = \sum_{p=0}^{m_\nu-1} d_\nu(p)\, Q_r^\nu(j - p n_\nu),$$

where $d_\nu = \mathcal{F}_{m_\nu}^{-1}(a_\nu)$ and the sequence $\{a_\nu(k)\}$ is defined by formula (3.10.7).

3.31 Prove that the coefficients $d_\nu(p)$ from the previous exercise can be represented as

$$d_\nu(p) = (-1)^{p+1} \sum_{l=-r}^{r} \binom{2r}{r-l} Q_{2r}^\nu\big((p + l + 1)n_\nu\big).$$

3.32 Calculate $\mathcal{F}_N(P_r^{\nu+1})$.

3.33 Prove that

$$\langle P_r^{\nu+1}(\cdot - l n_{\nu+1}),\ P_r^{\nu+1}(\cdot - l' n_{\nu+1})\rangle = \zeta_r^{\nu+1}(l - l'),$$

where $\zeta_r^{\nu+1}$ is DFT of order $m_{\nu+1}$ of the sequence $\{\|w_k^{\nu+1}\|^2\}$.

3.34 Prove that the spline $P_r^{\nu+1}(j - n_\nu)$ is even with respect to j.

3.35 Prove that a continuous periodic B-spline $B_\nu(x)$ introduced in par. 3.12.2 is an even function.

Comments

Discrete periodic Bernoulli functions are introduced in the paper [4]. Ibid the theorem about expansion of an arbitrary signal over the shifts of Bernoulli functions is proved. This theorem plays an important role in discrete harmonic analysis.

Discrete periodic splines and their numerical applications are the central point of the paper [29]. Piecewise polynomial nature of B-splines is investigated in [28].

Defining a spline as a linear combination of the shifts of a B-spline is a standard procedure. Less standard is an equivalent definition via a linear combination of the shifts of a Bernoulli function (Theorem 3.3.1). The latter definition is essentially used in devising a fundamental relation (3.3.10) which, in turn, is a basis of establishing the minimal norm property (Theorem 3.4.2). In continuous context the minimal norm property is peculiar to natural splines [27].

A solution of the discrete spline interpolation problem (along with the minimal norm property) is obtained in [29]. We note that discrete spline interpolation is used in construction of lifting schemes of wavelet decompositions of signals [33, 34, 53]. Hermite spline interpolation and its applications to computer aided geometric design are considered in [6]. Common approaches to wavelet processing of signals are presented in the monograph [18].

An analysis of the problem of discrete periodic data smoothing is performed within the framework of a common smoothing theory [42]. Along with that, to implement the common approach we utilize the techniques of discrete harmonic analysis to the full extent. We hope that a reader will experience an aesthetic enjoyment while examining this matter.

Formula (3.8.3) defining an orthogonal basis in a space of signals is the beginning of discrete spline harmonic analysis per se. Many problems are dealt with in terms of coefficients of an expansion over the orthogonal basis. In particular, it is these terms that are used to state a criterion of duality of two splines (Theorem 3.9.2). In practical terms, the spline dual to a B-spline helps solving a problem of spline processing of discrete periodic data with the least squares method.

Orthogonal splines are used to obtain a wavelet decomposition of the space of splines.

Sections 3.8–3.10 are written on the basis of the paper [13]. In continuous context a question of orthogonal periodic splines and their applications was considered in [40, 43, 52].

Limit properties of discrete periodic splines are investigated in the papers [19, 20]. The papers [7, 21] are devoted to application of discrete periodic splines to the problems of geometric modeling.

Some of the additional exercises are of interest on their own. For example, the problem 3.23 presents a so-called *calibration relation* for B-splines. A property of an interpolating spline noted in the problem 3.11 is referred to as the *best approximation property*. Problems 3.30–3.34 introduce a notion of B-wavelet and examine some of its properties.

Chapter 4
Fast Algorithms

4.1 Goertzel Algorithm

Let us consider a question of calculating *a single component* of a spectrum $X = \mathcal{F}_N(x)$. We fix $k \in 0 : N-1$ and write down

$$X(k) = \sum_{j=0}^{N-1} x(j)\, \omega_N^{-kj}$$
$$= x(0) + \sum_{j=1}^{N-1} x(j) \cos\Big(\frac{2\pi k}{N} j\Big) - i \sum_{j=1}^{N-1} x(j) \sin\Big(\frac{2\pi k}{N} j\Big).$$

Denote $\alpha = 2\pi k/N$, $c_j = \cos(\alpha j)$, $s_j = \sin(\alpha j)$, and

$$A(k) = \sum_{j=1}^{N-1} x(j)\, c_j, \qquad B(k) = \sum_{j=1}^{N-1} x(j)\, s_j.$$

Then

$$X(k) = x(0) + A(k) - i B(k). \tag{4.1.1}$$

Note that

$$c_j + c_{j-2} = \cos(\alpha j) + \cos\big(\alpha(j-2)\big) = 2\cos\big(\alpha(j-1)\big)\cos(\alpha) = 2\cos(\alpha)\, c_{j-1},$$

$$s_j + s_{j-2} = \sin(\alpha j) + \sin\big(\alpha(j-2)\big) = 2\sin\big(\alpha(j-1)\big)\cos(\alpha) = 2\cos(\alpha)\, s_{j-1}.$$

This induces the recurrent relations that serve as a basis for further transforms:

V. N. Malozemov and S. M. Masharsky, *Foundations of Discrete Harmonic Analysis*, Applied and Numerical Harmonic Analysis,
https://doi.org/10.1007/978-3-030-47048-7_4

$$\begin{aligned} c_j &= 2\cos(\alpha)\, c_{j-1} - c_{j-2}, \;\; j = 2, 3, \ldots, \\ c_0 &= 1, \;\; c_1 = \cos(\alpha); \end{aligned} \tag{4.1.2}$$

$$\begin{aligned} s_j &= 2\cos(\alpha)\, s_{j-1} - s_{j-2}, \;\; j = 2, 3, \ldots, \\ s_0 &= 0, \;\; s_1 = \sin(\alpha). \end{aligned} \tag{4.1.3}$$

In order to calculate $A(k)$ and $B(k)$ we construct a recurrent sequence $\{g_j\}$ basing on conditions

$$\begin{aligned} x(j) &= g_j - 2\cos(\alpha) g_{j+1} + g_{j+2}, \;\; j = N-1, N-2, \ldots, 1, \\ g_{N+1} &= g_N = 0. \end{aligned} \tag{4.1.4}$$

Such a construction is possible. Indeed, for $j = N-1$ we gain $g_{N-1} = x(N-1)$. The values $g_{N-2}, \ldots, g_1$ are determined sequentially from the formula

$$\begin{aligned} g_j &= x(j) + 2\cos(\alpha)\, g_{j+1} - g_{j+2}, \;\; j = N-2, \ldots, 1, \\ g_N &= 0, \;\; g_{N-1} = x(N-1). \end{aligned} \tag{4.1.5}$$

According to (4.1.4) and (4.1.2) we have

$$\begin{aligned} A(k) &= \sum_{j=1}^{N-1} x(j)\, c_j = \sum_{j=1}^{N-1} \big(g_j - 2\cos(\alpha)\, g_{j+1} + g_{j+2}\big)\, c_j \\ &= \sum_{j=1}^{N-1} g_j\, c_j - 2\cos(\alpha) \sum_{j=2}^{N} g_j\, c_{j-1} + \sum_{j=3}^{N+1} g_j\, c_{j-2} \\ &= g_1 c_1 + g_2 c_2 - 2\cos(\alpha)\, g_2 c_1 + \sum_{j=3}^{N-1} g_j \big(c_j - 2\cos(\alpha)\, c_{j-1} + c_{j-2}\big) \\ &= g_1\, c_1 + g_2 \big(c_2 - 2\cos(\alpha)\, c_1\big) = g_1\, c_1 - g_2\, c_0 = g_1 \cos(\alpha) - g_2. \end{aligned}$$

Similarly, with a reference to (4.1.4) and (4.1.3), we convert the expression for $B(k)$:

$$\begin{aligned} B(k) &= \sum_{j=1}^{N-1} x(j)\, s_j = \sum_{j=1}^{N-1} \big(g_j - 2\cos(\alpha)\, g_{j+1} + g_{j+2}\big)\, s_j \\ &= g_1\, s_1 + g_2\, s_2 - 2\cos(\alpha)\, g_2\, s_1 = g_1\, s_1 + g_2 \big(s_2 - 2\cos(\alpha)\, s_1\big) \\ &= g_1\, s_1 = g_1 \sin(\alpha). \end{aligned}$$

Now formula (4.1.1) gets a form

$$X(k) = x(0) + g_1 \cos(\alpha) - g_2 - i\, g_1 \sin(\alpha) = x(0) - g_2 + g_1\, \omega_N^{-k}. \qquad (4.1.6)$$

Calculation of $X(k)$ with a fixed k using formula (4.1.6) is referred to as *Goertzel algorithm*.

The key element of Goertzel algorithm is scheme (4.1.5) that describes construction of a sequence $\{g_j\}$. In fact, we do not need the whole sequence $\{g_j\}$ but only two its elements g_2 and g_1. Achievement of this goal is provided by the following group of operators:

Program Code

```
g := x(N − 1); g1 := 0; a := 2 ∗ cos(2 ∗ π ∗ k / N);
for j := N − 2 downto 1 do
begin g2 := g1; g1 := g;
    g := x(j) + a ∗ g1 − g2 end
```

As the output we obtain $g = g_1$ and $g1 = g_2$. The cycle uses $N - 2$ multiplications on a real number a and $2(N - 2)$ additions.

4.2 First Sequence of Orthogonal Bases

4.2.1 With the aid of Goertzel algorithm, it is possible to calculate the whole spectrum of a signal. However this is not the best way. More effective methods exist that are called *fast Fourier transforms* (FFTs). There are several FFT algorithms, and all of them depend on arithmetic properties of the length of the period N. We will focus on the case of $N = 2^s$.

Our approach to FFT is related to constructing a recurrent sequence of orthogonal bases in a space of signals. This matter is considered in the present section. The next section is devoted to a description of FFT.

4.2.2 In a space $\mathbb{C}_N$ with $N = 2^s$ we will construct a recurrent sequence of orthogonal bases $f_0, f_1, \ldots, f_s$. Here $f_\nu = \{f_\nu(k; j)\}_{k=0}^{N-1}$. A signal $f_\nu(k; j)$ as an element of a space $\mathbb{C}_N$ will be denoted as $f_\nu(k)$. We put $N_\nu = N/2^\nu$ and $\Delta_\nu = 2^{\nu-1}$. A sequence $f_\nu = \{f_\nu(k)\}_{k=0}^{N-1}$, $\nu = 0, 1, \ldots, s$, is defined as follows:

$$f_0(k) = \delta_N(\cdot - k), \quad k \in 0 : N - 1;$$

$$\begin{aligned} f_\nu(l + p\Delta_{\nu+1}) &= f_{\nu-1}(l + 2p\Delta_\nu) + \omega_{\Delta_{\nu+1}}^l\, f_{\nu-1}\big(l + (2p + 1)\Delta_\nu\big), \\ f_\nu(l + \Delta_\nu + p\Delta_{\nu+1}) &= f_{\nu-1}(l + 2p\Delta_\nu) - \omega_{\Delta_{\nu+1}}^l\, f_{\nu-1}\big(l + (2p + 1)\Delta_\nu\big), \end{aligned} \qquad (4.2.1)$$

$$p \in 0 : N_\nu - 1, \quad l \in 0 : \Delta_\nu - 1, \quad \nu = 1, \ldots, s.$$

An index k at $f_\nu(k)$ is represented in a form $k = p\Delta_{\nu+1} + r$, where $p \in 0 : N_\nu - 1$ and $r \in 0 : \Delta_{\nu+1} - 1$. In turn, $r = \sigma\Delta_\nu + l$, where $\sigma \in 0 : 1$ and $l \in 0 : \Delta_\nu - 1$. Thus, $k = p\Delta_{\nu+1} + \sigma\Delta_\nu + l$. Note that $\omega_{\Delta_{\nu+1}}^{\Delta_\nu} = \omega_2 = -1$. This allows writing down the recurrent relations (4.2.1) in a single line:

$$f_\nu(l + \sigma\Delta_\nu + p\Delta_{\nu+1}) = f_{\nu-1}(l + 2p\Delta_\nu) + \omega_{\Delta_{\nu+1}}^{l+\sigma\Delta_\nu} f_{\nu-1}\big(l + (2p+1)\Delta_\nu\big), \tag{4.2.2}$$

$$p \in 0 : N_\nu - 1, \quad l \in 0 : \Delta_\nu - 1, \quad \sigma \in 0 : 1, \quad \nu = 1, \ldots, s.$$

In particular, for $\nu = 1$ we gain

$$f_1(\sigma + 2p) = f_0(2p) + \omega_2^\sigma f_0(2p+1), \tag{4.2.3}$$

$$p \in 0 : N_1 - 1, \quad \sigma \in 0 : 1.$$

How does the basis f_s look like? Answering this question needs some additional preparation.

4.2.3 In Sect. 1.4 we introduced a permutation rev_ν. This permutation is defined on the set $\{0, 1, \ldots, 2^\nu - 1\}$; it maps a number $j = (j_{\nu-1}, j_{\nu-2}, \ldots, j_0)_2$ to a number $\mathrm{rev}_\nu(j) = (j_0, j_1, \ldots, j_{\nu-1})_2$ whose binary code equals to the reverted binary code of the number j. It is assumed by a definition that $\mathrm{rev}_0(0) = 0$.

The following lemma helps to clear up the explicit form of signals $f_\nu(k)$.

Lemma 4.2.1 *Given $q \in 0 : \Delta_\nu - 1$, valid are the equalities*

$$2\,\mathrm{rev}_{\nu-1}(q) = \mathrm{rev}_\nu(q),$$

$$2\,\mathrm{rev}_{\nu-1}(q) + 1 = \mathrm{rev}_\nu(\Delta_\nu + q).$$

Proof When $\nu = 1$, the assertion is trivial. Let $\nu \geq 2$. Then for $q \in 0 : \Delta_\nu - 1$ we have

$$q = q_{\nu-2}2^{\nu-2} + \cdots + q_0, \qquad \mathrm{rev}_{\nu-1}(q) = q_0 2^{\nu-2} + \cdots + q_{\nu-2},$$

$$2\,\mathrm{rev}_{\nu-1}(q) = q_0 2^{\nu-1} + \cdots + q_{\nu-2}\,2 + 0 = \mathrm{rev}_\nu(q),$$

$$2\,\mathrm{rev}_{\nu-1}(q) + 1 = q_0 2^{\nu-1} + \cdots + q_{\nu-2}\,2 + 1 = \mathrm{rev}_\nu(2^{\nu-1} + q) = \mathrm{rev}_\nu(\Delta_\nu + q).$$

The lemma is proved. □

4.2.4 Now we return to the recurrent relations (4.2.1) and (4.2.2).

Theorem 4.2.1 *Valid is the formula*

$$f_\nu(l + p\Delta_{\nu+1}) = \sum_{q=0}^{\Delta_{\nu+1}-1} \omega_{\Delta_{\nu+1}}^{l\,\mathrm{rev}_\nu(q)} f_0(q + p\Delta_{\nu+1}), \tag{4.2.4}$$

$$p \in 0 : N_\nu - 1, \quad l \in 0 : \Delta_{\nu+1} - 1, \quad \nu = 1, \ldots, s.$$

Proof When $\nu = 1$, formula (4.2.4) coincides with (4.2.3) if we replace σ by l there. We perform an induction step from $\nu - 1$ to ν.

Take $l \in 0 : \Delta_{\nu+1} - 1$ and represent it in a form $l = \sigma\Delta_\nu + l'$, where $l' \in 0 : \Delta_\nu - 1$ and $\sigma \in 0 : 1$. According to (4.2.2) and to the inductive hypothesis, for $p \in 0 : N_\nu - 1$ we have

$$\begin{aligned}
f_\nu(l + p\Delta_{\nu+1}) &= f_\nu(l' + \sigma\Delta_\nu + p\Delta_{\nu+1}) \\
&= f_{\nu-1}(l' + 2p\Delta_\nu) + \omega_{\Delta_{\nu+1}}^{l'+\sigma\Delta_\nu} f_{\nu-1}\big(l' + (2p+1)\Delta_\nu\big) \\
&= \sum_{q=0}^{\Delta_\nu-1} \omega_{\Delta_\nu}^{l'\mathrm{rev}_{\nu-1}(q)} f_0(q + 2p\Delta_\nu) \\
&\quad + \omega_{\Delta_{\nu+1}}^{l} \sum_{q=0}^{\Delta_\nu-1} \omega_{\Delta_\nu}^{l'\mathrm{rev}_{\nu-1}(q)} f_0\big(q + (2p+1)\Delta_\nu\big).
\end{aligned}$$

Let us examine the coefficients. Lemma 4.2.1 yields

$$\omega_{\Delta_\nu}^{l'\mathrm{rev}_{\nu-1}(q)} = \omega_{\Delta_\nu}^{(l'+\sigma\Delta_\nu)\mathrm{rev}_{\nu-1}(q)} = \omega_{\Delta_{\nu+1}}^{l(2\mathrm{rev}_{\nu-1}(q))} = \omega_{\Delta_{\nu+1}}^{l\,\mathrm{rev}_\nu(q)},$$

$$\omega_{\Delta_{\nu+1}}^{l}\omega_{\Delta_\nu}^{l'\mathrm{rev}_{\nu-1}(q)} = \omega_{\Delta_{\nu+1}}^{l(2\mathrm{rev}_{\nu-1}(q)+1)} = \omega_{\Delta_{\nu+1}}^{l\,\mathrm{rev}_\nu(q+\Delta_\nu)}.$$

Hence

$$\begin{aligned}
f_\nu(l + p\Delta_{\nu+1}) &= \sum_{q=0}^{\Delta_\nu-1} \omega_{\Delta_{\nu+1}}^{l\,\mathrm{rev}_\nu(q)} f_0(q + p\Delta_{\nu+1}) \\
&\quad + \sum_{q=0}^{\Delta_\nu-1} \omega_{\Delta_{\nu+1}}^{l\,\mathrm{rev}_\nu(q+\Delta_\nu)} f_0\big((q + \Delta_\nu) + p\Delta_{\nu+1}\big) \\
&= \sum_{q=0}^{\Delta_{\nu+1}-1} \omega_{\Delta_{\nu+1}}^{l\,\mathrm{rev}_\nu(q)} f_0(q + p\Delta_{\nu+1}).
\end{aligned}$$

The theorem is proved. □

When $\nu = s$, formula (4.2.4) takes a form

$$f_s(l;\, j) = \sum_{q=0}^{N-1} \omega_N^{l\,\mathrm{rev}_s(q)}\, \delta_N(j-q)$$
$$= \omega_N^{l\,\mathrm{rev}_s(j)} = u_l\big(\mathrm{rev}_s(j)\big), \quad l, j \in 0 : N-1. \tag{4.2.5}$$

We deduced that f_s is an exponential basis with a reverted argument.

4.2.5 Consider the signals $f_\nu(k)$ given a fixed ν.

Theorem 4.2.2 *For each $\nu \in 0 : s$ the system of signals*

$$f_\nu(0),\ f_\nu(1),\ \ldots,\ f_\nu(N-1) \tag{4.2.6}$$

is orthogonal and $\|f_\nu(k)\|^2 = 2^\nu$ holds for all $k \in 0 : N-1$.

Proof The assertion is known to be true for $\nu = 0$ (the corollary to Lemma 2.1.4); therefore, we assume that $\nu \in 1 : s$. We take $k, k' \in 0 : N-1$ and represent them in a form $k = l + p\Delta_{\nu+1}$, $k' = l' + p'\Delta_{\nu+1}$, where $l, l' \in 0 : \Delta_{\nu+1} - 1$ and $p, p' \in 0 : N_\nu - 1$. Bearing in mind formula (4.2.4), the definition of signals $f_0(k)$ and Lemma 2.1.4, we write

$$\langle f_\nu(k),\ f_\nu(k')\rangle = \langle f_\nu(l + p\Delta_{\nu+1}),\ f_\nu(l' + p'\Delta_{\nu+1})\rangle$$
$$= \Big\langle \sum_{q=0}^{\Delta_{\nu+1}-1} \omega_{\Delta_{\nu+1}}^{l\,\mathrm{rev}_\nu(q)}\, f_0(q + p\Delta_{\nu+1}),\ \sum_{q'=0}^{\Delta_{\nu+1}-1} \omega_{\Delta_{\nu+1}}^{l'\mathrm{rev}_\nu(q')}\, f_0(q' + p'\Delta_{\nu+1})\Big\rangle$$
$$= \sum_{q=0}^{\Delta_{\nu+1}-1} \sum_{q'=0}^{\Delta_{\nu+1}-1} \omega_{\Delta_{\nu+1}}^{l\,\mathrm{rev}_\nu(q) - l'\mathrm{rev}_\nu(q')}\, \delta_N\big(q - q' + (p-p')\Delta_{\nu+1}\big).$$

The argument of the unit pulse δ_N does not exceed $N-1$ in absolute value. When $p \neq p'$, it is other than zero for all $q, q' \in 0 : \Delta_{\nu+1} - 1$ because $|q - q'| \leq \Delta_{\nu+1} - 1$. Hence $\langle f_\nu(k),\ f_\nu(k')\rangle = 0$ for $p \neq p'$.

Let $p = p'$. Then

$$\langle f_\nu(k),\ f_\nu(k')\rangle = \sum_{q=0}^{\Delta_{\nu+1}-1} \omega_{\Delta_{\nu+1}}^{(l-l')\mathrm{rev}_\nu(q)}$$
$$= \sum_{q'=0}^{\Delta_{\nu+1}-1} \omega_{\Delta_{\nu+1}}^{(l-l')q'} = \Delta_{\nu+1}\, \delta_{\Delta_{\nu+1}}(l - l'). \tag{4.2.7}$$

We used formula (2.2.1) and the fact that the mapping $q \to \mathrm{rev}_\nu(q)$ is a permutation of the set $\{0, 1, \ldots, \Delta_{\nu+1} - 1\}$. On the basis of (4.2.7) we conclude that the scalar product $\langle f_\nu(k),\ f_\nu(k')\rangle$ is nonzero only when $p = p'$ and $l = l'$, i.e. only when

$k = k'$. In the latter case $\| f_\nu(k)\|^2 = \Delta_{\nu+1} = 2^\nu$ for all $k \in 0 : N-1$. The theorem is proved. □

Essentially, we ascertained that for each $\nu \in 0 : s$ the system of signals (4.2.6) forms an orthogonal basis in a space $\mathbb{C}_N$.

4.2.6 Let us show that the signals $f_\nu(l + p\Delta_{\nu+1})$ given some fixed ν, l and $p \in 0 : N_\nu - 1$ differ from $f_\nu(l)$ only by a shift of an argument.

Theorem 4.2.3 *Given* $l \in 0 : \Delta_{\nu+1} - 1$, *valid is the identity*

$$f_\nu(l + p\Delta_{\nu+1};\ j) \equiv f_\nu(l;\ j - p\Delta_{\nu+1}), \quad p \in 0 : N_\nu - 1. \tag{4.2.8}$$

Proof Let us write down formula (4.2.4) for $p = 0$:

$$f_\nu(l;\ j) = \sum_{q=0}^{\Delta_{\nu+1}-1} \omega_{\Delta_{\nu+1}}^{l\,\mathrm{rev}_\nu(q)}\, \delta_N(j - q).$$

Hence it follows that

$$\begin{aligned} f_\nu(l;\ j - p\Delta_{\nu+1}) &= \sum_{q=0}^{\Delta_{\nu+1}-1} \omega_{\Delta_{\nu+1}}^{l\,\mathrm{rev}_\nu(q)}\, \delta_N(j - p\Delta_{\nu+1} - q) \\ &= \sum_{q=0}^{\Delta_{\nu+1}-1} \omega_{\Delta_{\nu+1}}^{l\,\mathrm{rev}_\nu(q)}\, f_0(q + p\Delta_{\nu+1};\ j) = f_\nu(l + p\Delta_{\nu+1};\ j). \end{aligned}$$

The theorem is proved. □

4.3 Fast Fourier Transform

4.3.1 In the previous section we constructed $s + 1$ orthogonal bases $f_0, f_1, \ldots, f_s$ in a space $\mathbb{C}_N$ with $N = 2^s$. Let us take a signal $x \in \mathbb{C}_N$. It can be expanded over any of these bases. Bearing in mind our final goal, we will expand a signal $x_0(j) = x\big(\mathrm{rev}_s(j)\big)$, $j \in 0 : N-1$:

$$x_0 = \frac{1}{2^\nu} \sum_{k=0}^{N-1} x_\nu(k)\, f_\nu(k). \tag{4.3.1}$$

To determine the coefficients $x_\nu(k)$, we multiply both sides of (4.3.1) scalarly by $f_\nu(l)$. According to Theorem 4.2.2 we gain $\langle x_0,\ f_\nu(l)\rangle = x_\nu(l)$, so that

$$x_\nu(k) = \sum_{j=0}^{N-1} x_0(j)\,\overline{f_\nu}(k;\ j) = \sum_{j=0}^{N-1} x\big(\mathrm{rev}_s(j)\big)\,\overline{f_\nu}(k;\ j). \tag{4.3.2}$$

In particular,

$$x_0(k) = \sum_{j=0}^{N-1} x\big(\mathrm{rev}_s(j)\big)\,\delta_N(j-k) = x\big(\mathrm{rev}_s(k)\big).$$

The recurrent relation (4.2.1) yields

$$\begin{aligned} x_\nu(l + p\Delta_{\nu+1}) &= \langle x_0,\ f_\nu(l + p\Delta_{\nu+1})\rangle \\ &= \big\langle x_0,\ f_{\nu-1}(l + 2p\Delta_\nu) + \omega^{l}_{\Delta_{\nu+1}}\, f_{\nu-1}\big(l + (2p+1)\Delta_\nu\big)\big\rangle \\ &= x_{\nu-1}(l + 2p\Delta_\nu) + \omega^{-l}_{\Delta_{\nu+1}}\, x_{\nu-1}\big(l + (2p+1)\Delta_\nu\big). \end{aligned}$$

Similarly,

$$x_\nu(l + \Delta_\nu + p\Delta_{\nu+1}) = x_{\nu-1}(l + 2p\Delta_\nu) - \omega^{-l}_{\Delta_{\nu+1}}\, x_{\nu-1}\big(l + (2p+1)\Delta_\nu\big).$$

We come to a recurrent scheme

$$x_0(k) = x\big(\mathrm{rev}_s(k)\big), \quad k \in 0 : N-1;$$

$$\begin{aligned} x_\nu(l + p\Delta_{\nu+1}) &= x_{\nu-1}(l + 2p\Delta_\nu) + \omega^{-l}_{\Delta_{\nu+1}}\, x_{\nu-1}\big(l + (2p+1)\Delta_\nu\big), \\ x_\nu(l + \Delta_\nu + p\Delta_{\nu+1}) &= x_{\nu-1}(l + 2p\Delta_\nu) - \omega^{-l}_{\Delta_{\nu+1}}\, x_{\nu-1}\big(l + (2p+1)\Delta_\nu\big), \end{aligned} \tag{4.3.3}$$

$$p \in 0 : N_\nu - 1, \quad l \in 0 : \Delta_\nu - 1, \quad \nu = 1, \ldots, s.$$

Along this scheme we calculate the coefficients of expansion of the signal x_0 over all bases f_ν right up to f_s. Note that according to (4.3.2) and (4.2.5) we have

$$x_s(k) = \sum_{j=0}^{N-1} x\big(\mathrm{rev}_s(j)\big)\,\omega_N^{-k\,\mathrm{rev}_s(j)} = \sum_{j'=0}^{N-1} x(j')\,\omega_N^{-kj'} = X(k).$$

Thus, the coefficients $x_s(k)$ are nothing else but the spectral components of the signal x on the main period.

Calculations with formula (4.3.3) require $\sum_{\nu=1}^{s} N_\nu \Delta_\nu = \frac{1}{2}\sum_{\nu=1}^{s} N_\nu \Delta_{\nu+1} = \frac{1}{2}\, sN = \frac{1}{2}\, N \log_2 N$ multiplications and $2\sum_{\nu=1}^{s} N_\nu \Delta_\nu = N \log_2 N$ additions.

Scheme (4.3.3) is one of the versions of the fast Fourier transform for $N = 2^s$. It is referred to as the *decimation-in-time Cooley–Tukey algorithm*.

4.3.2 Formula (4.3.3) can be inverted:

$$x_s(k) = X(k), \quad k \in 0 : N-1;$$

$$\begin{aligned} x_{\nu-1}(l + 2p\Delta_\nu) &= \tfrac{1}{2}\big[x_\nu(l + p\Delta_{\nu+1}) + x_\nu(l + \Delta_\nu + p\Delta_{\nu+1})\big], \\ x_{\nu-1}\big(l + (2p+1)\Delta_\nu\big) &= \tfrac{1}{2}\,\omega^{l}_{\Delta_{\nu+1}}\big[x_\nu(l+p\Delta_{\nu+1}) - x_\nu(l+\Delta_\nu+p\Delta_{\nu+1})\big], \end{aligned} \tag{4.3.4}$$

$$p \in 0 : N_\nu - 1, \quad l \in 0 : \Delta_\nu - 1, \quad \nu = s, s-1, \ldots, 1.$$

Along formula (4.3.4) we descend down to $x_0(k) = x\big(\mathrm{rev}_s(k)\big)$. Replacing k by $\mathrm{rev}_s(k)$ we obtain

$$x(k) = x_0\big(\mathrm{rev}_s(k)\big), \quad k \in 0 : N-1.$$

Thereby we have pointed out the fast algorithm of reconstructing a signal x from its spectrum $X = \mathcal{F}_N(x)$ for $N = 2^s$.

4.4 Wavelet Bases

4.4.1 We rewrite formula (4.2.2) of transition from the basis $f_{\nu-1}$ to the basis f_ν:

$$f_\nu(l + \sigma\Delta_\nu + p\Delta_{\nu+1}) = f_{\nu-1}(l + 2p\Delta_\nu) + \omega_{\Delta_{\nu+1}}^{l+\sigma\Delta_\nu} f_{\nu-1}\big(l + (2p+1)\Delta_\nu\big), \tag{4.4.1}$$

$$p \in 0 : N_\nu - 1, \quad l \in 0 : \Delta_\nu - 1, \quad \sigma \in 0 : 1, \quad \nu = 1, \ldots, s.$$

Let us analyze the structure of this formula. It is convenient to assume that the basis $f_{\nu-1}$ is divided into Δ_ν blocks; the blocks are marked with an index l. Each block contains $N_{\nu-1}$ signals with inner indices $2p$ and $2p+1$ for $p \in 0 : N_\nu - 1$. According to (4.4.1) a block with an index l generates two blocks of the basis f_ν with indices l and $l + \Delta_\nu$, herein each block contains N_ν signals with an inner index p. The complete scheme of branching for $N = 2^3$ is presented in Fig. 4.1. By virtue of Theorem 4.2.2 all the bases $f_0, f_1, \ldots, f_s$ are orthogonal. According to Theorem 4.2.3 the signals of each block of the basis f_ν differ only by shifts of an argument; the length of a shift is a multiple of $\Delta_{\nu+1}$.

We can significantly increase the number of orthogonal bases in $\mathbb{C}_N$ with $N = 2^s$ if we use a vertical constituent. This is demonstrated in Fig. 4.2. The squares denote blocks. A number inside a square shows how many signals are there in this block. Hanging blocks are distinguished by double squares.

In all four variants of branching, unions of signals contained in hanging blocks form orthogonal bases.

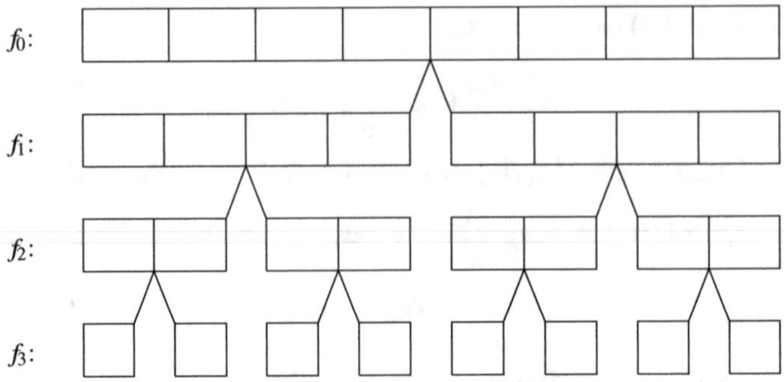

Fig. 4.1 Branching scheme for $N = 2^3$

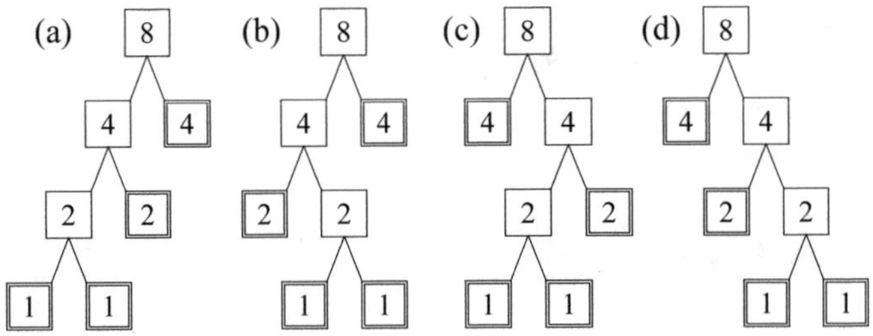

Fig. 4.2 Wavelet bases

Indeed, orthogonality of signals of the ν-th level is known. Signals of a hanging block of the $(\nu + 1)$-th level are linear combinations of signals of some block of the ν-th level, and therefore they are orthogonal to signals from another block of the ν-th level.

An orthogonal basis made of hanging blocks of all levels $\nu = 1, \ldots, s$ will be referred to as a *wavelet basis*, and the collection of all such bases will be referred to as a *wavelet packet*.

Below we will study in detail the wavelet basis generated by the branching scheme (a) in Fig. 4.2.

4.4.2 When $l = 0$, formula (4.2.1) gets a form

$$f_0(k) = \delta_N(\cdot - k), \quad k \in 0 : N - 1;$$

$$\begin{aligned} f_\nu(p\Delta_{\nu+1}) &= f_{\nu-1}(2p\Delta_\nu) + f_{\nu-1}\big((2p+1)\Delta_\nu\big), \\ f_\nu(\Delta_\nu + p\Delta_{\nu+1}) &= f_{\nu-1}(2p\Delta_\nu) - f_{\nu-1}\big((2p+1)\Delta_\nu\big), \end{aligned} \tag{4.4.2}$$

$$p \in 0 : N_\nu - 1, \quad \nu = 1, \ldots, s.$$

The signals $\{f_\nu(\Delta_\nu + p\Delta_{\nu+1})\}$ will enter a wavelet basis, while the signals $\{f_\nu(p\Delta_{\nu+1})\}$ will participate in further branching. The recurrent relations (4.4.2) can be written down in a single line:

$$f_\nu(\sigma\Delta_\nu + p\Delta_{\nu+1}) = f_{\nu-1}(2p\Delta_\nu) + (-1)^\sigma f_{\nu-1}\big((2p+1)\Delta_\nu\big), \tag{4.4.3}$$

$$p \in 0 : N_\nu - 1, \quad \sigma \in 0 : 1, \quad \nu = 1, \dots, s.$$

We introduce linear spans

$$V_\nu = \operatorname{lin}\Big(\{f_\nu(p\Delta_{\nu+1})\}_{p=0}^{N_\nu-1}\Big), \quad \nu = 0, 1, \dots, s;$$

$$W_\nu = \operatorname{lin}\Big(\{f_\nu(\Delta_\nu + p\Delta_{\nu+1})\}_{p=0}^{N_\nu-1}\Big), \quad \nu = 1, \dots, s.$$

It is evident that $V_0 = \operatorname{lin}\Big(\{\delta_N(\cdot - p)\}_{p=0}^{N-1}\Big) = \mathbb{C}_N$. On the strength of (4.4.2) we have $V_\nu \subset V_{\nu-1}$ and $W_\nu \subset V_{\nu-1}$. Since the signals

$$\{f_\nu(p\Delta_{\nu+1})\}_{p=0}^{N_\nu-1}, \quad \{f_\nu(\Delta_\nu + p\Delta_{\nu+1})\}_{p=0}^{N_\nu-1}$$

belong to $V_{\nu-1}$ and are pairwise orthogonal, and their total amount coincides with the dimension of $V_{\nu-1}$, we conclude that they form an orthogonal basis of $V_{\nu-1}$. Moreover, $V_{\nu-1}$ is an orthogonal sum of V_ν and W_ν, i.e.

$$V_{\nu-1} = V_\nu \oplus W_\nu, \quad \nu = 1, \dots, s. \tag{4.4.4}$$

This formula corresponds to a branching step. Consequently applying (4.4.4) we come to an orthogonal decomposition of the space $\mathbb{C}_N$:

$$\begin{aligned} \mathbb{C}_N = V_0 &= V_1 \oplus W_1 = (V_2 \oplus W_2) \oplus W_1 = \dots \\ &= V_s \oplus W_s \oplus W_{s-1} \oplus \dots \oplus W_2 \oplus W_1. \end{aligned} \tag{4.4.5}$$

Here $V_s = \operatorname{lin}\big(f_s(0)\big)$. According to (4.2.5) there holds $f_s(0; j) \equiv 1$, so V_s is a subspace of signals that are identically equal to a complex constant.

Subspaces W_ν are referred to as *wavelet subspaces*. Identity (4.2.8) yields

$$f_\nu(\Delta_\nu + p\Delta_{\nu+1};\ j) = f_\nu(\Delta_\nu;\ j - p\Delta_{\nu+1}),$$

$$p \in 0 : N_\nu - 1, \quad \nu = 1, \dots, s.$$

This means that the basis of W_ν consists of shifts of the signal $f_\nu(\Delta_\nu;\ j)$; the shifts are multiples of $\Delta_{\nu+1}$.

Theorem 4.4.1 *Given $\nu \in 1 : s$, valid is the formula*

$$f_\nu(\Delta_\nu;\ j) = \begin{cases} 1 & \text{for} \quad j \in 0 : \Delta_\nu - 1, \\ -1 & \text{for} \quad j \in \Delta_\nu : \Delta_{\nu+1} - 1, \\ 0 & \text{for} \quad j \in \Delta_{\nu+1} : N - 1. \end{cases} \tag{4.4.6}$$

Proof On the basis of (4.2.4) we write

$$\begin{aligned} f_\nu(\Delta_\nu;\ j) &= \sum_{q=0}^{\Delta_{\nu+1}-1} \omega_{\Delta_{\nu+1}}^{\Delta_\nu \mathrm{rev}_\nu(q)}\, \delta_N(j-q) \\ &= \sum_{q=0}^{\Delta_\nu - 1} \omega_2^{\mathrm{rev}_\nu(q)}\, \delta_N(j-q) + \sum_{q=\Delta_\nu}^{\Delta_{\nu+1}-1} \omega_2^{\mathrm{rev}_\nu(q)}\, \delta_N(j-q). \end{aligned} \tag{4.4.7}$$

For $\nu = 1$ we have $f_1(\Delta_1;\ j) = \delta_N(j) - \delta_N(j-1)$, which corresponds to (4.4.6).

Let $\nu \geq 2$. When $q \in 0 : \Delta_\nu - 1$, Lemma 4.2.1 yields $\mathrm{rev}_\nu(q) = 2\,\mathrm{rev}_{\nu-1}(q)$, therefore $\omega_2^{\mathrm{rev}_\nu(q)} = (-1)^{2\,\mathrm{rev}_{\nu-1}(q)} = 1$. When $q \in \Delta_\nu : \Delta_{\nu+1} - 1$, it can be represented in a form $q = \Delta_\nu + q'$, where $q' \in 0 : \Delta_\nu - 1$. Lemma 4.2.1 now yields $\mathrm{rev}_\nu(q) = \mathrm{rev}_\nu(\Delta_\nu + q') = 2\,\mathrm{rev}_{\nu-1}(q') + 1$, so that $\omega_2^{\mathrm{rev}_\nu(q)} = (-1)^{2\,\mathrm{rev}_{\nu-1}(q')+1} = -1$. Substituting the obtained expressions for the coefficients into (4.4.7) we gain

$$f_\nu(\Delta_\nu;\ j) = \sum_{q=0}^{\Delta_\nu - 1} \delta_N(j-q) - \sum_{q=\Delta_\nu}^{\Delta_{\nu+1}-1} \delta_N(j-q). \tag{4.4.8}$$

This conforms to (4.4.6). The theorem is proved. □

4.5 Haar Basis. Fast Haar Transform

4.5.1 According to (4.4.5) the signals

$$f_s(0);\ \ f_\nu(\Delta_\nu + p\Delta_{\nu+1}),\ \ p \in 0 : N_\nu - 1,\ \ \nu = s, s-1, \ldots, 1 \tag{4.5.1}$$

form an orthogonal basis in a space $\mathbb{C}_N$ with $N = 2^s$. It is referred to as the *discrete Haar basis related to decimation in time*. Figure 4.3 depicts the Haar basis for $N = 2^3$.

Any signal $x \in \mathbb{C}_N$ can be expanded over basis (4.5.1):

$$x = 2^{-s}\, x_s(0)\, f_s(0) + \sum_{\nu=1}^{s} 2^{-\nu} \sum_{p=0}^{N_\nu - 1} x_\nu(\Delta_\nu + p\Delta_{\nu+1})\, f_\nu(\Delta_\nu + p\Delta_{\nu+1}). \tag{4.5.2}$$

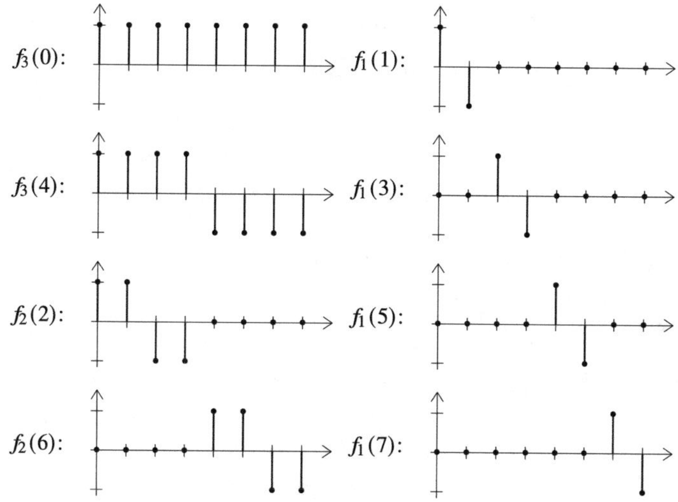

Fig. 4.3 Haar basis related to decimation in time for $N = 2^3$

In order to simplify the indexing we introduce the following notations:

$$\varphi_0(p) = f_0(p) = \delta_N(\cdot - p), \quad p \in 0 : N - 1;$$

$$\varphi_\nu(p + \sigma N_\nu) = f_\nu(\sigma \Delta_\nu + p\Delta_{\nu+1}),$$

$$p \in 0 : N_\nu - 1, \quad \sigma \in 0 : 1, \quad \nu = 1, \ldots, s.$$

In particular, $\varphi_s(\sigma) = f_s(\sigma \Delta_s)$ for $\sigma \in 0 : 1$. For $s = 3$ signals (4.5.1)

$$f_3(0),\ f_3(4),\ f_2(2),\ f_2(6),\ f_1(1),\ f_1(3),\ f_1(5),\ f_1(7)$$

shown in Fig. 4.3 coincide with the signals

$$\varphi_3(0),\ \varphi_3(1),\ \varphi_2(2),\ \varphi_2(3),\ \varphi_1(4),\ \varphi_1(5),\ \varphi_1(6),\ \varphi_1(7).$$

According to (4.4.3) we have

$$\varphi_\nu(p + \sigma N_\nu) = \varphi_{\nu-1}(2p) + (-1)^\sigma\, \varphi_{\nu-1}(2p + 1).$$

We come to the recurrent relations

$$\varphi_0(p) = \delta_N(\cdot - p), \quad p \in 0 : N - 1;$$

$$\begin{aligned}\varphi_\nu(p) &= \varphi_{\nu-1}(2p) + \varphi_{\nu-1}(2p+1),\\ \varphi_\nu(p+N_\nu) &= \varphi_{\nu-1}(2p) - \varphi_{\nu-1}(2p+1),\end{aligned} \tag{4.5.3}$$

$$p \in 0 : N_\nu - 1, \quad \nu = 1, \ldots, s.$$

In our new notations, Haar basis (4.5.1) will be constituted of the signals

$$\varphi_s(0);\ \varphi_\nu(p+N_\nu),\ \ p \in 0 : N_\nu - 1,\ \ \nu = s, s-1, \ldots, 1.$$

We put $\xi_\nu(k) = \langle x,\ \varphi_\nu(k)\rangle$. In particular,

$$\xi_0(p) = \langle x,\ \varphi_0(p)\rangle = \langle x,\ f_0(p)\rangle = x(p),$$

$$\xi_s(0) = \langle x,\ \varphi_s(0)\rangle = \langle x,\ f_s(0)\rangle = x_s(0),$$

$$\xi_\nu(p+N_\nu) = \langle x,\ \varphi_\nu(p+N_\nu)\rangle = \langle x,\ f_\nu(\Delta_\nu + p\Delta_{\nu+1})\rangle = x_\nu(\Delta_\nu + p\Delta_{\nu+1}).$$

On the basis of (4.5.3) we gain

$$\xi_0(p) = x(p), \quad p \in 0 : N-1;$$

$$\begin{aligned}\xi_\nu(p) &= \xi_{\nu-1}(2p) + \xi_{\nu-1}(2p+1),\\ \xi_\nu(p+N_\nu) &= \xi_{\nu-1}(2p) - \xi_{\nu-1}(2p+1),\end{aligned} \tag{4.5.4}$$

$$p \in 0 : N_\nu - 1, \quad \nu = 1, \ldots, s.$$

Formula (4.5.2) takes a form

$$x = 2^{-s}\,\xi_s(0)\,\varphi_s(0) + \sum_{\nu=1}^{s} 2^{-\nu} \sum_{p=0}^{N_\nu-1} \xi_\nu(p+N_\nu)\,\varphi_\nu(p+N_\nu). \tag{4.5.5}$$

Along scheme (4.5.4), for every ν we calculate N_ν coefficients $\xi_\nu(p+N_\nu)$ of the wavelet expansion (4.5.5) and N_ν coefficients $\xi_\nu(p)$ that will be utilized with the next ν.

4.5.2 We will give an example of expanding a signal over Haar basis. Let $N = 2^3$ and the signal x be defined by its samples on the main period as $x = (1, -1, -1, 1, 1, 1, -1, -1)$. Calculations performed along formula (4.5.4) are presented in the Table 4.1.

According to (4.5.5) we obtain the expansion

$$x = \tfrac{1}{4}\,4\,\varphi_2(3) + \tfrac{1}{2}\,2\,\varphi_1(4) - \tfrac{1}{2}\,2\,\varphi_1(5) = f_2(6) + f_1(1) - f_1(3).$$

Table 4.1 Calculation of Haar coefficients

ξ_0	1	–1	–1	1	1	1	–1	–1
ξ_1	0	0	2	–2	2	–2	0	0
ξ_2	0	0	0	4				
ξ_3	0	0						

This result can be verified immediately taking into account the form of Haar basic functions shown in Fig. 4.3.

4.5.3 Scheme (4.5.4) of calculation of the coefficients of expansion (4.5.5) is referred to as the *decimation-in-time fast Haar transform*. This transform requires only additions; the number of additions is

$$2\sum_{\nu=1}^{s} N_\nu = 2\,(2^{s-1} + 2^{s-2} + \cdots + 2 + 1) = 2(N-1).$$

4.5.4 Formula (4.5.4) can be inverted:

$$\xi_{\nu-1}(2p) = \tfrac{1}{2}\big[\xi_\nu(p) + \xi_\nu(p+N_\nu)\big],$$
$$\xi_{\nu-1}(2p+1) = \tfrac{1}{2}\big[\xi_\nu(p) - \xi_\nu(p+N_\nu)\big],$$
$$p \in 0 : N_\nu - 1, \quad \nu = s, s-1, \ldots, 1.$$

Herein $x(p) = \xi_0(p)$, $p \in 0 : N-1$. We have derived the fast algorithm of reconstructing the samples of a signal x given in form (4.5.5). The reconstruction is performed on the main period.

4.6 Decimation in Frequency

4.6.1 If we take an orthogonal system of signals and perform the same permutation of an argument of each signal, the transformed signals will remain pairwise orthogonal. This simple idea allows us to construct new orthogonal bases in a space $\mathbb{C}_N$.

Let $N = 2^s$ and $f_0, f_1, \ldots, f_s$ be orthogonal bases in $\mathbb{C}_N$ defined in par. 4.2.2. We put

$$g_\nu(k;\, j) = f_\nu\big(\mathrm{rev}_s(k);\, \mathrm{rev}_s(j)\big), \quad \nu \in 0 : s.$$

In particular,

$$g_0(k;\ j) = \delta_N\big(\mathrm{rev}_s(j) - \mathrm{rev}_s(k)\big) = \delta_N(j-k).$$

According to (4.2.5) we have

$$g_s(k;\ j) = \omega_N^{\mathrm{rev}_s(k)\, j}. \tag{4.6.1}$$

It is evident that for each $\nu \in 0 : s$ the signals $g_\nu(0), g_\nu(1), \ldots, g_\nu(N-1)$ are pairwise orthogonal and there holds $\|g_\nu(k)\|^2 = 2^\nu$, $k \in 0 : N-1$.

Theorem 4.6.1 *There hold the recurrent relations*

$$g_0(k) = \delta_N(\cdot - k), \quad k \in 0 : N-1;$$

$$\begin{aligned} g_\nu(2lN_\nu + p) &= g_{\nu-1}(lN_{\nu-1}+p) + \omega_N^{\mathrm{rev}_s(2l)} g_{\nu-1}(lN_{\nu-1}+N_\nu+p), \\ g_\nu\big((2l+1)N_\nu + p\big) &= g_{\nu-1}(lN_{\nu-1}+p) - \omega_N^{\mathrm{rev}_s(2l)} g_{\nu-1}(lN_{\nu-1}+N_\nu+p), \end{aligned} \tag{4.6.2}$$

$$p \in 0 : N_\nu - 1, \quad l \in 0 : \Delta_\nu - 1, \quad \nu = 1, \ldots, s.$$

Proof As a preliminary, we will ascertain that

$$\mathrm{rev}_s\big((2l+\sigma)N_\nu + p\big) = \mathrm{rev}_{\nu-1}(l) + \sigma\Delta_\nu + \Delta_{\nu+1}\,\mathrm{rev}_{s-\nu}(p) \tag{4.6.3}$$

for $p \in 0 : N_\nu - 1$, $l \in 0 : \Delta_\nu - 1$, $\sigma \in 0 : 1$, $\nu = 1, \ldots, s$. Recall that $\mathrm{rev}_0(0) = 0$ by a definition, so formula (4.6.3) for $\nu = 1$ takes a form

$$\mathrm{rev}_s(\sigma N_1 + p) = \sigma + 2\,\mathrm{rev}_{s-1}(p), \quad p \in 0 : N_1 - 1, \quad \sigma \in 0 : 1.$$

The latter equality can be easily verified:

$$\begin{aligned} \mathrm{rev}_s(\sigma N_1 + p) &= \mathrm{rev}_s(\sigma 2^{s-1} + p_{s-2}2^{s-2} + \cdots + p_0) \\ &= p_0 2^{s-1} + \cdots + p_{s-2}2 + \sigma = 2\,\mathrm{rev}_{s-1}(p) + \sigma. \end{aligned}$$

Let $\nu \geq 2$. Then

$$\begin{aligned} \mathrm{rev}_s\big((2l+\sigma)N_\nu + p\big) &= \mathrm{rev}_s(l_{\nu-2}2^{s-1} + \cdots + l_0 2^{s-\nu+1} + \sigma 2^{s-\nu} \\ &\quad + p_{s-\nu-1}2^{s-\nu-1} + \cdots + p_0) \\ &= p_0 2^{s-1} + \cdots + p_{s-\nu-1}2^\nu + \sigma 2^{\nu-1} + l_0 2^{\nu-2} + \cdots + l_{\nu-2} \\ &= \Delta_{\nu+1}\,\mathrm{rev}_{s-\nu}(p) + \sigma\Delta_\nu + \mathrm{rev}_{\nu-1}(l), \end{aligned}$$

as it was to be ascertained.

The recurrent relations (4.6.2) can be written down in a single line

$$g_\nu\big((2l+\sigma)N_\nu+p\big) = g_{\nu-1}(lN_{\nu-1}+p) + (-1)^\sigma \omega_N^{\mathrm{rev}_s(2l)} g_{\nu-1}(lN_{\nu-1}+N_\nu+p), \tag{4.6.4}$$

where $\sigma \in 0:1$. To verify (4.6.4) we use formulae (4.6.3) and (4.2.2). We gain

$$\begin{aligned}
&g_\nu\big((2l+\sigma)N_\nu + p;\ j\big) = f_\nu\big(\mathrm{rev}_s((2l+\sigma)N_\nu + p);\ \mathrm{rev}_s(j)\big)\\
&= f_\nu\big(\mathrm{rev}_{\nu-1}(l) + \sigma\Delta_\nu + \Delta_{\nu+1}\,\mathrm{rev}_{s-\nu}(p);\ \mathrm{rev}_s(j)\big)\\
&= f_{\nu-1}\big(\mathrm{rev}_{\nu-1}(l) + 2\,\mathrm{rev}_{s-\nu}(p)\Delta_\nu;\ \mathrm{rev}_s(j)\big)\\
&+(-1)^\sigma \omega_{\Delta_{\nu+1}}^{\mathrm{rev}_{\nu-1}(l)} f_{\nu-1}\big(\mathrm{rev}_{\nu-1}(l) + (2\mathrm{rev}_{s-\nu}(p)+1)\Delta_\nu;\ \mathrm{rev}_s(j)\big)\\
&= f_{\nu-1}\big(\mathrm{rev}_s(lN_{\nu-1}+p);\ \mathrm{rev}_s(j)\big)\\
&+(-1)^\sigma\ \omega_{\Delta_{\nu+1}}^{\mathrm{rev}_{\nu-1}(l)}\ f_{\nu-1}\big(\mathrm{rev}_s(lN_{\nu-1}+N_\nu+p);\ \mathrm{rev}_s(j)\big)\\
&= g_{\nu-1}(lN_{\nu-1}+p;\ j) + (-1)^\sigma\ \omega_{\Delta_{\nu+1}}^{\mathrm{rev}_{\nu-1}(l)}\ g_{\nu-1}(lN_{\nu-1}+N_\nu+p;\ j).
\end{aligned}$$

It is remaining to check that $\omega_{\Delta_{\nu+1}}^{\mathrm{rev}_{\nu-1}(l)} = \omega_N^{\mathrm{rev}_s(2l)}$ for $l \in 0:\Delta_\nu - 1$. For $\nu = 1$ this is obvious, and for $\nu \geq 2$ this is a consequence of the equality

$$2^{s-\nu}\,\mathrm{rev}_{\nu-1}(l) = l_0 2^{s-2} + \cdots + l_{\nu-2}2^{s-\nu} = \mathrm{rev}_s(2l).$$

The theorem is proved. □

Theorem 4.6.2 *Valid is the equality*

$$g_\nu(lN_\nu + p) = \sum_{q=0}^{\Delta_{\nu+1}-1} \omega_N^{q\,\mathrm{rev}_s(l)}\ g_0(qN_\nu + p), \tag{4.6.5}$$

$$p \in 0:N_\nu - 1, \quad l \in 0:\Delta_{\nu+1} - 1, \quad \nu = 1,\ldots,s.$$

Proof At first we note that

$$\begin{aligned}
\mathrm{rev}_s\big(\mathrm{rev}_\nu(q)N_\nu + p\big) &= p_0 2^{s-1} + \cdots + p_{s-\nu-1}2^\nu + q_{\nu-1}2^{\nu-1} + \cdots + q_0\\
&= q + \Delta_{\nu+1}\,\mathrm{rev}_{s-\nu}(p).
\end{aligned} \tag{4.6.6}$$

According to (4.2.4) and (4.6.6) we have

$$\begin{aligned}
&g_\nu(lN_\nu + p;\ j)\\
&= f_\nu\big(\mathrm{rev}_s(l_{\nu-1}2^{s-1} + \cdots + l_0 2^{s-\nu} + p_{s-\nu-1}2^{s-\nu-1} + \cdots + p_0);\ \mathrm{rev}_s(j)\big)\\
&= f_\nu\big(\mathrm{rev}_\nu(l) + \Delta_{\nu+1}\,\mathrm{rev}_{s-\nu}(p);\ \mathrm{rev}_s(j)\big)\\
&= \sum_{q=0}^{\Delta_{\nu+1}-1} \omega_{\Delta_{\nu+1}}^{\mathrm{rev}_\nu(l)\,\mathrm{rev}_\nu(q)}\ f_0\big(q + \Delta_{\nu+1}\,\mathrm{rev}_{s-\nu}(p);\ \mathrm{rev}_s(j)\big)
\end{aligned}$$

$$= \sum_{q=0}^{\Delta_{\nu+1}-1} \omega_{\Delta_{\nu+1}}^{\mathrm{rev}_\nu(l)\,\mathrm{rev}_\nu(q)} f_0\big(\mathrm{rev}_s(\mathrm{rev}_\nu(q)N_\nu + p);\ \mathrm{rev}_s(j)\big)$$

$$= \sum_{q=0}^{\Delta_{\nu+1}-1} \omega_{\Delta_{\nu+1}}^{\mathrm{rev}_\nu(l)\,q}\ g_0(qN_\nu + p);\ j\big).$$

It is remaining to take into account that $\omega_{\Delta_{\nu+1}}^{\mathrm{rev}_\nu(l)} = \omega_N^{\mathrm{rev}_s(l)}$ for $l \in 0 : \Delta_{\nu+1} - 1$. The theorem is proved. □

Theorem 4.6.3 *Valid is the identity*

$$g_\nu(lN_\nu + p;\ j) \equiv g_\nu(lN_\nu;\ j - p), \tag{4.6.7}$$

$$p \in 0 : N_\nu - 1, \quad l \in 0 : \Delta_{\nu+1} - 1, \quad \nu = 1, \ldots, s.$$

Proof Equality (4.6.5) yields

$$g_\nu(lN_\nu;\ j) = \sum_{q=0}^{\Delta_{\nu+1}-1} \omega_N^{q\,\mathrm{rev}_s(l)}\ \delta_N(j - qN_\nu).$$

Therefore

$$g_\nu(lN_\nu;\ j - p) = \sum_{q=0}^{\Delta_{\nu+1}-1} \omega_N^{q\,\mathrm{rev}_s(l)}\ \delta_N\big(j - (qN_\nu + p)\big)$$

$$= \sum_{q=0}^{\Delta_{\nu+1}-1} \omega_N^{q\,\mathrm{rev}_s(l)}\ g_0(qN_\nu + p;\ j) = g_\nu(lN_\nu + p;\ j).$$

The theorem is proved. □

4.6.2 A signal $y \in \mathbb{C}_N$ with $N = 2^s$ can be expanded over any basis g_ν:

$$y = \frac{1}{2^\nu} \sum_{k=0}^{N-1} y_\nu(k)\, g_\nu(k). \tag{4.6.8}$$

Here $y_\nu(k) = \langle y,\ g_\nu(k)\rangle$. Relying on (4.6.2), in a usual way we come to recurrent relations for coefficients $y_\nu(k)$ of expansion (4.6.8):

$$y_0(k) = y(k), \quad k \in 0 : N - 1;$$

$$\begin{aligned} y_\nu(2lN_\nu + p) &= y_{\nu-1}(lN_{\nu-1}+p) + \omega_N^{-\mathrm{rev}_s(2l)} y_{\nu-1}(lN_{\nu-1}+N_\nu+p), \\ y_\nu\big((2l+1)N_\nu + p\big) &= y_{\nu-1}(lN_{\nu-1}+p) - \omega_N^{-\mathrm{rev}_s(2l)} y_{\nu-1}(lN_{\nu-1}+N_\nu+p), \end{aligned} \tag{4.6.9}$$

$$p \in 0 : N_\nu - 1, \quad l \in 0 : \Delta_\nu - 1, \quad \nu = 1, \ldots, s.$$

For $\nu = s$ according to (4.6.1) we obtain

$$y_s(k) = \sum_{j=0}^{N-1} y(j)\, \overline{g_s}(k;\, j) = \sum_{j=0}^{N-1} y(j)\, \omega_N^{-\mathrm{rev}_s(k)\, j} = Y\big(\mathrm{rev}_s(k)\big).$$

Hence it follows that the components of Fourier spectrum Y of a signal y are determined by the formula

$$Y(k) = y_s\big(\mathrm{rev}_s(k)\big), \quad k \in 0 : N - 1.$$

Scheme (4.6.9) is referred to as the *decimation-in-frequency Cooley–Tukey algorithm* for calculation of the discrete Fourier transform.

Formula (4.6.9) can be inverted:

$$y_s(k) = Y\big(\mathrm{rev}_s(k)\big), \quad k \in 0 : N - 1;$$

$$y_{\nu-1}(lN_{\nu-1} + p) = \tfrac{1}{2}\big[y_\nu(2lN_\nu + p) + y_\nu\big((2l+1)N_\nu + p\big)\big],$$

$$y_{\nu-1}(lN_{\nu-1} + N_\nu + p) = \tfrac{1}{2}\, \omega_N^{\mathrm{rev}_s(2l)}\big[y_\nu(2lN_\nu + p) - y_\nu\big((2l+1)N_\nu + p\big)\big],$$

$$p \in 0 : N_\nu - 1, \quad l \in 0 : \Delta_\nu - 1, \quad \nu = s, s-1, \ldots, 1.$$

Herein $y(k) = y_0(k)$, $k \in 0 : N - 1$.

4.6.3 The structure of formula (4.6.4) is similar to the structure of formula (4.4.1). It is convenient to assume that the basis $g_{\nu-1}$ is divided into Δ_ν blocks; the blocks are marked by an index l. Each block contains $N_{\nu-1}$ signals with inner indices p and $p + N_\nu$ for $p \in 0 : N_\nu - 1$. According to (4.6.4) a block with an index l generates two blocks of the basis g_ν with inner indices $2l$ and $2l + 1$, herein each block contains N_ν signals with an index p. On the strength of (4.6.7) the signals inside a block differ only by a shift of an argument. The complete scheme of branching is the same as it was shown in Fig. 4.1, par. 4.4.1.

Wavelet bases are formed of blocks of different levels. This is done in the same way as in par. 4.4.1.

4.6.4 We will investigate the branching scheme (a) in Fig. 4.2. It corresponds to putting $l = 0$ in (4.6.2). We derive the recurrent relations

$$\begin{aligned} g_0(k) &= \delta_N(\cdot - k), \quad k \in 0 : N - 1; \\ g_\nu(p) &= g_{\nu-1}(p) + g_{\nu-1}(p + N_\nu), \\ g_\nu(p + N_\nu) &= g_{\nu-1}(p) - g_{\nu-1}(p + N_\nu), \\ p &\in 0 : N_\nu - 1, \quad \nu = 1, \ldots, s. \end{aligned} \tag{4.6.10}$$

The signals $\{g_\nu(p+N_\nu)\}$ will enter a wavelet basis and the signals $\{g_\nu(p)\}$ will participate in further branching.

The signals

$$g_s(0);\ g_\nu(p+N_\nu),\ p \in 0: N_\nu - 1,\ \nu = s, s-1, \ldots, 1, \tag{4.6.11}$$

form an orthogonal basis in a space $\mathbb{C}_N$ with $N = 2^s$. It is referred to as the *discrete Haar basis related to decimation in frequency.*

According to (4.6.1) we have $g_s(0;\ j) \equiv 1$. Identity (4.6.7) yields

$$g_\nu(p+N_\nu;\ j) \equiv g_\nu(N_\nu;\ j-p), \quad p \in 0: N_\nu - 1.$$

Theorem 4.6.4 *For $\nu \in 1: s$ there holds*

$$g_\nu(N_\nu;\ j) = \delta_{N_{\nu-1}}(j) - \delta_{N_{\nu-1}}(j - N_\nu), \quad j \in \mathbb{Z}.$$

Proof Using formula (4.6.5) with $l = 1$ and $p = 0$ we gain

$$\begin{aligned} g_\nu(N_\nu;\ j) &= \sum_{q=0}^{\Delta_{\nu+1}-1} \omega_N^{q\,\mathrm{rev}_s(1)}\, \delta_N(j - qN_\nu) \\ &= \sum_{q=0}^{\Delta_{\nu+1}-1} (-1)^q\, \delta_N(j - qN_\nu) \\ &= \sum_{q=0}^{\Delta_\nu-1} \delta_N(j - qN_{\nu-1}) - \sum_{q=0}^{\Delta_\nu-1} \delta_N(j - N_\nu - qN_{\nu-1}) \\ &= \delta_{N_{\nu-1}}(j) - \delta_{N_{\nu-1}}(j - N_\nu). \end{aligned}$$

The theorem is proved. □

Figure 4.4 depicts basis (4.6.11) for $N = 2^3$.

4.6.5 Any signal $y \in \mathbb{C}_N$ can be expanded over basis (4.6.11):

$$y = 2^{-s}\, y_s(0)\, g_s(0) + \sum_{\nu=1}^{s} 2^{-\nu} \sum_{p=0}^{N_\nu-1} y_\nu(p+N_\nu)\, g_\nu(p+N_\nu). \tag{4.6.12}$$

Here $y_\nu(k) = \langle y,\ g_\nu(k)\rangle$. Bearing in mind (4.6.10) we deduce recurrent relations for the coefficients of expansion (4.6.12):

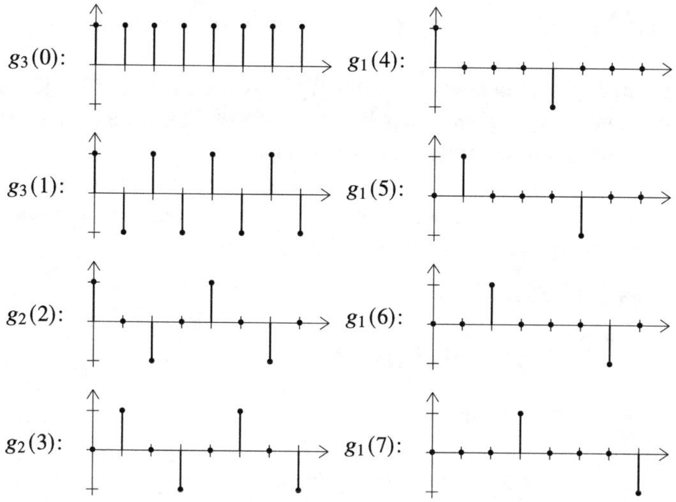

Fig. 4.4 Haar basis related to decimation in frequency for $N = 2^3$

$$\begin{aligned} y_0(k) &= y(k), \quad k \in 0 : N-1; \\ y_\nu(p) &= y_{\nu-1}(p) + y_{\nu-1}(p + N_\nu), \\ y_\nu(p + N_\nu) &= y_{\nu-1}(p) - y_{\nu-1}(p + N_\nu), \\ p &\in 0 : N_\nu - 1, \quad \nu = 1, \ldots, s. \end{aligned} \tag{4.6.13}$$

Along this scheme, for every ν we calculate N_ν coefficients $y_\nu(p + N_\nu)$ of the wavelet expansion (4.6.12) and N_ν coefficients $y_\nu(p)$ that will be utilized with the next ν.

As an example we will expand the signal from par. 4.5.2 over basis (4.6.11). It is convenient to rename this signal as y in lieu of x. Calculations performed along formula (4.6.13) are presented in Table 4.2.

According to (4.6.12) we obtain the expansion

$$y = g_2(2) - g_1(5) + g_1(7).$$

Table 4.2 Calculation of Haar coefficients

y_0	1	–1	–1	1	1	1	–1	–1
y_1	2	0	–2	0	0	–2	0	2
y_2	0	0	4	0				
y_3	0	0						

This result can be verified immediately taking into account the form of Haar basic functions shown in Fig. 4.4.

Scheme (4.6.13) of calculation of the coefficients of expansion (4.6.12) is referred to as the *decimation-in-frequency fast Haar transform*. This transform requires only additions; the number of operations is $2(N-1)$.

Note that the coefficients of expansion (4.6.12) are contained in the table $\{y_\nu(k)\}$ constructed along formula (4.6.9). Thus, in a process of calculating Fourier coefficients we incidentally calculate Haar coefficients as well.

4.6.6 Formula (4.6.13) can be inverted:

$$y_{\nu-1}(p) = \tfrac{1}{2}[y_\nu(p) + y_\nu(p+N_\nu)],$$

$$y_{\nu-1}(p+N_\nu) = \tfrac{1}{2}[y_\nu(p) - y_\nu(p+N_\nu)],$$

$$p \in 0 : N_\nu - 1, \quad \nu = s, s-1, \dots, 1.$$

Herein $y(p) = y_0(p)$, $p \in 0 : N-1$. We have derived the fast algorithm of reconstructing the samples of a signal y given in form (4.6.12). The reconstruction is performed on the main period.

4.7 Sampling Theorem in Haar Bases

4.7.1 We begin with Haar basis related to decimation in time.

Lemma 4.7.1 *Given a signal $x \in \mathbb{C}_N$, for each $k \in 1 : s$ there holds an equality*

$$x = 2^{-k} \sum_{p=0}^{N_k-1} \xi_k(p)\varphi_k(p) + \sum_{\nu=1}^{k} 2^{-\nu} \sum_{p=0}^{N_\nu-1} \xi_\nu(p+N_\nu)\varphi_\nu(p+N_\nu). \tag{4.7.1}$$

In case of $k = s$ we obtain a wavelet expansion (4.5.5).

Proof Let $k = 1$. We have

$$x = \sum_{p=0}^{N-1} x(p)\,\delta_N(\cdot - p) = \sum_{p=0}^{N-1} \xi_0(p)\,\varphi_0(p)$$

$$= \sum_{p=0}^{N_1-1} \big[\xi_0(2p)\,\varphi_0(2p) + \xi_0(2p+1)\,\varphi_0(2p+1)\big].$$

We put

$$\widetilde{x} = \sum_{p=0}^{N_1-1} \big[\xi_0(2p+1)\,\varphi_0(2p) + \xi_0(2p)\,\varphi_0(2p+1)\big].$$

Then

$$x = \tfrac{1}{2}(x + \widetilde{x}) + \tfrac{1}{2}(x - \widetilde{x}) =: \tfrac{1}{2}\, v_1 + \tfrac{1}{2}\, w_1. \tag{4.7.2}$$

On the basis of (4.5.3) and (4.5.4) we gain

$$\begin{aligned} v_1 &= \sum_{p=0}^{N_1-1} \big[\xi_0(2p) + \xi_0(2p+1)\big]\big[\varphi_0(2p) + \varphi_0(2p+1)\big] \\ &= \sum_{p=0}^{N_1-1} \xi_1(p)\, \varphi_1(p), \end{aligned}$$

$$\begin{aligned} w_1 &= \sum_{p=0}^{N_1-1} \big[\xi_0(2p) - \xi_0(2p+1)\big]\big[\varphi_0(2p) - \varphi_0(2p+1)\big] \\ &= \sum_{p=0}^{N_1-1} \xi_1(p+N_1)\, \varphi_1(p+N_1). \end{aligned}$$

Now formula (4.7.2) conforms to (4.7.1) for $k = 1$.

We perform an induction step from k to $k + 1$. We denote

$$\begin{aligned} v_k &= \sum_{p=0}^{N_k-1} \xi_k(p)\, \varphi_k(p) \\ &= \sum_{p=0}^{N_{k+1}-1} \big[\xi_k(2p)\, \varphi_k(2p) + \xi_k(2p+1)\, \varphi_k(2p+1)\big] \end{aligned}$$

and introduce a signal

$$\widetilde{v}_k = \sum_{p=0}^{N_{k+1}-1} \big[\xi_k(2p+1)\, \varphi_k(2p) + \xi_k(2p)\, \varphi_k(2p+1)\big].$$

Then

$$v_k = \tfrac{1}{2}(v_k + \widetilde{v}_k) + \tfrac{1}{2}(v_k - \widetilde{v}_k) =: \tfrac{1}{2}\, v_{k+1} + \tfrac{1}{2}\, w_{k+1}. \tag{4.7.3}$$

Here

$$v_{k+1} = \sum_{p=0}^{N_{k+1}-1} \xi_{k+1}(p)\, \varphi_{k+1}(p),$$

$$w_{k+1} = \sum_{p=0}^{N_{k+1}-1} \xi_{k+1}(p + N_{k+1})\, \varphi_{k+1}(p + N_{k+1}).$$

Combining (4.7.1) and (4.7.3) we gain

$$x = 2^{-k-1} \sum_{p=0}^{N_{k+1}-1} \xi_{k+1}(p)\varphi_{k+1}(p) + \sum_{\nu=1}^{k+1} 2^{-\nu} \sum_{p=0}^{N_\nu-1} \xi_\nu(p + N_\nu)\varphi_\nu(p + N_\nu).$$

The lemma is proved. □

Theorem 4.7.1 (Sampling Theorem) *Let $x \in \mathbb{C}_N$ be a signal such that for some $k \in 1 : s$ there holds $\xi_\nu(p + N_\nu) = 0$ for all $p \in 0 : N_\nu - 1$ and $\nu = 1, \ldots, k$. Then*

$$x = \sum_{p=0}^{N_k-1} x(2^k p)\, \varphi_k(p). \tag{4.7.4}$$

Proof By virtue of (4.7.1) it is sufficient to verify that $\xi_k(p) = 2^k\, x(2^k p)$ for $p \in 0 : N_k - 1$. On the basis of (4.5.4) we have

$$\xi_{\nu-1}(2p) = \tfrac{1}{2}\big[\xi_\nu(p) + \xi_\nu(p + N_\nu)\big].$$

Taking into account the hypothesis of the theorem we gain $\xi_\nu(p) = 2\,\xi_{\nu-1}(2p)$ for all $p \in 0 : N_\nu - 1$ and $\nu = 1, \ldots, k$. Hence for $p \in 0 : N_k - 1$ there holds

$$\xi_k(p) = 2\,\xi_{k-1}(2p) = 2^2\,\xi_{k-2}(2^2 p) = \cdots = 2^k\,\xi_0(2^k p) = 2^k\, x(2^k p).$$

The theorem is proved. □

We note that according to (4.2.4) there holds

$$\varphi_k(0;\ j) = f_k(0;\ j) = \sum_{q=0}^{2^k-1} \delta_N(j - q),$$

i.e. $\varphi_k(0;\ j)$ is a periodic step that equals to unity for $j = 0, 1, \ldots, 2^k - 1$ and equals to zero for $j = 2^k, \ldots, N - 1$. By virtue of (4.2.8), for $p \in 0 : N_k - 1$ we have

$$\varphi_k(p;\ j) = f_k(p\Delta_{k+1};\ j) = f_k(0;\ j - p\Delta_{k+1}) = \varphi_k(0;\ j - 2^k p).$$

This observation lets us rewrite (4.7.4) in a form

$$x(j) = \sum_{p=0}^{N_k-1} x(2^k p)\, \varphi_k(0;\ j - 2^k p).$$

The latter formula shows that in the premises of Theorem 4.7.1, the signal x is a step-function defined by equalities $x(j) = x(2^k p)$ for $j \in \{2^k p,\ 2^k p + 1, \ldots, 2^k(p+1) - 1\}$, $p = 0, 1, \ldots, N_k - 1$.

4.7.2 Now we turn to Haar basis related to decimation in frequency.

Lemma 4.7.2 *Given a signal $y \in \mathbb{C}_N$, for each $k \in 1 : s$ there holds an equality*

$$y = 2^{-k} \sum_{p=0}^{N_k-1} y_k(p) g_k(p) + \sum_{\nu=1}^{k} 2^{-\nu} \sum_{p=0}^{N_\nu-1} y_\nu(p + N_\nu) g_\nu(p + N_\nu). \tag{4.7.5}$$

In case of $k = s$ we obtain a wavelet expansion (4.6.12).

Proof Let $k = 1$. We have

$$\begin{aligned} y(j) &= \tfrac{1}{2}\big[y(j) + y(j - N_1)\big] + \tfrac{1}{2}\big[y(j) - y(j - N_1)\big] \\ &=: \tfrac{1}{2}\, v_1(j) + \tfrac{1}{2}\, w_1(j). \end{aligned} \tag{4.7.6}$$

According to (4.6.10) and (4.6.13) we write

$$\begin{aligned} v_1 &= \sum_{p=0}^{N-1} y(p)\big[\delta_N(\cdot - p) + \delta_N(\cdot - \langle p + N_1\rangle_N)\big] \\ &= \sum_{p=0}^{N-1} y_0(p)\big[g_0(p) + g_0(\langle p + N_1\rangle_N)\big] \\ &= \sum_{p=0}^{N_1-1} y_0(p)\, g_1(p) + \sum_{p=0}^{N_1-1} y_0(p + N_1)\, g_1(p) \\ &= \sum_{p=0}^{N_1-1} y_1(p)\, g_1(p), \end{aligned}$$

$$\begin{aligned} w_1 &= \sum_{p=0}^{N_1-1} y_0(p)\, g_1(p + N_1) - \sum_{p=0}^{N_1-1} y_0(p + N_1)\, g_1(p + N_1) \\ &= \sum_{p=0}^{N_1-1} y_1(p + N_1)\, g_1(p + N_1). \end{aligned}$$

Now formula (4.7.6) corresponds to (4.7.5) for $k = 1$.

We perform an induction step from k to $k+1$. We denote

$$v_k = \sum_{p=0}^{N_k-1} y_k(p)\, g_k(p)$$

and write down

$$\begin{aligned} v_k &= \frac{1}{2}\sum_{p=0}^{N_k-1} y_k(p)\big[g_k(p) + g_k(\langle p + N_{k+1}\rangle_{N_k})\big] \\ &+ \frac{1}{2}\sum_{p=0}^{N_k-1} y_k(p)\big[g_k(p) - g_k(\langle p + N_{k+1}\rangle_{N_k})\big] \\ &=: \tfrac{1}{2}\, v_{k+1} + \tfrac{1}{2}\, w_{k+1}. \end{aligned} \tag{4.7.7}$$

Here

$$\begin{aligned} v_{k+1} &= \sum_{p=0}^{N_{k+1}-1} y_k(p)\, g_{k+1}(p) + \sum_{p=0}^{N_{k+1}-1} y_k(p + N_{k+1})\, g_{k+1}(p) \\ &= \sum_{p=0}^{N_{k+1}-1} y_{k+1}(p)\, g_{k+1}(p) \end{aligned}$$

and

$$\begin{aligned} w_{k+1} &= \sum_{p=0}^{N_{k+1}-1} y_k(p)\, g_{k+1}(p+N_{k+1}) - \sum_{p=0}^{N_{k+1}-1} y_k(p+N_{k+1})\, g_{k+1}(p+N_{k+1}) \\ &= \sum_{p=0}^{N_{k+1}-1} y_{k+1}(p + N_{k+1})\, g_{k+1}(p + N_{k+1}). \end{aligned}$$

Combining (4.7.5) and (4.7.7) we gain

$$y = 2^{-k-1}\sum_{p=0}^{N_{k+1}-1} y_{k+1}(p)\, g_{k+1}(p) + \sum_{\nu=1}^{k+1} 2^{-\nu} \sum_{p=0}^{N_\nu-1} y_\nu(p + N_\nu)\, g_\nu(p + N_\nu).$$

The lemma is proved. □

Theorem 4.7.2 (Sampling Theorem) *Let $y \in \mathbb{C}_N$ be a signal such that for some $k \in 1 : s$ there holds $y_\nu(p + N_\nu) = 0$ for all $p \in 0 : N_\nu - 1$ and $\nu = 1, \ldots, k$. Then*

$$y = \sum_{p=0}^{N_k-1} y(p)\, g_k(p). \tag{4.7.8}$$

Proof By virtue of (4.7.5) it is sufficient to verify that $y_k(p) = 2^k\, y(p)$ for $p \in 0 : N_k - 1$. On the basis of (4.6.13) we have

$$y_{\nu-1}(p) = \tfrac{1}{2}\big[y_\nu(p) + y_\nu(p + N_\nu)\big].$$

Taking into account the hypothesis of the theorem we gain $y_\nu(p) = 2\, y_{\nu-1}(p)$ for all $p \in 0 : N_\nu - 1$ and $\nu = 1, \ldots, k$. Hence for $p \in 0 : N_k - 1$ there holds

$$y_k(p) = 2\, y_{k-1}(p) = 4\, y_{k-2}(p) = \cdots = 2^k\, y_0(p) = 2^k\, y(p).$$

The theorem is proved. □

Note that according to (4.6.5) there holds

$$g_k(0;\ j) = \sum_{q=0}^{\Delta_{k+1}-1} \delta_N(j - qN_k) = \delta_{N_k}(j).$$

On the strength of (4.6.7), for $p \in 0 : N_k - 1$ we have

$$g_k(p;\ j) = g_k(0;\ j - p) = \delta_{N_k}(j - p). \tag{4.7.9}$$

This lets us rewrite (4.7.8) in a form

$$y(j) = \sum_{p=0}^{N_k-1} y(p)\, \delta_{N_k}(j - p).$$

The latter formula shows that in the premises of Theorem 4.7.2, the signal y is N_k-periodic.

Essentially, we ascertained that with the aid of expanding a signal over Haar basis related to decimation in frequency one can detect its hidden periodicity.

4.8 Convolution Theorem in Haar Bases

4.8.1 Formula (4.6.12) gives an expansion of a signal $y \in \mathbb{C}_N$ over the discrete Haar basis related to decimation in frequency. As it was mentioned in par. 4.6.4, the basic signals satisfy to the identity

$$g_\nu(p + N_\nu;\ j) \equiv g_\nu(N_\nu;\ j - p),$$

$$p \in 0 : N_\nu - 1, \quad \nu = 1, \ldots, s - 1.$$

In the other words, every basic signal of the ν-th level is a shift of a single signal $g_\nu(N_\nu)$. We introduce a notation $\psi_\nu(j) = g_\nu(N_\nu;\ j)$. According to Theorem 4.6.4 we have

$$\psi_\nu(j) = \delta_{N_{\nu-1}}(j) - \delta_{N_{\nu-1}}(j - N_\nu), \quad \nu = 1, \ldots, s. \tag{4.8.1}$$

In particular, $\psi_\nu(-j) = \psi_\nu(j)$, $\psi_\nu(j - N_\nu) = -\psi_\nu(j)$. To simplify formula (4.6.12) we put

$$\beta = y_s(0); \quad \widehat{y}_\nu(p) = y_\nu(p + N_\nu), \;\; p \in 0 : N_\nu - 1, \;\; \nu = 1, \ldots, s.$$

In the new notations formula (4.6.12) takes a form

$$y(j) = 2^{-s}\beta + \sum_{\nu=1}^{s} 2^{-\nu} \sum_{p=0}^{N_\nu - 1} \widehat{y}_\nu(p)\, \psi_\nu(j - p), \quad j \in \mathbb{Z}. \tag{4.8.2}$$

We took advantage of the fact that $g_s(0;\ j) \equiv 1$.

We will examine in more detail the properties of the signal ψ_ν.

Lemma 4.8.1 *Valid is the formula*

$$\psi_\nu(j) = (-1)^{\lfloor j/N_\nu \rfloor}\, \delta_{N_\nu}(j), \quad j \in \mathbb{Z}. \tag{4.8.3}$$

Proof By virtue of (4.8.1) the signal ψ_ν is $N_{\nu-1}$-periodic. Since $N_{\nu-1} = 2N_\nu$, of the same property is the signal in the right side of (4.8.3). Hence it is sufficient to verify equality (4.8.3) on the main period $0 : N_{\nu-1} - 1$. When $j = 0$ or $j = N_\nu$, it is a consequence of (4.8.1). When $j \in 1 : N_\nu - 1$ or $j \in N_\nu + 1 : N_{\nu-1} - 1$, both sides of (4.8.3) are equal to zero. The lemma is proved. □

Lemma 4.8.2 *Given $p \in 0 : N_\nu - 1$, valid is the equality*

$$\psi_\nu(j - p) = (-1)^{\lfloor j/N_\nu \rfloor}\, \delta_{N_\nu}(j - p), \quad j \in \mathbb{Z}. \tag{4.8.4}$$

Proof Let us show that

$$(-1)^{\lfloor (j-p)/N_\nu \rfloor}\, \delta_{N_\nu}(j - p) = (-1)^{\lfloor j/N_\nu \rfloor}\, \delta_{N_\nu}(j - p). \tag{4.8.5}$$

As long as both sides of this equality are $N_{\nu-1}$-periodic (in terms of j) signals, it is sufficient to verify it on the period $p : p + N_{\nu-1} - 1$. When $j = p$ or $j = p + N_\nu$, equality (4.8.5) is true. It is also true for any other j from the given period because in this case both sides of (4.8.5) are equal to zero.

Now (4.8.4) follows from (4.8.3) and (4.8.5). □

Lemma 4.8.3 *For any integers j and p there holds*

$$\psi_\nu(j-p) = (-1)^{\lfloor p/N_\nu \rfloor}\, \psi_\nu\big(j - \langle p\rangle_{N_\nu}\big). \tag{4.8.6}$$

Proof According to (4.8.3), for each integer j and l we have

$$\psi_\nu(j - lN_\nu) = (-1)^l\, \psi_\nu(j).$$

Taking into account that $j - p = \big(j - \langle p\rangle_{N_\nu}\big) - \lfloor p/N_\nu\rfloor\, N_\nu$ we come to (4.8.6). □

4.8.2 We will investigate how the decimation-in-frequency discrete Haar transform acts on a cyclic convolution. Recall that a cyclic convolution of signals x and y from $\mathbb{C}_N$ is a signal $u = x * y$ with samples

$$u(j) = \sum_{k=0}^{N-1} x(k)\, y(j-k).$$

Theorem 4.8.1 (Convolution Theorem) *Let u be a cyclic convolution of signals x and y. If along with* (4.8.2) *we have expansions*

$$x(j) = 2^{-s}\alpha + \sum_{\nu=1}^{s} 2^{-\nu} \sum_{p=0}^{N_\nu-1} \widehat{x}_\nu(p)\, \psi_\nu(j-p), \tag{4.8.7}$$

$$u(j) = 2^{-s}\gamma + \sum_{\nu=1}^{s} 2^{-\nu} \sum_{p=0}^{N_\nu-1} \widehat{u}_\nu(p)\, \psi_\nu(j-p), \tag{4.8.8}$$

then it is necessary that $\gamma = \alpha\beta$, $\widehat{u}_s(0) = \widehat{x}_s(0)\, \widehat{y}_s(0)$ and

$$\widehat{u}_\nu(p) = \sum_{q=0}^{p} \widehat{x}_\nu(q)\, \widehat{y}_\nu(p-q) - \sum_{q=p+1}^{N_\nu-1} \widehat{x}_\nu(q)\, \widehat{y}_\nu(p-q+N_\nu), \tag{4.8.9}$$

$$p \in 0 : N_\nu - 1, \quad \nu = s-1, s-2, \ldots, 1.$$

Proof Let us transform a formula

$$y(j-k) = 2^{-s}\beta + \sum_{\nu=1}^{s} 2^{-\nu} \sum_{p=0}^{N_\nu-1} \widehat{y}_\nu(p)\, \psi_\nu(j-k-p).$$

As it was mentioned in par. 4.8.1, there holds $\psi_\nu(-j) = \psi_\nu(j)$. Along with (4.8.6) it gives us

$$\psi_\nu(j-k-p) = \psi_\nu\big(k - (j-p)\big) = (-1)^{\lfloor (j-p)/N_\nu \rfloor}\, \psi_\nu\big(k - \langle j-p\rangle_{N_\nu}\big).$$

We have

$$y(j-k) = 2^{-s}\beta + \sum_{\nu=1}^{s} 2^{-\nu} \sum_{p=0}^{N_\nu - 1} (-1)^{\lfloor (j-p)/N_\nu \rfloor} \widehat{y}_\nu(p)\, \psi_\nu\big(k - \langle j-p\rangle_{N_\nu}\big).$$

We change the variables: $q = \langle j-p\rangle_{N_\nu}$. Then $p = \langle j-q\rangle_{N_\nu}$ and

$$\left\lfloor \frac{j-p}{N_\nu} \right\rfloor = \left\lfloor \frac{j - \langle j-q\rangle_{N_\nu}}{N_\nu} \right\rfloor = \left\lfloor \frac{j-q-\langle j-q\rangle_{N_\nu} + q}{N_\nu} \right\rfloor = \left\lfloor \frac{j-q}{N_\nu} \right\rfloor.$$

We come to a formula

$$y(j-k) = 2^{-s}\beta + \sum_{\nu=1}^{s} 2^{-\nu} \sum_{q=0}^{N_\nu - 1} (-1)^{\lfloor (j-q)/N_\nu \rfloor} \widehat{y}_\nu\big(\langle j-q\rangle_{N_\nu}\big)\psi_\nu(k-q). \quad (4.8.10)$$

Let us substitute (4.8.7) and (4.8.10) into a convolution formula. Bearing in mind reality, orthogonality and norming of the basic functions we gain

$$u(j) = 2^{-s}\alpha\beta + \sum_{\nu=1}^{s} 2^{-\nu} \sum_{q=0}^{N_\nu - 1} (-1)^{\lfloor (j-q)/N_\nu \rfloor} \widehat{x}_\nu(q)\, \widehat{y}_\nu\big(\langle j-q\rangle_{N_\nu}\big).$$

We write

$$\begin{aligned}
&(-1)^{\lfloor (j-q)/N_\nu \rfloor} \widehat{y}_\nu\big(\langle j-q\rangle_{N_\nu}\big) \\
&= (-1)^{\lfloor j/N_\nu \rfloor + \lfloor (\langle j\rangle_{N_\nu} - q)/N_\nu \rfloor} \widehat{y}_\nu\big(\langle \langle j\rangle_{N_\nu} - q\rangle_{N_\nu}\big) \\
&= (-1)^{\lfloor j/N_\nu \rfloor} \sum_{p=0}^{N_\nu - 1} (-1)^{\lfloor (p-q)/N_\nu \rfloor} \widehat{y}_\nu\big(\langle p-q\rangle_{N_\nu}\big)\, \delta_{N_\nu}\big(\langle j\rangle_{N_\nu} - p\big).
\end{aligned}$$

According to (4.8.4) we have

$$\delta_{N_\nu}\big(\langle j\rangle_{N_\nu} - p\big) = \delta_{N_\nu}(j-p) = (-1)^{\lfloor j/N_\nu \rfloor} \psi_\nu(j-p),$$

therefore

$$u(j) = 2^{-s}\alpha\beta + \sum_{\nu=1}^{s} 2^{-\nu} \sum_{p=0}^{N_\nu - 1} \left\{ \sum_{q=0}^{N_\nu - 1} (-1)^{\lfloor (p-q)/N_\nu \rfloor} \widehat{x}_\nu(q) \widehat{y}_\nu\big(\langle p-q\rangle_{N_\nu}\big) \right\} \psi_\nu(j-p).$$

But we already have representation (4.8.8) for the signal u. By virtue of uniqueness of expansion over an orthogonal basis we conclude that $\gamma = \alpha\beta$ and the sum in braces is nothing else but $\widehat{u}_\nu(p)$. It is remaining to note that

$$\sum_{q=0}^{N_\nu-1}(-1)^{\lfloor (p-q)/N_\nu \rfloor}\,\widehat{x}_\nu(q)\,\widehat{y}_\nu\big(\langle p-q\rangle_{N_\nu}\big) = \sum_{q=0}^{p}\widehat{x}_\nu(q)\,\widehat{y}_\nu(p-q) - \sum_{q=p+1}^{N_\nu-1}\widehat{x}_\nu(q)\,\widehat{y}_\nu(p-q+N_\nu).$$

The theorem is proved. □

The expression in the right side of formula (4.8.9) is referred to as a *skew-cyclic convolution* of signals $\widehat{x}_\nu$ and $\widehat{y}_\nu$. Thus, coefficients of the ν-th level in the expansion of a cyclic convolution u over Haar basis related to decimation in frequency are obtained as a result of a skew-cyclic convolution of expansion coefficients of the ν-th level of the signals x and y.

4.8.3 Now we turn to the discrete Haar basis related to decimation in time (see Sects. 4.4.2 and 4.5.1). We have

$$\begin{aligned}\varphi_\nu(p+N_\nu;\ j) &= f_\nu(\Delta_\nu + p\Delta_{\nu+1};\ j) = f_\nu(\Delta_\nu;\ j - p\Delta_{\nu+1})\\ &= \varphi_\nu(N_\nu;\ j - p\Delta_{\nu+1}).\end{aligned}$$

For the sake of simplicity we will write $\varphi_\nu(j)$ instead of $\varphi_\nu(N_\nu;\ j)$. Formula (4.4.8) yields

$$\varphi_\nu(j) = \sum_{q=0}^{\Delta_\nu-1}\delta_N(j-q) - \sum_{q=\Delta_\nu}^{\Delta_{\nu+1}-1}\delta_N(j-q). \tag{4.8.11}$$

We take expansion (4.5.5) and rename $\alpha := \xi_s(0)$ and $\widehat{\xi}_\nu(p) := \xi_\nu(p+N_\nu)$. Formula (4.5.5) takes a form

$$x(j) = 2^{-s}\alpha + \sum_{\nu=1}^{s}2^{-\nu}\sum_{p=0}^{N_\nu-1}\widehat{\xi}_\nu(p)\,\varphi_\nu(j-p\Delta_{\nu+1}). \tag{4.8.12}$$

We took advantage of the fact that $\varphi_s(0;\ j) \equiv f_s(0;\ j) \equiv 1$.

We will examine in more detail the properties of the signal φ_ν.

Lemma 4.8.4 *Valid is the equality*

$$\varphi_\nu(j) = (-1)^{\lfloor j/\Delta_\nu \rfloor}\,\delta_{N_\nu}\big(\lfloor j/\Delta_{\nu+1}\rfloor\big), \quad j \in \mathbb{Z}. \tag{4.8.13}$$

Proof The right side of (4.8.13) is N-periodic because $N = 2N_\nu\Delta_\nu = N_\nu\Delta_{\nu+1}$; the left side is N-periodic as well. Hence it is sufficient to verify equality (4.8.13) for $j \in 0 : N-1$.

The right side of (4.8.13) equals to 1 for $j \in 0 : \Delta_\nu - 1$, equals to -1 for $j \in \Delta_\nu : \Delta_{\nu+1} - 1$, and equals to zero for $j \in \Delta_{\nu+1} : N - 1$ (take into account that $N = N_\nu \Delta_{\nu+1}$). According to (4.8.12), $\varphi_\nu(j)$ has the same values in the indicated nodes j. The lemma is proved. □

Substituting $j - p\Delta_{\nu+1}$ instead of j in (4.8.13) we gain

$$\varphi_\nu(j - p\Delta_{\nu+1}) = (-1)^{\lfloor j/\Delta_\nu \rfloor}\, \delta_{N_\nu}\big(\lfloor j/\Delta_{\nu+1} \rfloor - p\big). \tag{4.8.14}$$

In case of $j \in 0 : N - 1$ the coefficient $(-1)^{\lfloor j/\Delta_\nu \rfloor}$ can be represented in another way. Let $j = (j_{s-1}, \ldots, j_0)_2$. Then

$$\left\lfloor \frac{j}{\Delta_\nu} \right\rfloor = \frac{j_{s-1}2^{s-1} + \cdots + j_\nu 2^\nu + j_{\nu-1}2^{\nu-1}}{2^{\nu-1}} = j_{s-1}2^{s-\nu} + \cdots + j_\nu 2 + j_{\nu-1},$$

so

$$(-1)^{\lfloor j/\Delta_\nu \rfloor} = (-1)^{j_{\nu-1}}. \tag{4.8.15}$$

In the following lemma we will use the operation $\oplus$ of bitwise summation modulo 2 (see Sect. 1.5).

Lemma 4.8.5 *Given* $k \in 0 : N - 1$, $k = (k_{s-1}, \ldots, k_0)_2$, *valid is the equality*

$$\varphi_\nu(j \oplus k) = (-1)^{k_{\nu-1}}\, \varphi_\nu\big(j - \lfloor k/\Delta_{\nu+1} \rfloor \Delta_{\nu+1}\big), \tag{4.8.16}$$

$$j \in 0 : N - 1.$$

Proof We write

$$j = \lfloor j/\Delta_{\nu+1} \rfloor \Delta_{\nu+1} + j_{\nu-1}\Delta_\nu + \langle j \rangle_{\Delta_\nu},$$

$$k = \lfloor k/\Delta_{\nu+1} \rfloor \Delta_{\nu+1} + k_{\nu-1}\Delta_\nu + \langle k \rangle_{\Delta_\nu}.$$

It is clear that

$$j \oplus k = \big(\lfloor j/\Delta_{\nu+1} \rfloor \oplus \lfloor k/\Delta_{\nu+1} \rfloor\big)\Delta_{\nu+1} + \langle j_{\nu-1} + k_{\nu-1} \rangle_2\, \Delta_\nu + \big(\langle j \rangle_{\Delta_\nu} \oplus \langle k \rangle_{\Delta_\nu}\big).$$

According to (4.8.13) and (4.8.15) we have

$$\varphi_\nu(j \oplus k) = (-1)^{j_{\nu-1}+k_{\nu-1}}\, \delta_{N_\nu}\big(\lfloor j/\Delta_{\nu+1} \rfloor \oplus \lfloor k/\Delta_{\nu+1} \rfloor\big).$$

Let us use the equality $\delta_{N_\nu}(a \oplus b) = \delta_{N_\nu}(a - b)$, where $a, b \in 0 : N_\nu - 1$ (it is true both for $a = b$ and for $a \neq b$). We gain

$$\varphi_\nu(j \oplus k) = (-1)^{j_{\nu-1}+k_{\nu-1}}\, \delta_{N_\nu}\big(\lfloor j/\Delta_{\nu+1} \rfloor - \lfloor k/\Delta_{\nu+1} \rfloor\big). \tag{4.8.17}$$

Formulae (4.8.14) and (4.8.15) yield

$$(-1)^{j_{\nu-1}}\,\delta_{N_\nu}\big(\lfloor j/\Delta_{\nu+1}\rfloor - \lfloor k/\Delta_{\nu+1}\rfloor\big) = \varphi_\nu\big(j - \lfloor k/\Delta_{\nu+1}\rfloor\,\Delta_{\nu+1}\big). \tag{4.8.18}$$

Combining (4.8.17) and (4.8.18) we come to (4.8.16). The lemma is proved. □

Corollary 4.8.1 *For $p \in 0 : N_\nu - 1$ we have*

$$\varphi_\nu(j \oplus p\Delta_{\nu+1}) = \varphi_\nu(j - p\Delta_{\nu+1}), \quad j \in 0 : N - 1. \tag{4.8.19}$$

4.8.4 Application of the decimation-in-time discrete Haar transform to a cyclic convolution does not produce a satisfactory result. More efficient is application of this transform to a dyadic convolution.

Let x and y be signals of $\mathbb{C}_N$. A signal $z \in \mathbb{C}_N$ with samples

$$z(j) = \sum_{k=0}^{N-1} x(k)\,y(j \oplus k), \quad j \in 0 : N - 1 \tag{4.8.20}$$

is referred to as a *dyadic convolution* of signals x and y.

Theorem 4.8.2 (Dyadic Convolution Theorem) *Let z be a dyadic convolution of signals x and y. If along with* (4.8.12) *we have expansions*

$$y(j) = 2^{-s}\beta + \sum_{\nu=1}^{s} 2^{-\nu} \sum_{p=0}^{N_\nu - 1} \widehat{\eta}_\nu(p)\,\varphi_\nu(j - p\Delta_{\nu+1}),$$

$$z(j) = 2^{-s}\gamma + \sum_{\nu=1}^{s} 2^{-\nu} \sum_{p=0}^{N_\nu - 1} \widehat{\zeta}_\nu(p)\,\varphi_\nu(j - p\Delta_{\nu+1}),$$

then it is necessary that $\gamma = \alpha\beta$ and

$$\widehat{\zeta}_\nu(p) = \sum_{q=0}^{N_\nu - 1} \widehat{\xi}_\nu(q)\,\widehat{\eta}_\nu(p \oplus q), \tag{4.8.21}$$

$$p \in 0 : N_\nu - 1, \quad \nu = 1, \ldots, s.$$

Proof We fix $j \in 0 : N - 1$. According to (4.8.19) we have

$$y(j \oplus k) = 2^{-s}\beta + \sum_{\nu=1}^{s} 2^{-\nu} \sum_{p=0}^{N_\nu - 1} \widehat{\eta}_\nu(p)\,\varphi_\nu\big((j \oplus k) \oplus p\Delta_{\nu+1}\big).$$

Since

$$(j \oplus k) \oplus p\Delta_{\nu+1} = k \oplus (j \oplus p\Delta_{\nu+1})$$
$$= k \oplus \big((\lfloor j/\Delta_{\nu+1}\rfloor \oplus p)\,\Delta_{\nu+1} + j_{\nu-1}\,\Delta_\nu + \langle j\rangle_{\Delta_\nu}\big),$$

equality (4.8.16) yields

$$y(j \oplus k) = 2^{-s}\beta + \sum_{\nu=1}^{s} 2^{-\nu} \sum_{p=0}^{N_\nu-1} (-1)^{j_{\nu-1}}\, \widehat{\eta}_\nu(p)\, \varphi_\nu\big(k - (\lfloor j/\Delta_{\nu+1}\rfloor \oplus p)\Delta_{\nu+1}\big).$$

Performing a change of variables $q = p \oplus \lfloor j/\Delta_{\nu+1}\rfloor$ we come to a formula

$$y(j \oplus k) = 2^{-s}\beta + \sum_{\nu=1}^{s} 2^{-\nu} \sum_{q=0}^{N_\nu-1} (-1)^{j_{\nu-1}}\, \widehat{\eta}_\nu\big(q \oplus \lfloor j/\Delta_{\nu+1}\rfloor\big)\, \varphi_\nu(k - q\Delta_{\nu+1}). \tag{4.8.22}$$

Let us substitute (4.8.12) and (4.8.22) into (4.8.20). Bearing in mind reality, orthogonality, and norming of the basic signals we gain

$$z(j) = 2^{-s}\alpha\beta + \sum_{\nu=1}^{s} 2^{-\nu} \sum_{q=0}^{N_\nu-1} (-1)^{j_{\nu-1}}\, \widehat{\xi}_\nu(q)\, \widehat{\eta}_\nu\big(q \oplus \lfloor j/\Delta_{\nu+1}\rfloor\big). \tag{4.8.23}$$

The following transform is based on formulae (4.8.14) and (4.8.15):

$$(-1)^{j_{\nu-1}}\, \widehat{\eta}_\nu\big(q \oplus \lfloor j/\Delta_{\nu+1}\rfloor\big) = \sum_{p=0}^{N_\nu-1} \widehat{\eta}_\nu(q \oplus p)\big[(-1)^{\lfloor j/\Delta_\nu\rfloor}\, \delta_{N_\nu}\big(\lfloor j/\Delta_{\nu+1}\rfloor - p\big)\big]$$
$$= \sum_{p=0}^{N_\nu-1} \widehat{\eta}_\nu(p \oplus q)\, \varphi_\nu(j - p\Delta_{\nu+1}). \tag{4.8.24}$$

Substituting (4.8.24) into (4.8.23) we come to the expansion

$$z(j) = 2^{-s}\alpha\beta + \sum_{\nu=1}^{s} 2^{-\nu} \sum_{p=0}^{N_\nu-1} \left\{ \sum_{q=0}^{N_\nu-1} \widehat{\xi}_\nu(q)\, \widehat{\eta}_\nu(p \oplus q) \right\} \varphi_\nu(j - p\Delta_{\nu+1}).$$

Uniqueness of expansion over an orthogonal basis guarantees that $\gamma = \alpha\beta$ holds, and that the sum in braces is equal to $\widehat{\zeta}_\nu(p)$. The theorem is proved. □

Formula (4.8.21) shows that coefficients of the ν-th level in the expansion of a dyadic convolution z over Haar basis related to decimation in time are obtained as a result of a dyadic convolution of expansion coefficients of the ν-th level of the signals x and y.

4.9 Second Sequence of Orthogonal Bases

4.9.1 We retain the notations $N = 2^s$, $N_\nu = N/2^\nu$, $\Delta_\nu = 2^{\nu-1}$. We will construct another sequence of orthogonal bases $w_\nu = \{w_\nu(k;\ j)\}_{k=0}^{N-1}$, $\nu = 0, 1, \dots, s$, with the aid of recurrent relations

$$w_0(k) = \delta_N(\cdot - k), \quad k \in 0 : N-1;$$

$$\begin{aligned} w_\nu(l + p\Delta_{\nu+1}) &= w_{\nu-1}(l + 2p\Delta_\nu) + w_{\nu-1}\big(l + (2p+1)\Delta_\nu\big), \\ w_\nu(l + \Delta_\nu + p\Delta_{\nu+1}) &= w_{\nu-1}(l + 2p\Delta_\nu) - w_{\nu-1}\big(l + (2p+1)\Delta_\nu\big), \end{aligned} \tag{4.9.1}$$

$$p \in 0 : N_\nu - 1, \quad l \in 0 : \Delta_\nu - 1, \quad \nu = 1, \dots, s.$$

These formulae differ from (4.2.1) only by the coefficient $\omega^l_{\Delta_{\nu+1}}$ being replaced with unity. A transition from the basis $w_{\nu-1}$ to the basis w_ν can be written in a single line:

$$w_\nu(l + \sigma\Delta_\nu + p\Delta_{\nu+1}) = w_{\nu-1}(l + 2p\Delta_\nu) + (-1)^\sigma\, w_{\nu-1}\big(l + (2p+1)\Delta_\nu\big), \tag{4.9.2}$$

$$p \in 0 : N_\nu - 1, \quad l \in 0 : \Delta_\nu - 1, \quad \sigma \in 0 : 1, \quad \nu = 1, \dots, s.$$

In particular, for $\nu = 1$ we gain

$$w_1(\sigma + 2p) = w_0(2p) + (-1)^\sigma\, w_0(2p+1), \tag{4.9.3}$$

$$p \in 0 : N_1 - 1, \quad \sigma \in 0 : 1.$$

How do the signals $w_s(k;\ j)$ look like? Answering this question needs some additional preparation.

4.9.2 We introduce a sequence of matrices

$$A_1 = \begin{bmatrix} 1 & 1 \\ 1 & -1 \end{bmatrix}; \quad A_\nu = \begin{bmatrix} A_{\nu-1} & A_{\nu-1} \\ A_{\nu-1} & -A_{\nu-1} \end{bmatrix}, \quad \nu = 2, \dots, s. \tag{4.9.4}$$

A matrix A_ν is referred to as a *Hadamard matrix*. It is a square matrix of order $\Delta_{\nu+1}$. We will suppose that the indices of its rows and columns vary from 0 to $\Delta_{\nu+1} - 1$.

Let $k, j \in 0 : \Delta_{\nu+1} - 1$, $k = (k_{\nu-1}, k_{\nu-2}, \dots, k_0)_2$ and $j = (j_{\nu-1}, j_{\nu-2}, \dots, j_0)_2$. We put

$$\{k, j\}_\nu = \sum_{\alpha=0}^{\nu-1} k_\alpha j_\alpha.$$

Theorem 4.9.1 *Elements of a Hadamard matrix satisfy to the formula*

$$A_\nu[k, j] = (-1)^{\{k,j\}_\nu}, \quad k, j \in 0 : \Delta_{\nu+1} - 1. \tag{4.9.5}$$

Proof When $\nu = 1$, the assertion is obvious because $A_1[k, j] = (-1)^{kj}$ holds for $k, j \in 0:1$. We perform an induction step from $\nu - 1$ to ν.

We take $k, j \in 0 : \Delta_{\nu+1} - 1$ and represent them in the following manner: $k = k_{\nu-1}2^{\nu-1} + l$, $j = j_{\nu-1}2^{\nu-1} + q$. Here $k_{\nu-1}, j_{\nu-1} \in 0:1$ and $l, q \in 0 : \Delta_\nu - 1$. According to (4.9.4) we have

$$A_\nu[k, j] = (-1)^{k_{\nu-1} j_{\nu-1}} A_{\nu-1}[l, q]. \tag{4.9.6}$$

Taking into account the inductive hypothesis we gain

$$A_\nu[k, j] = (-1)^{k_{\nu-1} j_{\nu-1} + \{l,q\}_{\nu-1}} = (-1)^{\{k,j\}_\nu}.$$

The theorem is proved. □

Later on we will need formula (4.9.6) in a form

$$A_\nu[l + \sigma\Delta_\nu, q + \tau\Delta_\nu] = (-1)^{\sigma\tau} A_{\nu-1}[l, q], \tag{4.9.7}$$

$$l, q \in 0 : \Delta_\nu - 1, \quad \sigma, \tau \in 0 : 1.$$

4.9.3 Let us get an explicit expression for the signals $w_\nu(k)$.

Theorem 4.9.2 *There holds a representation*

$$w_\nu(l + p\Delta_{\nu+1}) = \sum_{q=0}^{\Delta_{\nu+1}-1} A_\nu[l, q]\, w_0(q + p\Delta_{\nu+1}), \tag{4.9.8}$$

$$p \in 0 : N_\nu - 1, \quad l \in 0 : \Delta_{\nu+1} - 1, \quad \nu = 1, \ldots, s.$$

Proof When $\nu = 1$, formula (4.9.8) coincides with (4.9.3) if we replace σ by l in the latter one. We perform an induction step from $\nu - 1$ to ν.

We represent an index $l \in 0 : \Delta_{\nu+1} - 1$ in a form $l = \sigma\Delta_\nu + l'$, where $l' \in 0 : \Delta_\nu - 1$ and $\sigma \in 0:1$. On the basis of (4.9.2) and the inductive hypothesis we write

$$\begin{aligned}
w_\nu(l + p\Delta_{\nu+1}) &= w_\nu(l' + \sigma\Delta_\nu + p\Delta_{\nu+1}) \\
&= w_{\nu-1}(l' + 2p\Delta_\nu) + (-1)^\sigma\, w_{\nu-1}\big(l' + (2p+1)\Delta_\nu\big) \\
&= \sum_{q=0}^{\Delta_\nu - 1} A_{\nu-1}[l', q]\, w_0(q + 2p\Delta_\nu) \\
&\quad + (-1)^\sigma \sum_{q=0}^{\Delta_\nu - 1} A_{\nu-1}[l', q]\, w_0\big(q + (2p+1)\Delta_\nu\big).
\end{aligned}$$

Formula (4.9.7) yields

$$A_{\nu-1}[l', q] = A_\nu[l' + \sigma\Delta_\nu,\ q] = A_\nu[l, q],$$

$$(-1)^\sigma A_{\nu-1}[l', q] = A_\nu[l' + \sigma\Delta_\nu,\ q + \Delta_\nu] = A_\nu[l,\ q + \Delta_\nu].$$

Taking this into account we gain

$$\begin{aligned} w_\nu(l + p\Delta_{\nu+1}) &= \sum_{q=0}^{\Delta_\nu - 1} A_\nu[l, q]\, w_0(q + p\Delta_{\nu+1}) \\ &\quad + \sum_{q=0}^{\Delta_\nu - 1} A_\nu[l, q + \Delta_\nu]\, w_0(q + \Delta_\nu + p\Delta_{\nu+1}) \\ &= \sum_{q=0}^{\Delta_{\nu+1} - 1} A_\nu[l, q]\, w_0(q + p\Delta_{\nu+1}). \end{aligned}$$

The theorem is proved. □

When $\nu = s$, formula (4.9.8) takes a form

$$w_s(l;\ j) = \sum_{q=0}^{N-1} A_s[l, q]\, \delta_N(j - q) = A_s[l, j] = (-1)^{\{l, j\}_s},$$

$$l, j \in 0 : N - 1.$$

The functions

$$v_k(j) = (-1)^{\{k, j\}_s}, \quad k, j \in 0 : N - 1 \tag{4.9.9}$$

are referred to as the *discrete Walsh functions*. Thus,

$$w_s(k;\ j) = v_k(j), \quad k, j \in 0 : N - 1. \tag{4.9.10}$$

Figure 4.5 depicts Walsh functions for $N = 8$.

4.9.4 Consider the signals $w_\nu(k)$ given a fixed ν.

Theorem 4.9.3 *For each* $\nu \in 0 : s$ *the signals*

$$w_\nu(0),\ w_\nu(1),\ \ldots,\ w_\nu(N - 1) \tag{4.9.11}$$

are pairwise orthogonal and $\|w_\nu(k)\|^2 = 2^\nu$ *holds for all* $k \in 0 : N - 1$.

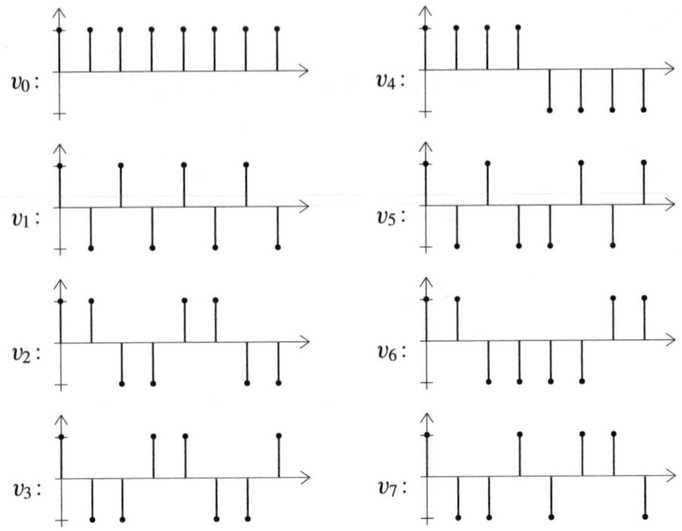

Fig. 4.5 Walsh functions for $N = 8$

Proof The assertion is known to be true for $\nu = 0$. Let $\nu \in 1 : s$. We take $k,\ k' \in 0 : N - 1$ and represent them in a form $k = l + p\Delta_{\nu+1},\ k' = l' + p'\Delta_{\nu+1}$, where $l,\ l' \in 0 : \Delta_{\nu+1} - 1$ and $p,\ p' \in 0 : N_\nu - 1$. According to (4.9.8) and Lemma 2.1.4 we have

$$\langle w_\nu(k),\ w_\nu(k')\rangle = \sum_{q,q'=0}^{\Delta_{\nu+1}-1} A_\nu[l, q]\, A_\nu[l', q']\, \delta_N\big(q - q' + (p - p')\Delta_{\nu+1}\big).$$

If $p \neq p'$ then $\langle w_\nu(k),\ w_\nu(k')\rangle = 0$. Let $p = p'$. In this case we have

$$\langle w_\nu(k),\ w_\nu(k')\rangle = \sum_{q=0}^{\Delta_{\nu+1}-1} A_\nu[l, q]\, A_\nu[l', q].$$

Note that by virtue of (4.9.5) the matrix A_ν is symmetric. Furthermore,

$$A_\nu A_\nu = 2^\nu I_{2^\nu}, \quad \nu = 1, 2, \ldots, \tag{4.9.12}$$

where I_{2^ν} is an identity matrix of order 2^ν. For $\nu = 1$ this is evident. We perform an induction step from $\nu - 1$ to ν. According to (4.9.4)

$$A_\nu A_\nu = \begin{bmatrix} A_{\nu-1} & A_{\nu-1} \\ A_{\nu-1} & -A_{\nu-1} \end{bmatrix} \begin{bmatrix} A_{\nu-1} & A_{\nu-1} \\ A_{\nu-1} & -A_{\nu-1} \end{bmatrix}$$
$$= \begin{bmatrix} 2A_{\nu-1}A_{\nu-1} & \mathbb{O} \\ \mathbb{O} & 2A_{\nu-1}A_{\nu-1} \end{bmatrix}$$
$$= \begin{bmatrix} 2^\nu I_{2^{\nu-1}} & \mathbb{O} \\ \mathbb{O} & 2^\nu I_{2^{\nu-1}} \end{bmatrix} = 2^\nu I_{2^\nu}.$$

Validity of equality (4.9.12) is ascertained. On the basis of the stated properties of a Hadamard matrix A_ν we gain

$$\langle w_\nu(k),\, w_\nu(k')\rangle = \sum_{q=0}^{\Delta_{\nu+1}-1} A_\nu[l,q]\, A_\nu[q,l'] = (A_\nu A_\nu)[l,l'] = 2^\nu I_{2^\nu}[l,l'].$$

Now we can make a conclusion that the scalar product $\langle w_\nu(k),\, w_\nu(k')\rangle$ is nonzero only when $p = p'$ and $l = l'$, i.e. only when $k = k'$. In the latter case $\|w_\nu(k)\|^2 = 2^\nu$ for all $k \in 0 : N-1$.

The theorem is proved. □

Essentially, we ascertained that for each $\nu \in 0 : s$ signals (4.9.11) form an orthogonal basis in a space $\mathbb{C}_N$. In particular, the Walsh functions $v_0,\, v_1,\, \ldots,\, v_{N-1}$ of form (4.9.9) constitute an orthogonal basis. This basis is referred to as *Walsh–Hadamard basis* or, more often, *Walsh basis*. Note that $\|v_k\|^2 = N$ holds for all $k \in 0 : N-1$.

4.10 Fast Walsh Transform

4.10.1 Any signal $x \in \mathbb{C}_N$ can be expanded over Walsh basis:

$$x = \frac{1}{N} \sum_{k=0}^{N-1} x_s(k)\, v_k. \tag{4.10.1}$$

Here $x_s(k) = \langle x, v_k\rangle$ or, in the explicit form,

$$x_s(k) = \sum_{j=0}^{N-1} x(j)\, v_k(j), \quad k \in 0 : N-1. \tag{4.10.2}$$

The transform $\mathcal{W}_N$ that maps a signal $x \in \mathbb{C}_N$ to a signal $x_s = \mathcal{W}_N(x)$ with components (4.10.2) is referred to as the *discrete Walsh transform* (DWT). By analogy with DFT, a signal x_s is referred to as a *Walsh spectrum* of a signal x. Formula (4.10.1) can be interpreted as the inversion formula for DWT.

We will consider a question of fast calculation of a Walsh spectrum.

4.10.2 For each $\nu \in 0 : s$ a signal $x \in \mathbb{C}_N$ can be expanded over the basis (4.9.11):

$$x = \frac{1}{2^\nu} \sum_{k=0}^{N-1} x_\nu(k)\, w_\nu(k). \tag{4.10.3}$$

Here $x_\nu(k) = \langle x,\ w_\nu(k)\rangle$. In case of $\nu = s$ formula (4.10.3) coincides with (4.10.1). It follows from (4.9.10).

The recurrent relations (4.9.1) for the basic functions generate the recurrent relations for the coefficients of expansion (4.10.3):

$$x_0(k) = x(k), \quad k \in 0 : N-1;$$

$$\begin{aligned} x_\nu(l + p\Delta_{\nu+1}) &= x_{\nu-1}(l + 2p\Delta_\nu) + x_{\nu-1}\big(l + (2p+1)\Delta_\nu\big), \\ x_\nu(l + \Delta_\nu + p\Delta_{\nu+1}) &= x_{\nu-1}(l + 2p\Delta_\nu) - x_{\nu-1}\big(l + (2p+1)\Delta_\nu\big), \end{aligned} \tag{4.10.4}$$

$$p \in 0 : N_\nu - 1, \quad l \in 0 : \Delta_\nu - 1, \quad \nu = 1, \ldots, s.$$

When $\nu = s$, we obtain the Walsh spectrum x_s of the signal x.

Scheme (4.10.4) is referred to as the *decimation-in-time fast Walsh transform*. This scheme requires only additions; the number of operations is $N \log_2 N$.

Relation (4.10.4) can be inverted:

$$\begin{aligned} x_{\nu-1}(l + 2p\Delta_\nu) &= \tfrac{1}{2}\big[x_\nu(l + p\Delta_{\nu+1}) + x_\nu(l + \Delta_\nu + p\Delta_{\nu+1})\big], \\ x_{\nu-1}\big(l + (2p+1)\Delta_\nu\big) &= \tfrac{1}{2}\big[x_\nu(l + p\Delta_{\nu+1}) - x_\nu(l + \Delta_\nu + p\Delta_{\nu+1})\big], \end{aligned} \tag{4.10.5}$$

$$p \in 0 : N_\nu - 1, \quad l \in 0 : \Delta_\nu - 1, \quad \nu = s, s-1, \ldots, 1.$$

Herein $x(k) = x_0(k)$, $k \in 0 : N-1$. Scheme (4.10.5) makes it possible to quickly reconstruct the samples of a signal x given in form (4.10.1). The reconstruction is performed on the main period.

4.10.3 The sequence of orthogonal bases $\{w_\nu(k;\ j)\}_{k=0}^{N-1}$, $\nu = 0, 1, \ldots, s$, generates wavelet bases and a wavelet packet. This is done in the same way as in case of the sequence $\{f_\nu(k;\ j)\}_{k=0}^{N-1}$, $\nu = 0, 1, \ldots, s$ (see par. 4.4.1). The block structure of the bases w_ν is utilized; this structure follows from the formula analogous to (4.2.8):

$$w_\nu(l + p\Delta_{\nu+1};\ j) \equiv w_\nu(l;\ j - p\Delta_{\nu+1}),$$

$$p \in 0 : N_\nu - 1, \quad l \in 0 : \Delta_{\nu+1} - 1, \quad \nu = 0, 1, \ldots s.$$

We will not go into details. We just note that the branching scheme (a) in Fig. 4.2 in this case also generates Haar basis related to decimation in time.

4.11 Ordering of Walsh Functions

4.11.1 Signals of the exponential basis are ordered by frequency. Index k in a notation

$$u_k(j) = \omega_N^{kj} = \exp\big(i\tfrac{2\pi k}{N} j\big), \quad k = 0, 1, \dots, N-1$$

is a *frequency* of the signal $u_k(j)$ in the following sense: while j varies from 0 to N (to the beginning of the next period), the argument of a complex number $u_k(j)$ being equal to $\frac{2\pi k}{N} j$ monotonically increases from 0 to $2\pi k$, i.e. a point $u_k(j)$ runs around the unit circle of the complex plane k times.

We will consider Walsh functions

$$v_k(j) = (-1)^{\{k,j\}_s} = \exp\big(i\pi\{k, j\}_s\big), \quad k = 0, 1, \dots, N-1,$$

from this point of view. Unfortunately, the value $\{k, j\}_s$ does not increase monotonically together with j. An example of this value's behavior for $N = 2^3$ and $k = 2 = (0, 1, 0)_2$ is presented in Table 4.3.

To attain monotonicity, we represent $v_k(j)$ in a form

$$v_k(j) = (-1)^{\sum_{\alpha=0}^{s-1} k_\alpha (j_\alpha + j_{\alpha+1}2 + \cdots + j_{s-1}2^{s-1-\alpha})}.$$

Note that

$$j/2^\alpha = j_{s-1}2^{s-1-\alpha} + \cdots + j_{\alpha+1}2 + j_\alpha + j_{\alpha-1}2^{-1} + \cdots + j_0 2^{-\alpha},$$

so

$$\lfloor j/2^\alpha \rfloor = j_{s-1}2^{s-1-\alpha} + \cdots + j_{\alpha+1}2 + j_\alpha.$$

We gain

$$v_k(j) = (-1)^{\sum_{\alpha=0}^{s-1} k_\alpha \lfloor j/2^\alpha \rfloor}, \quad j \in 0 : N-1.$$

Introducing a notation

$$\theta_k(j) = \sum_{\alpha=0}^{s-1} k_\alpha \lfloor j/2^\alpha \rfloor$$

we come to a representation

Table 4.3 Behavior of the value $\{k, j\}_3$

j	0	1	2	3	4	5	6	7
$\{k, j\}_3$	0	0	1	1	0	0	1	1

$$v_k(j) = (-1)^{\theta_k(j)}, \quad j \in 0 : N-1. \tag{4.11.1}$$

Formula (4.11.1) is valid for $j = N = 2^s$ as well. In this case we have

$$\theta_k(N) = \sum_{\alpha=0}^{s-1} k_\alpha 2^{s-\alpha} = 2 \sum_{\alpha=0}^{s-1} k_\alpha 2^{s-1-\alpha} = 2\,\mathrm{rev}_s(k)$$

(a definition of the permutation rev_s can be found in Sect. 1.4). Therefore, the right side of (4.11.1) for $j = N$ is equal to unity. By virtue of N-periodicity, the left side of (4.11.1) equals to unity too. Indeed, $v_k(N) = v_k(0) = 1$. Thus, equality (4.11.1) holds for $j \in 0 : N$.

We rewrite (4.11.1) in a form

$$v_k(j) = \exp\big(i\pi\theta_k(j)\big), \quad j \in 0 : N.$$

It is evident that the function $\theta_k(j)$ varies from 0 to $2\,\mathrm{rev}_s(k)$ *monotonically non-decreasing* while j increases from 0 to N. As a consequence, the argument $\pi\theta_k(j)$ of a complex number $v_k(j)$ varies from 0 to $2\pi\,\mathrm{rev}_s(k)$ monotonically non-decreasing while j increases from 0 to N. Hence a point $v_k(j)$ runs around the unit circle of the complex plane $\mathrm{rev}_s(k)$ times. Thus a number $\mathrm{rev}_s(k)$ is treated as a frequency of a function v_k.

We denote $\widehat{v}_k = v_{\mathrm{rev}_s(k)}$. A function $\widehat{v}_k$ has a frequency of k because $\mathrm{rev}_s\big(\mathrm{rev}_s(k)\big) = k$. It can be represented in a form

$$\widehat{v}_k(j) = (-1)^{\sum_{\alpha=0}^{s-1} k_{s-1-\alpha} j_\alpha}, \quad k, j \in 0 : N-1.$$

Walsh functions $\widehat{v}_0, \widehat{v}_1, \ldots, \widehat{v}_{N-1}$ are ordered by frequency. They comprise a *Walsh–Paley basis* of the space $\mathbb{C}_N$.

Figure 4.6 depicts the functions $\widehat{v}_k(j)$ for $N = 8$.

4.11.2 There exists another ordering of Walsh functions—by the number of sign changes on the main period. To clarify this matter we need to return to Hadamard matrices (see par. 4.9.2).

We denote by $\mathrm{wal}_\nu(k)$ the number of sign changes in a row of Hadamard matrix A_ν with the index $k \in 0 : \Delta_{\nu+1} - 1$. According to (4.9.4) we have $\mathrm{wal}_1(0) = 0$ and $\mathrm{wal}_1(1) = 1$.

Theorem 4.11.1 *The following recurrent relations hold:*

$$\mathrm{wal}_1(k) = k, \quad k \in 0 : 1;$$

$$\mathrm{wal}_\nu(2k) = \mathrm{wal}_{\nu-1}(k), \tag{4.11.2}$$

$$\mathrm{wal}_\nu(2k+1) = 2^\nu - 1 - \mathrm{wal}_{\nu-1}(k), \tag{4.11.3}$$

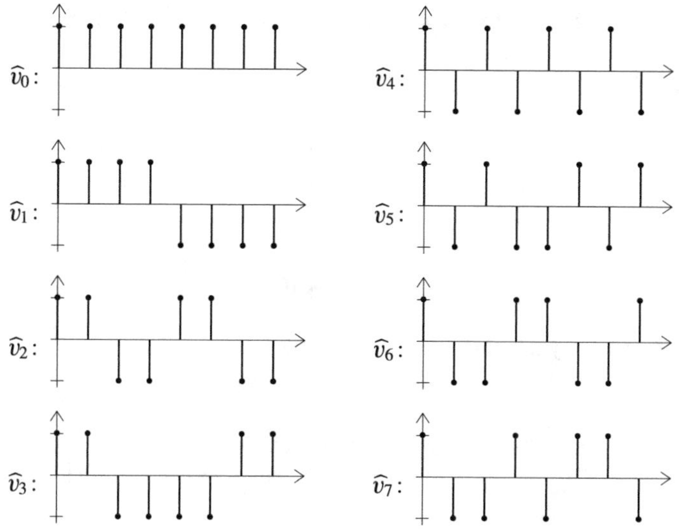

Fig. 4.6 Walsh functions ordered by frequency, for $N = 8$

$$k \in 0 : \Delta_\nu - 1, \quad \nu = 2, \ldots, s.$$

Proof The first relation is true. We will verify (4.11.2) and (4.11.3). Recall that $A_\nu[k, j] = (-1)^{\{k,j\}_\nu}$. Hence it follows that for $k, j \in 0 : \Delta_\nu - 1$ there hold

$$A_\nu[2k, 2j] = A_\nu[2k, 2j + 1] = A_\nu[2k + 1, 2j] = A_{\nu-1}[k, j],$$

$$A_\nu[2k + 1, 2j + 1] = -A_{\nu-1}[k, j].$$

At first, let us show that (4.11.2) holds. Since $A_\nu[2k, 2j] = A_\nu[2k, 2j + 1]$, we may not take into account the elements $A_\nu[2k, 2j + 1]$ while determining the number of sign changes. The remaining elements are $A_\nu[2k, 2j] = A_{\nu-1}[k, j]$; they have $\mathrm{wal}_{\nu-1}(k)$ sign changes by a definition. Relation (4.11.2) is ascertained.

Let us rewrite equality (4.11.3) in a form

$$\mathrm{wal}_\nu(2k) + \mathrm{wal}_\nu(2k + 1) = 2^\nu - 1. \tag{4.11.4}$$

We introduce submatrices of order two

$$G_j = \begin{bmatrix} A_\nu[2k, j - 1] & A_\nu[2k, j] \\ A_\nu[2k + 1, j - 1] & A_\nu[2k + 1, j] \end{bmatrix},$$

$$j = 1, \ldots, \Delta_{\nu+1} - 1.$$

We will show that one of the rows of G_j has a sign change while another one does not. Let us consider two cases.

(a) $j = 2j' + 1,\ j' \in 0 : \Delta_\nu - 1$. We write

$$G_j = \begin{bmatrix} A_\nu[2k, 2j'] & A_\nu[2k, 2j' + 1] \\ A_\nu[2k + 1, 2j'] & A_\nu[2k + 1, 2j' + 1] \end{bmatrix} = \begin{bmatrix} A_{\nu-1}[k, j'] & A_{\nu-1}[k, j'] \\ A_{\nu-1}[k, j'] & -A_{\nu-1}[k, j'] \end{bmatrix}.$$

It is obvious that only the second row of G_j has a sign change.

(b) $j = 2j',\ j' \in 1 : \Delta_\nu - 1$. In this case we have

$$G_j = \begin{bmatrix} A_\nu[2k, 2(j' - 1) + 1] & A_\nu[2k, 2j'] \\ A_\nu[2k + 1, 2(j' - 1) + 1] & A_\nu[2k + 1, 2j'] \end{bmatrix} = \begin{bmatrix} A_{\nu-1}[k, j' - 1] & A_{\nu-1}[k, j'] \\ -A_{\nu-1}[k, j' - 1] & A_{\nu-1}[k, j'] \end{bmatrix}.$$

We see that G_j has only one sign change either in the first or in the second row.

The sequence $G_1, \ldots, G_{\Delta_{\nu+1}-1}$ accumulates $\Delta_{\nu+1} - 1$ sign changes in the rows of the matrix A_ν with indices $2k$ and $2k + 1$, which conforms to (4.11.4).

The theorem is proved. □

Corollary 4.11.1 *The mapping* $k \to \mathrm{wal}_\nu(k)$ *is a permutation of the set* $\{0, 1, \ldots, \Delta_{\nu+1} - 1\}$.

Indeed, this is evident for $\nu = 1$. If it is true for $\nu - 1$ then it is true for ν as well, because by virtue of (4.11.2) the even indices $\mathrm{wal}_\nu(2k)$ contain numbers from 0 to $\Delta_\nu - 1$, and the odd indices $\mathrm{wal}_\nu(2k + 1)$ in accordance with (4.11.3) contain numbers from Δ_ν to $\Delta_{\nu+1} - 1$.

4.11.3 Recall that $A_s[k, j] = v_k(j)$ for $k = 0, 1, \ldots, N - 1$ (see par. 4.9.3). Therefore the value $\mathrm{wal}_s(k)$ is the number of sign changes of the Walsh function $v_k(j)$ on the main period. As it was ascertained above, a number of sign changes differs for different Walsh functions and varies from 0 to $N - 1$.

We denote $\widetilde{v}_k = v_{\mathrm{wal}_s^{-1}(k)}$. As long as $\mathrm{wal}_s\big(\mathrm{wal}_s^{-1}(k)\big) = k$, a function $\widetilde{v}_k$ has k sign changes on the main period. Walsh functions $\widetilde{v}_0, \widetilde{v}_1, \ldots, \widetilde{v}_{N-1}$ are ordered by the number of sign changes.

Figure 4.7 shows functions $\widetilde{v}_k(j)$ for $N = 8$.

4.11.4 With the aid of the permutations rev_s and wal_s we defined frequency and number of sign changes of Walsh functions. It turns out that these permutations are bound with each other through the permutation grey_s (see Sect. 1.4).

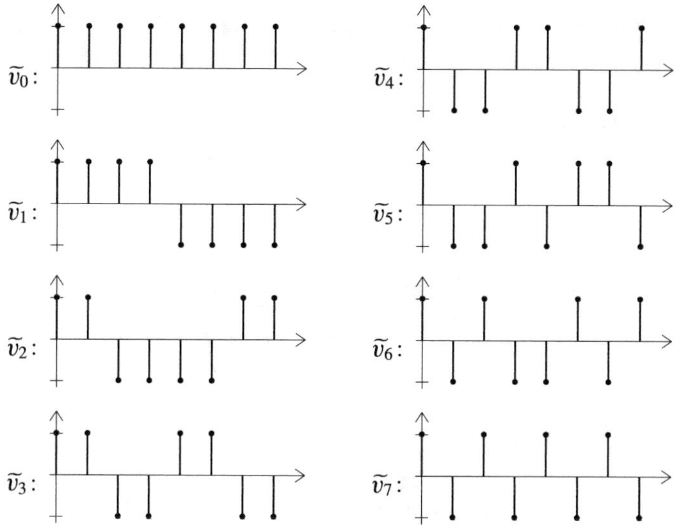

Fig. 4.7 Walsh functions ordered by sign changes, for $N = 8$

Theorem 4.11.2 *For each* $\nu \in 1 : s$ *there holds an equality*

$$\mathrm{grey}_\nu\big(\mathrm{wal}_\nu(k)\big) = \mathrm{rev}_\nu(k), \quad k \in 0 : \Delta_{\nu+1} - 1. \tag{4.11.5}$$

Proof Let us remind the recurrent relations for the permutations rev_ν and grey_ν:

$$\mathrm{rev}_1(k) = k, \quad k \in 0 : 1;$$

$$\mathrm{rev}_\nu(2k) = \mathrm{rev}_{\nu-1}(k),$$

$$\mathrm{rev}_\nu(2k + 1) = 2^{\nu-1} + \mathrm{rev}_{\nu-1}(k),$$

$$k \in 0 : \Delta_\nu - 1, \quad \nu = 2, 3, \ldots ;$$

$$\mathrm{grey}_1(k) = k, \quad k \in 0 : 1;$$

$$\mathrm{grey}_\nu(k) = \mathrm{grey}_{\nu-1}(k),$$

$$\mathrm{grey}_\nu(2^\nu - 1 - k) = 2^{\nu-1} + \mathrm{grey}_{\nu-1}(k),$$

$$k \in 0 : \Delta_\nu - 1, \quad \nu = 2, 3, \ldots$$

For $\nu = 1$ formula (4.11.5) is valid. We perform an induction step from $\nu - 1$ to ν.

We will consider the cases of even and odd k separately. Let $k = 2k'$, $k' \in 0 : \Delta_\nu - 1$. According to Theorem 4.11.1 and to the inductive hypothesis we have

$$\begin{aligned}\mathrm{grey}_\nu\big(\mathrm{wal}_\nu(2k')\big) &= \mathrm{grey}_\nu\big(\mathrm{wal}_{\nu-1}(k')\big)\\ &= \mathrm{grey}_{\nu-1}\big(\mathrm{wal}_{\nu-1}(k')\big) = \mathrm{rev}_{\nu-1}(k') = \mathrm{rev}_\nu(2k').\end{aligned}$$

Similarly, for $k = 2k' + 1$, $k' \in 0 : \Delta_\nu - 1$, we gain

$$\begin{aligned}\mathrm{grey}_\nu\big(\mathrm{wal}_\nu(2k'+1)\big) &= \mathrm{grey}_\nu\big(2^\nu - 1 - \mathrm{wal}_{\nu-1}(k')\big)\\ &= 2^{\nu-1} + \mathrm{grey}_{\nu-1}\big(\mathrm{wal}_{\nu-1}(k')\big)\\ &= 2^{\nu-1} + \mathrm{rev}_{\nu-1}(k') = \mathrm{rev}_\nu(2k'+1).\end{aligned}$$

The theorem is proved. □

Corollary 4.11.2 *Valid is the equality*

$$\mathrm{wal}_s^{-1}(k) = \mathrm{rev}_s\big(\mathrm{grey}_s(k)\big), \quad k \in 0 : N-1.$$

Indeed, according to (4.11.5) we have

$$\mathrm{grey}_s\big(\mathrm{wal}_s(k)\big) = \mathrm{rev}_s(k), \quad k \in 0 : N-1.$$

Now we pass to the inverse mappings:

$$\mathrm{wal}_s^{-1}\big(\mathrm{grey}_s^{-1}(k)\big) = \mathrm{rev}_s^{-1}(k), \quad k \in 0 : N-1.$$

Replacing k by $\mathrm{grey}_s(k)$ in the latter relation we gain

$$\mathrm{wal}_s^{-1}(k) = \mathrm{rev}_s^{-1}\big(\mathrm{grey}_s(k)\big), \quad k \in 0 : N-1.$$

It is remaining to take into account that $\mathrm{rev}_s^{-1}(l) = \mathrm{rev}_s(l)$ for $l \in 0 : N-1$.

4.12 Sampling Theorem in Walsh Basis

4.12.1 Let us write down the expansion of a signal $x \in \mathbb{C}_N$ over the Walsh basis ordered by frequency:

$$x = \frac{1}{N}\sum_{k=0}^{N-1} \xi(k)\,\widehat{v}_k, \tag{4.12.1}$$

where $\xi(k) = \langle x, \widehat{v}_k\rangle$. We are interested in a case when $\xi(k) = 0$ for $k \in \Delta_{\nu+1} : N-1$, $\nu \in 0 : s-1$.

We denote

$$h_\nu(j) = \sum_{q=0}^{N_\nu - 1} \delta_N(j-q).$$

The function $h_\nu(j)$ is an N-periodic step. On the main period, it is equal to unity for $j \in 0 : N_\nu - 1$ and is equal to zero for $j \in N_\nu : N-1$.

Theorem 4.12.1 *Consider expansion* (4.12.1). *If there holds* $\xi(k) = 0$ *for* $k \in \Delta_{\nu+1} : N-1,\ \nu \in 0 : s-1$, *then*

$$x(j) = \sum_{l=0}^{\Delta_{\nu+1}-1} x(lN_\nu)\, h_\nu(j - lN_\nu), \quad j \in 0 : N-1. \tag{4.12.2}$$

Formula (4.12.2) shows that in the premises of the theorem the signal $x(j)$ is a step-function. It equals to $x(lN_\nu)$ for $j \in lN_\nu : (l+1)N_\nu - 1$, $l = 0, 1, \ldots, \Delta_{\nu+1} - 1$.

4.12.2 We will prepend the proof of the theorem with a few auxiliary assertions.

Lemma 4.12.1 *Valid is the formula*

$$\delta_{\Delta_{\nu+1}}\big(\mathrm{rev}_s(j)\big) = \sum_{q=0}^{N_\nu-1} \delta_N(j-q), \quad j \in 0 : N-1. \tag{4.12.3}$$

Proof If $j = (j_{s-1}, j_{s-2}, \ldots, j_0)_2$ then there holds

$$\mathrm{rev}_s(j) = j_0 2^{s-1} + \cdots + j_{s-\nu-1} 2^\nu + j_{s-\nu} 2^{\nu-1} + \cdots + j_{s-1}.$$

Denote $p = j_0 2^{s-\nu-1} + \cdots + j_{s-\nu-1}$ and $r = j_{s-\nu} 2^{\nu-1} + \cdots + j_{s-1}$. Then $\mathrm{rev}_s(j) = p\Delta_{\nu+1} + r$, herein $r \in 0 : \Delta_{\nu+1} - 1$. On the strength of $\Delta_{\nu+1}$-periodicity of the unit pulse $\delta_{\Delta_{\nu+1}}$ we have

$$\delta_{\Delta_{\nu+1}}\big(\mathrm{rev}_s(j)\big) = \delta_{\Delta_{\nu+1}}(r). \tag{4.12.4}$$

Let $j \in 0 : N_\nu - 1$. In this case $j_{s-1} = j_{s-2} = \cdots = j_{s-\nu} = 0$. In particular, $r = 0$. According to equality (4.12.4) we get $\delta_{\Delta_{\nu+1}}\big(\mathrm{rev}_s(j)\big) = 1$. And the right side of (4.12.3) also equals to unity for the given indices j.

If $j \in N_\nu : N-1$ then at least one of the digits $j_{s-1}, j_{s-2}, \ldots, j_{s-\nu}$ is nonzero. As a consequence we gain $r \neq 0$, i.e. $r \in 1 : \Delta_{\nu+1} - 1$. Formula (4.12.4) yields $\delta_{\Delta_{\nu+1}}\big(\mathrm{rev}_s(j)\big) = 0$. The right side of (4.12.3) also equals to zero for the mentioned indices j.

The lemma is proved. □

Formula (4.12.3) can be compactly rewritten in a form $\delta_{\Delta_{\nu+1}}\big(\mathrm{rev}_s(j)\big) = h_\nu(j)$ whence it follows that

$$h_\nu\big(\mathrm{rev}_s(j)\big) = \delta_{\Delta_{\nu+1}}(j), \quad j \in 0 : N-1. \tag{4.12.5}$$

Lemma 4.12.2 *For* $j = (j_{s-1}, j_{s-2}, \dots, j_0)_2$ *there holds*

$$\delta_{\Delta_{\nu+1}}(j) = \prod_{\alpha=0}^{\nu-1} \delta_2(j_\alpha). \tag{4.12.6}$$

Proof We have

$$j = j_{s-1}2^{s-1} + \cdots + j_\nu 2^\nu + j_{\nu-1}2^{\nu-1} + \cdots + j_0.$$

Denote $p' = j_{s-1}2^{s-\nu-1} + \cdots + j_\nu$ and $r' = j_{\nu-1}2^{\nu-1} + \cdots + j_0$. Then $j = p'\Delta_{\nu+1} + r'$ and

$$\delta_{\Delta_{\nu+1}}(j) = \delta_{\Delta_{\nu+1}}(r'). \tag{4.12.7}$$

Two cases are possible.

(1) $j_0 = j_1 = \cdots = j_{\nu-1} = 0$. In this case $r' = 0$. According to (4.12.7) we have $\delta_{\Delta_{\nu+1}}(j) = 1$. It is obvious that the right side of (4.12.6) equals to unity as well.

(2) At least one of the digits $j_0, j_1, \dots, j_{\nu-1}$ is nonzero. Let $j_{\alpha'} = 1$ for some $\alpha' \in 0 : \nu - 1$. Then the factor $\delta_2(j_{\alpha'})$ in the right side of (4.12.6) is equal to zero, so the whole product equals to zero. Since $r' \in 1 : \Delta_{\nu+1} - 1$ in this case, equality (4.12.7) yields $\delta_{\Delta_{\nu+1}}(j) = 0$.

The lemma is proved. □

Lemma 4.12.3 *The following expansion holds*

$$h_\nu(j) = \frac{1}{\Delta_{\nu+1}} \sum_{k=0}^{\Delta_{\nu+1}-1} \widehat{v}_k(j), \quad j \in 0 : N - 1. \tag{4.12.8}$$

Proof Denote the right side of (4.12.8) as $f_\nu(j)$. We will show that

$$f_\nu\big(\mathrm{rev}_s(j)\big) = \delta_{\Delta_{\nu+1}}(j). \tag{4.12.9}$$

According to the definition of Walsh functions $\widehat{v}_k$ we have

$$\begin{aligned} f_\nu\big(\mathrm{rev}_s(j)\big) &= \frac{1}{\Delta_{\nu+1}} \sum_{k=0}^{\Delta_{\nu+1}-1} (-1)^{\{\mathrm{rev}_s(k),\, \mathrm{rev}_s(j)\}_s} \\ &= \frac{1}{\Delta_{\nu+1}} \sum_{k=0}^{\Delta_{\nu+1}-1} \prod_{\alpha=0}^{s-1} (-1)^{k_\alpha j_\alpha}. \end{aligned}$$

Since $k = (k_{\nu-1}, k_{\nu-2}, \ldots, k_0)_2$, we write

$$f_\nu\big(\mathrm{rev}_s(j)\big) = \frac{1}{\Delta_{\nu+1}} \sum_{k_{\nu-1}=0}^{1} \cdots \sum_{k_0=0}^{1} \prod_{\alpha=0}^{\nu-1} (-1)^{k_\alpha j_\alpha}$$
$$= \frac{1}{\Delta_{\nu+1}} \prod_{\alpha=0}^{\nu-1} \sum_{k_\alpha=0}^{1} (-1)^{k_\alpha j_\alpha}.$$

Let us use the formula

$$\frac{1}{2} \sum_{k_\alpha=0}^{1} (-1)^{k_\alpha j_\alpha} = \frac{1}{2} \sum_{k_\alpha=0}^{1} \omega_2^{k_\alpha j_\alpha} = \delta_2(j_\alpha).$$

We gain

$$f_\nu\big(\mathrm{rev}_s(j)\big) = \prod_{\alpha=0}^{\nu-1} \delta_2(j_\alpha).$$

Now (4.12.9) follows from (4.12.6).

Relations (4.12.9) and (4.12.5) yield $f_\nu\big(\mathrm{rev}_s(j)\big) = h_\nu\big(\mathrm{rev}_s(j)\big)$. Replacing j by $\mathrm{rev}_s(j)$ in the latter equality we come to (4.12.8).

The lemma is proved. □

Formula (4.12.8) is an expansion of the step h_ν over the Walsh basis ordered by frequency.

Lemma 4.12.4 *For all k and l from the set* $0 : \Delta_{\nu+1} - 1$ *there holds*

$$\{\mathrm{rev}_\nu(l), k\}_\nu = \{\mathrm{rev}_s(k), l N_\nu\}_s. \tag{4.12.10}$$

Proof Let $l = (l_{\nu-1}, l_{\nu-2}, \ldots, l_0)_2$ and $k = (k_{\nu-1}, k_{\nu-2}, \ldots, k_0)_2$. Then $\mathrm{rev}_\nu(l) = (l_0, l_1, \ldots, l_{\nu-1})_2$ and

$$\{\mathrm{rev}_\nu(l), k\}_\nu = l_{\nu-1}k_0 + l_{\nu-2}k_1 + \cdots + l_0 k_{\nu-1}. \tag{4.12.11}$$

Moreover,

$$\mathrm{rev}_s(k) = k_0 2^{s-1} + k_1 2^{s-2} + \cdots + k_{\nu-1} 2^{s-\nu},$$

$$l N_\nu = l_{\nu-1} 2^{s-1} + l_{\nu-2} 2^{s-2} + \cdots + l_0 2^{s-\nu},$$

so that

$$\{\mathrm{rev}_s(k), l N_\nu\}_s = k_0 l_{\nu-1} + k_1 l_{\nu-2} + \cdots + k_{\nu-1} l_0. \tag{4.12.12}$$

Comparing (4.12.11) and (4.12.12) we come to (4.12.10). The lemma is proved. □

Lemma 4.12.5 *For $j \in 0 : N-1$ and $l \in 0 : \Delta_{\nu+1}-1$ there holds*

$$h_\nu(j \oplus lN_\nu) = h_\nu(j - lN_\nu). \tag{4.12.13}$$

Proof Let us use the fact that $\delta_N(j \oplus k) = \delta_N(j-k)$ for all j and k from $0 : N-1$. The definition of h_ν yields

$$\begin{aligned} h_\nu(j \oplus lN_\nu) &= \sum_{q=0}^{N_\nu-1} \delta_N\big((j \oplus lN_\nu) - q\big) = \sum_{q=0}^{N_\nu-1} \delta_N\big((j \oplus lN_\nu) \oplus q\big) \\ &= \sum_{q=0}^{N_\nu-1} \delta_N\big(j \oplus (lN_\nu \oplus q)\big). \end{aligned}$$

Since $lN_\nu = l_{\nu-1}2^{s-1} + l_{\nu-2}2^{s-2} + \cdots + l_0 2^{s-\nu}$ and $q = q_{s-\nu-1}2^{s-\nu-1} + \cdots + q_0$, we have $lN_\nu \oplus q = lN_\nu + q$. Hence

$$\begin{aligned} h_\nu(j \oplus lN_\nu) &= \sum_{q=0}^{N_\nu-1} \delta_N\big(j \oplus (lN_\nu + q)\big) = \sum_{q=0}^{N_\nu-1} \delta_N\big(j - (lN_\nu + q)\big) \\ &= \sum_{q=0}^{N_\nu-1} \delta_N\big((j - lN_\nu) - q\big) = h_\nu(j - lN_\nu). \end{aligned}$$

The lemma is proved. □

4.12.3 Now we turn to the proof of the theorem. On the basis of the theorem's hypothesis and formula (4.12.1) we write

$$x(j) = \frac{1}{N} \sum_{k=0}^{\Delta_{\nu+1}-1} \xi(k)\, \widehat{v}_k(j), \quad j \in 0 : N-1. \tag{4.12.14}$$

We fix an integer $j \in 0 : N-1$, introduce a function $d(k) = \widehat{v}_k(j)$, $k \in 0 : \Delta_{\nu+1} - 1$, and expand d over the Walsh basis that is defined on the set $\{0, 1, \ldots, \Delta_{\nu+1}-1\}$ and ordered by frequency. To do that we calculate Fourier–Walsh coefficients of the function d. According to (4.12.10) we have

$$\begin{aligned} D(l) &= \sum_{k=0}^{\Delta_{\nu+1}-1} d(k)\, (-1)^{\{\mathrm{rev}_\nu(l),\, k\}_\nu} \\ &= \sum_{k=0}^{\Delta_{\nu+1}-1} \widehat{v}_k(j)\, (-1)^{\{\mathrm{rev}_s(k),\, lN_\nu\}_s} = \sum_{k=0}^{\Delta_{\nu+1}-1} \widehat{v}_k(j)\, \widehat{v}_k(lN_\nu), \end{aligned}$$

$$l \in 0 : \Delta_{\nu+1} - 1.$$

We note that

$$\widehat{v}_k(j)\,\widehat{v}_k(j') = \widehat{v}_k(j \oplus j'), \quad j, j' \in 0 : N-1. \tag{4.12.15}$$

Indeed,

$$\begin{aligned}\widehat{v}_k(j)\,\widehat{v}_k(j') &= (-1)^{\{\mathrm{rev}_s(k),\, j\}_s + \{\mathrm{rev}_s(k),\, j'\}_s} \\ &= \prod_{\alpha=0}^{s-1} (-1)^{k_{s-1-\alpha}(j_\alpha + j'_\alpha)_2} = \prod_{\alpha=0}^{s-1} (-1)^{k_{s-1-\alpha}(j \oplus j')_\alpha} = \widehat{v}_k(j \oplus j').\end{aligned}$$

Taking into account (4.12.15), (4.12.8) and (4.12.13), we gain

$$D(l) = \sum_{k=0}^{\Delta_{\nu+1}-1} \widehat{v}_k(j \oplus lN_\nu) = \Delta_{\nu+1}\, h_\nu(j \oplus lN_\nu) = \Delta_{\nu+1}\, h_\nu(j - lN_\nu).$$

With the aid of the DWT inversion formula and equality (4.12.10) we can reconstruct the values $\widehat{v}_k(j)$ for $k \in 0 : \Delta_{\nu+1} - 1$:

$$\begin{aligned}\widehat{v}_k(j) = d(k) &= \frac{1}{\Delta_{\nu+1}} \sum_{l=0}^{\Delta_{\nu+1}-1} D(l)\,(-1)^{\{\mathrm{rev}_\nu(l),\, k\}_\nu} \\ &= \sum_{l=0}^{\Delta_{\nu+1}-1} h_\nu(j - lN_\nu)\,(-1)^{\{\mathrm{rev}_s(k),\, lN_\nu\}_s} \\ &= \sum_{l=0}^{\Delta_{\nu+1}-1} h_\nu(j - lN_\nu)\,\widehat{v}_k(lN_\nu).\end{aligned} \tag{4.12.16}$$

It is remaining to substitute (4.12.16) into (4.12.14). This yields

$$\begin{aligned}x(j) &= \frac{1}{N} \sum_{k=0}^{\Delta_{\nu+1}-1} \xi(k) \sum_{l=0}^{\Delta_{\nu+1}-1} h_\nu(j - lN_\nu)\,\widehat{v}_k(lN_\nu) \\ &= \sum_{l=0}^{\Delta_{\nu+1}-1} h_\nu(j - lN_\nu) \left\{ \frac{1}{N} \sum_{k=0}^{\Delta_{\nu+1}-1} \xi(k)\,\widehat{v}_k(lN_\nu) \right\} \\ &= \sum_{l=0}^{\Delta_{\nu+1}-1} h_\nu(j - lN_\nu)\, x(lN_\nu).\end{aligned}$$

The theorem is proved. □

4.12.4 Theorem 4.12.1 can be inverted. To be more precise, the following assertion is true.

Theorem 4.12.2 *Let x be a step-function of a form*

$$x(j) = \sum_{l=0}^{\Delta_{\nu+1}-1} a(l)\, h_\nu(j - lN_\nu). \tag{4.12.17}$$

Then its Fourier–Walsh coefficients $\xi(k)$ defined by the formula

$$\xi(k) = \sum_{j=0}^{N-1} x(j)\, \widehat{v}_k(j)$$

are equal to zero for $k \in \Delta_{\nu+1} : N-1$.

Proof On the basis of (4.12.13), (4.12.8), and (4.12.15) we have

$$\begin{aligned}
\xi(k) &= \sum_{l=0}^{\Delta_{\nu+1}-1} a(l) \sum_{j=0}^{N-1} h_\nu(j \oplus lN_\nu)\, \widehat{v}_k(j) \\
&= \frac{1}{\Delta_{\nu+1}} \sum_{l=0}^{\Delta_{\nu+1}-1} a(l) \sum_{j=0}^{N-1} \widehat{v}_k(j) \sum_{p=0}^{\Delta_{\nu+1}-1} \widehat{v}_p(j)\, \widehat{v}_p(lN_\nu) \\
&= \frac{1}{\Delta_{\nu+1}} \sum_{l=0}^{\Delta_{\nu+1}-1} a(l) \sum_{p=0}^{\Delta_{\nu+1}-1} \widehat{v}_p(lN_\nu) \sum_{j=0}^{N-1} \widehat{v}_k(j)\, \widehat{v}_p(j).
\end{aligned}$$

For each $k \in \Delta_{\nu+1} : N-1$ and $p \in 0 : \Delta_{\nu+1} - 1$, Walsh functions $\widehat{v}_k$ and $\widehat{v}_p$ are orthogonal. This guarantees that the coefficients $\xi(k)$ are equal to zero for $k \in \Delta_{\nu+1} : N-1$. The theorem is proved. □

4.13 Ahmed–Rao Bases

4.13.1 We still presume that $N = 2^s$. Let us take arbitrary nonzero complex numbers $t(0), t(1), \ldots, t(N/2-1)$ and construct one more sequence of bases in $\mathbb{C}_N$:

$$g_0(k) = \delta_N(\cdot - k), \quad k \in 0 : N-1;$$

$$\begin{aligned}
g_\nu(2lN_\nu + p) &= g_{\nu-1}(lN_{\nu-1} + p) + t(l)\, g_{\nu-1}(lN_{\nu-1} + N_\nu + p), \\
g_\nu\big((2l+1)N_\nu + p\big) &= g_{\nu-1}(lN_{\nu-1} + p) - t(l)\, g_{\nu-1}(lN_{\nu-1} + N_\nu + p),
\end{aligned} \tag{4.13.1}$$

$$p \in 0 : N_\nu - 1, \quad l \in 0 : \Delta_\nu - 1, \quad \nu = 1, \ldots, s.$$

Formula (4.13.1) differs from (4.6.2) only by the coefficient $\omega_N^{\mathrm{rev}_s(2l)}$ being replaced with $t(l)$. A transition from the basis $g_{\nu-1}$ to the basis g_ν can be written in a single line:

$$g_\nu\big((2l+\sigma)N_\nu + p\big) = \sum_{\tau=0}^{1}(-1)^{\sigma\tau}\,[t(l)]^\tau\, g_{\nu-1}(lN_{\nu-1} + \tau N_\nu + p), \qquad (4.13.2)$$

where $\sigma \in 0:1$. In particular, for $\nu = 1$ we have

$$g_1(\sigma N_1 + p) = \sum_{\tau=0}^{1}(-1)^{\sigma\tau}\,[t(0)]^\tau\, g_0(\tau N_1 + p), \qquad (4.13.3)$$

$$p \in 0 : N_1 - 1, \quad \sigma \in 0 : 1.$$

Let us express signals of the ν-th level through signals of the zero level. In order to do this we introduce a sequence of matrices $T_1,\ T_2,\ \ldots,\ T_s$ by the rule

$$T_1 = \begin{bmatrix} 1 & t(0) \\ 1 & -t(0) \end{bmatrix};$$

$$T_\nu[2l+\sigma, 2q+\tau] = (-1)^{\sigma\tau}\,[t(l)]^\tau\, T_{\nu-1}[l,q], \qquad (4.13.4)$$

$$l, q \in 0 : \Delta_\nu - 1, \quad \sigma, \tau \in 0 : 1, \quad \nu = 2, \ldots, s.$$

Theorem 4.13.1 *The following representation is valid:*

$$g_\nu(lN_\nu + p) = \sum_{q=0}^{\Delta_{\nu+1}-1} T_\nu[l,q]\, g_0(qN_\nu + p), \qquad (4.13.5)$$

$$p \in 0 : N_\nu - 1, \quad l \in 0 : \Delta_{\nu+1} - 1, \quad \nu = 1, \ldots, s.$$

Proof When $\nu = 1$, formula (4.13.5) coincides with (4.13.3) up to notations. We perform an induction step from $\nu - 1$ to ν, $\nu \geq 2$.

We take an index $l \in 0 : \Delta_{\nu+1} - 1$ and represent it in a form $l = 2l' + \sigma$, $\sigma \in 0 : 1$, $l' \in 0 : \Delta_\nu - 1$. On the basis of (4.13.2), (4.13.4) and the inductive hypothesis we write

$$\begin{aligned} g_\nu(lN_\nu + p) &= g_\nu\big((2l'+\sigma)N_\nu + p\big) \\ &= \sum_{\tau=0}^{1}(-1)^{\sigma\tau}\,[t(l')]^\tau\, g_{\nu-1}(l'N_{\nu-1} + \tau N_\nu + p) \end{aligned}$$

$$= \sum_{\tau=0}^{1} (-1)^{\sigma\tau} [t(l')]^{\tau} \sum_{q'=0}^{\Delta_\nu - 1} T_{\nu-1}[l', q'] \, g_0(q' N_{\nu-1} + \tau N_\nu + p)$$

$$= \sum_{\tau=0}^{1} \sum_{q'=0}^{\Delta_\nu - 1} T_\nu[2l' + \sigma, 2q' + \tau] \, g_0\big((2q' + \tau) N_\nu + p\big)$$

$$= \sum_{q=0}^{\Delta_{\nu+1}-1} T_\nu[l, q] \, g_0(q N_\nu + p).$$

The theorem is proved. □

From (4.13.5) with $\nu = s$ it follows that

$$g_s(l;\ j) = T_s[l, j], \quad l, j \in 0 : N - 1. \tag{4.13.6}$$

4.13.2 Let us find an explicit expression for the elements of the matrix T_ν.

Theorem 4.13.2 *Given $\nu \in 1 : s$, for $l,\ q \in 0 : \Delta_{\nu+1} - 1$, $l = (l_{\nu-1}, \ldots, l_0)_2$, $q = (q_{\nu-1}, \ldots, q_0)_2$, valid is the formula*

$$T_\nu[l, q] = (-1)^{\{l,q\}_\nu} \prod_{\alpha=0}^{\nu-1} \big[t(\lfloor l/2^{\alpha+1} \rfloor)\big]^{q_\alpha}. \tag{4.13.7}$$

Proof When $\nu = 1$, formula (4.13.7) takes a form

$$T_1[l, q] = (-1)^{lq} \, [t(0)]^q, \quad l, q \in 0 : 1.$$

The validity of this equality is obvious. We perform an induction step from $\nu - 1$ to ν, $\nu \geq 2$.

Let $l = 2l' + \sigma$ and $q = 2q' + \tau$, where $\sigma, \tau \in 0 : 1$ and $l', q' \in 0 : \Delta_\nu - 1$. Binary digits of the numbers l and l' and q and q' are bound with the relations

$$l_{\alpha+1} = l'_\alpha, \quad q_{\alpha+1} = q'_\alpha, \quad l_0 = \sigma, \quad q_0 = \tau.$$

Bearing in mind the inductive hypothesis and equality (4.13.4) we gain

$$T_\nu[l, q] = T_\nu[2l' + \sigma, 2q' + \tau] = (-1)^{\sigma\tau} \, [t(l')]^{\tau} \, T_{\nu-1}[l', q']$$

$$= (-1)^{\sigma\tau} \, [t(l')]^{\tau} \, (-1)^{\{l',q'\}_{\nu-1}} \prod_{\alpha=0}^{\nu-2} \big[t(\lfloor l'/2^{\alpha+1} \rfloor)\big]^{q'_\alpha}$$

$$= (-1)^{\{l,q\}_\nu} \, [t(\lfloor l/2 \rfloor)]^{q_0} \prod_{\alpha=1}^{\nu-1} \big[t(\lfloor l'/2^{\alpha} \rfloor)\big]^{q_\alpha}.$$

It is remaining to verify that for $\alpha \in 1 : \nu - 1$ there holds

$$\lfloor l'/2^\alpha \rfloor = \lfloor l/2^{\alpha+1} \rfloor.$$

When $\alpha = \nu - 1$, the latter equality turns to an undoubtedly true assertion $0 = 0$. Let $\alpha \in 1 : \nu - 2$. If $l' = l'_{\nu-2}2^{\nu-2} + \cdots + l'_0$ then $l = 2l' + \sigma = l'_{\nu-2}2^{\nu-1} + \cdots + l'_0 2 + \sigma$ and

$$\lfloor l'/2^\alpha \rfloor = l'_{\nu-2}2^{\nu-2-\alpha} + \cdots + l'_\alpha = \lfloor l/2^{\alpha+1} \rfloor.$$

The theorem is proved. □

On the basis of (4.13.6) and (4.13.7) we come to a representation

$$g_s(k;\ j) = (-1)^{\{k,j\}_s} \prod_{\alpha=0}^{s-1} \left[t(\lfloor k/2^{\alpha+1} \rfloor)\right]^{j_\alpha}$$
$$= v_k(j) \prod_{\alpha=0}^{s-1} \left[t(\lfloor k/2^{\alpha+1} \rfloor)\right]^{j_\alpha}, \quad k, j \in 0 : N-1, \qquad (4.13.8)$$

where $v_k(j)$ is the Walsh function of order k.

4.13.3 Let us clarify the conditions when the family g_ν consists of pairwise orthogonal signals.

Lemma 4.13.1 *If* $|t(l)| = 1$ *for* $l \in 0 : \Delta_\nu - 1$ *then for any* $k, k' \in 0 : \Delta_{\nu+1} - 1$ *there holds*

$$\sum_{j=0}^{\Delta_{\nu+1}-1} T_\nu[k, j]\, \overline{T_\nu[k', j]} = 2^\nu \delta_{\Delta_{\nu+1}}(k - k'). \qquad (4.13.9)$$

Proof For $\nu = 1$ we have $|t(0)| = 1$, therefore

$$\sum_{j=0}^{1} T_1[k, j]\, \overline{T_1[k', j]} = \sum_{j=0}^{1} (-1)^{kj} \left[t(0)\right]^j (-1)^{k'j} \left[\overline{t(0)}\right]^j$$
$$= \sum_{j=0}^{1} (-1)^{(k-k')j} = \sum_{j=0}^{1} \omega_2^{(k-k')j} = 2\delta_2(k - k').$$

We perform an induction step from $\nu - 1$ to ν, $\nu \geq 2$.

Let $k = 2l + \sigma$, $k' = 2l' + \sigma'$, and $j = 2q + \tau$, where $\sigma, \sigma', \tau \in 0 : 1$ and $l, l', q \in 0 : \Delta_\nu - 1$. On the basis of (4.13.4) and the inductive hypothesis we write

$$\sum_{j=0}^{\Delta_{\nu+1}-1} T_\nu[k,j]\,\overline{T_\nu[k',j]}$$
$$= \sum_{q=0}^{\Delta_\nu-1}\sum_{\tau=0}^{1} T_\nu[2l+\sigma, 2q+\tau]\,\overline{T_\nu[2l'+\sigma', 2q+\tau]}$$
$$= \sum_{q=0}^{\Delta_\nu-1}\sum_{\tau=0}^{1} (-1)^{\sigma\tau}\,[t(l)]^\tau\, T_{\nu-1}[l,q]\,(-1)^{\sigma'\tau}\,\overline{[t(l')]^\tau}\;\overline{T_{\nu-1}[l',q]}$$
$$= \Big(\sum_{q=0}^{\Delta_\nu-1} T_{\nu-1}[l,q]\,\overline{T_{\nu-1}[l',q]}\Big)\Big(\sum_{\tau=0}^{1}(-1)^{(\sigma-\sigma')\tau}\,[t(l)]^\tau\,\overline{[t(l')]^\tau}\Big)$$
$$= 2^{\nu-1}\delta_{\Delta_\nu}(l-l')\sum_{\tau=0}^{1}(-1)^{(\sigma-\sigma')\tau}\,[t(l)]^\tau\,\overline{[t(l')]}^\tau.$$

If $l \neq l'$ then the obtained expression equals to zero. Let $l = l'$. By virtue of the lemma's hypothesis $|t(l)| = 1$ for $l \in 0 : \Delta_\nu - 1$, hence

$$\sum_{j=0}^{\Delta_{\nu+1}-1} T_\nu[k,j]\,\overline{T_\nu[k',j]} = 2^{\nu-1}\sum_{\tau=0}^{1}(-1)^{(\sigma-\sigma')\tau} = 2^\nu\delta_2(\sigma-\sigma').$$

We see that the left side of (4.13.9) is nonzero only when $l = l'$ and $\sigma = \sigma'$, i.e. only when $k = k'$. In the latter case we have

$$\sum_{j=0}^{\Delta_{\nu+1}-1} |T_\nu[k,j]|^2 = 2^\nu, \quad k \in 0 : \Delta_{\nu+1} - 1.$$

The lemma is proved. □

Actually, Lemma 4.13.1 states that, provided $|t(l)| = 1$ for $l \in 0 : \Delta_\nu - 1$, the rows of the matrix T_ν are pairwise orthogonal and the squared norm of each row is equal to 2^ν.

Theorem 4.13.3 *Provided* $|t(l)| = 1$ *for* $l \in 0 : \Delta_\nu - 1$, *the signals*

$$g_\nu(0),\ g_\nu(1),\ \ldots,\ g_\nu(N-1) \tag{4.13.10}$$

are pairwise orthogonal and $\|g_\nu(k)\|^2 = 2^\nu$ *holds for all* $k \in 0 : N - 1$.

Proof Let $k = lN_\nu + p$ and $k' = l'N_\nu + p'$, where $p, p' \in 0 : N_\nu - 1$ and $l, l' \in 0 : \Delta_{\nu+1} - 1$. According to (4.13.5) and Lemma 2.1.4 we have

$$\begin{aligned}\langle g_\nu(k), g_\nu(k')\rangle &= \langle g_\nu(lN_\nu + p),\ g_\nu(l'N_\nu + p')\rangle \\ &= \sum_{q=0}^{\Delta_{\nu+1}-1} \sum_{q'=0}^{\Delta_{\nu+1}-1} T_\nu[l,q]\,\overline{T_\nu[l',q']}\,\delta_N\big((q-q')N_\nu + (p-p')\big).\end{aligned}$$

When $q \neq q'$, the corresponding terms in the double sum are equal to zero. Taking into account (4.13.9) we gain

$$\begin{aligned}\langle g_\nu(k), g_\nu(k')\rangle &= \delta_N(p-p') \sum_{q=0}^{\Delta_{\nu+1}-1} T_\nu[l,q]\,\overline{T_\nu[l',q]} \\ &= 2^\nu \delta_N(p-p')\,\delta_{\Delta_{\nu+1}}(l-l').\end{aligned}$$

We see that the scalar product $\langle g_\nu(k), g_\nu(k')\rangle$ is nonzero only when $p = p'$ and $l = l'$, i.e. only when $k = k'$. In the latter case $\|g_\nu(k)\|^2 = 2^\nu$ holds for all $k \in 0 : N-1$. The theorem is proved. □

Corollary 4.13.1 *Provided* $|t(l)| = 1$ *for* $l \in 0 : N/2 - 1$, *for each* $\nu \in 1 : s$ *signals* (4.13.10) *form an orthogonal basis in a space* $\mathbb{C}_N$. *The squared norm of each signal is equal to* 2^ν.

4.13.4 We will consider the particular case of choosing the coefficients $t(l)$ whose moduli are equal to unity. Let us fix $r \in 1 : s$ and put

$$t^{(r)}(l) = \begin{cases} \omega_N^{\mathrm{rev}_s(2l)} & \text{for } l \in 0 : \Delta_r - 1, \\ 1 & \text{for } l \in \Delta_r : \Delta_s - 1. \end{cases} \tag{4.13.11}$$

Consider the signals

$$g_\nu^{(r)}(0),\ g_\nu^{(r)}(1),\ \ldots,\ g_\nu^{(r)}(N-1) \tag{4.13.12}$$

defined by the recurrent relations (4.13.1) with the coefficients $t(l) = t^{(r)}(l)$. Since $|t^{(r)}(l)| \equiv 1$, Theorem 4.13.3 yields that signals (4.13.12) form an orthogonal basis in a space $\mathbb{C}_N$ for all $\nu \in 1 : s$ and $r \in 1 : s$. A collection of signals (4.13.12) with $\nu = s$ is referred to as *Ahmed–Rao basis with an index* r.

When $r = s$, formula (4.13.11) takes a form $t^{(s)}(l) = \omega_N^{\mathrm{rev}_s(2l)}$, $l \in 0 : \Delta_s - 1$. In this case the recurrent relations (4.13.1) with $t(l) = t^{(s)}(l)$ coincide with (4.6.2) and therefore generate a sequence of bases leading to the exponential basis. An explicit expression for the signals of this sequence is presented in Theorem 4.6.2. In particular, formula (4.6.1) yields

$$g_s^{(s)}(k;\ j) = \omega_N^{\mathrm{rev}_s(k)j}, \quad k, j \in 0 : N-1.$$

When $r = 1$, we have $t^{(1)}(l) = 1$, $l \in 0 : \Delta_s - 1$. On the basis of (4.13.8) we come to a formula $g_s^{(1)}(k;\ j) = v_k(j)$, where $v_k(j)$ is a Walsh function of order k.

Note that Walsh basis is obtained both by sequence (4.9.1) and sequence (4.13.1) with $t(l) \equiv 1$.

4.13.5 We will investigate in more detail Ahmed–Rao bases with an index $r \in 2 : s-1$. To do that we turn to the matrices $T_\nu^{(r)}$ whose elements are calculated by formula (4.13.7) with $t(l) = t^{(r)}(l)$, $l \in 0 : \Delta_s - 1$.

Lemma 4.13.2 *The following equality is valid for* $\nu \le r$*:*

$$T_\nu^{(r)}[l, q] = \omega_N^{\mathrm{rev}_s(l) q}, \quad l, q \in 0 : \Delta_{\nu+1} - 1. \tag{4.13.13}$$

Proof Let $l = (l_{\nu-1}, \ldots, l_0)_2$ and $q = (q_{\nu-1}, \ldots, q_0)_2$. For $\nu \le r$ we have $\lfloor l/2^{\alpha+1} \rfloor \in 0 : \Delta_r - 1$ and

$$t^{(r)}(\lfloor l/2^{\alpha+1} \rfloor) = \omega_N^{\mathrm{rev}_s(2\lfloor l/2^{\alpha+1} \rfloor)}, \quad \alpha \in 0 : \nu - 1. \tag{4.13.14}$$

We will show that

$$t^{(r)}(\lfloor l/2^{\alpha+1} \rfloor) = (-1)^{-l_\alpha}\, \omega_N^{2^\alpha \mathrm{rev}_s(l)}, \quad \alpha \in 0 : \nu - 1. \tag{4.13.15}$$

When $\alpha = \nu - 1$, the left side of (4.13.15) equals to unity. Since

$$\begin{aligned} 2^{\nu-1} \mathrm{rev}_s(l) &= 2^{\nu-1}(l_0 2^{s-1} + \cdots + l_{\nu-1} 2^{s-\nu}) \\ &= l_0 2^{s+\nu-2} + \cdots + l_{\nu-1} 2^{s-1}, \end{aligned}$$

the right side of (4.13.15) also equals to unity when $\alpha = \nu - 1$.

Let us take $\alpha \in 0 : \nu - 2$. We write

$$\begin{aligned} \mathrm{rev}_s(2\lfloor l/2^{\alpha+1} \rfloor) &= \mathrm{rev}_s\big((l_{\nu-1}, \ldots, l_{\alpha+1}, 0)_2\big) \\ &= l_{\alpha+1} 2^{s-2} + \cdots + l_{\nu-1} 2^{s-\nu+\alpha} \\ &= 2^\alpha [l_0 2^{s-1} + \cdots + l_\alpha 2^{s-\alpha-1} + l_{\alpha+1} 2^{s-\alpha-2} + \cdots \\ &\quad + l_{\nu-1} 2^{s-\nu} - (l_0 2^{s-1} + \cdots + l_\alpha 2^{s-\alpha-1})] \\ &= 2^\alpha \mathrm{rev}_s(l) - (l_0 2^{s+\alpha-1} + \cdots + l_\alpha 2^{s-1}). \end{aligned}$$

From here and from (4.13.14) follows formula (4.13.15).

Formula (4.13.15) yields

$$\begin{aligned} \prod_{\alpha=0}^{\nu-1} \big[t^{(r)}(\lfloor l/2^{\alpha+1} \rfloor)\big]^{q_\alpha} &= \prod_{\alpha=0}^{\nu-1} (-1)^{-l_\alpha q_\alpha}\, \omega_N^{2^\alpha \mathrm{rev}_s(l) q_\alpha} \\ &= (-1)^{\{l,q\}_\nu} \omega_N^{\mathrm{rev}_s(l) \sum_{\alpha=0}^{\nu-1} q_\alpha 2^\alpha} \\ &= (-1)^{\{l,q\}_\nu} \omega_N^{\mathrm{rev}_s(l) q}. \end{aligned} \tag{4.13.16}$$

Combining formula (4.13.7) with $t(l) = t^{(r)}(l)$ and (4.13.16), we come to (4.13.13). The lemma is proved. □

Now we will consider the case of $\nu \geq r+1$. Note that an integer $l \in \Delta_{r+1} : \Delta_{\nu+1} - 1$ can be represented in a form

$$l = (1, l_{p-1}, \ldots, l_{r-1}, \ldots, l_0)_2, \quad p \in r : \nu - 1. \tag{4.13.17}$$

Here p is the index of the most significant nonzero digit in a binary code of l.

We introduce one more notation. If $q = (q_{\nu-1}, q_{\nu-2}, \ldots, q_0)_2$ then we put

$$[q]_\alpha = (q_{\nu-1}, q_{\nu-2}, \ldots, q_\alpha, 0, \ldots, 0)_2, \quad \alpha \in 1 : \nu - 1.$$

Lemma 4.13.3 *Given $\nu \geq r+1$ and $q \in 0 : \Delta_{\nu+1} - 1$, valid is the equality*

$$T_\nu^{(r)}[l, q] = \begin{cases} \omega_N^{\mathrm{rev}_s(l)q}, & l \in 0 : \Delta_{r+1} - 1; \\ (-1)^{\{l,q\}_{p-r+1}} \omega_N^{\mathrm{rev}_s(l)[q]_{p-r+1}}, & \\ \qquad l = (1, l_{p-1}, \ldots, l_0)_2, \ p \in r : \nu - 1. & \end{cases} \tag{4.13.18}$$

Proof For $l \in 0 : \Delta_{r+1} - 1$ we have $\lfloor l/2^{\alpha+1} \rfloor \in 0 : \Delta_r - 1$, therefore formula (4.13.14) holds. It is remaining to transit from (4.13.14) to (4.13.15). We will show how to perform such a transition in this case.

Let $l = (l_{r-1}, \ldots, l_0)_2$, so that $l_{\nu-1} = \cdots = l_r = 0$ and

$$\begin{aligned} 2^\alpha \mathrm{rev}_s(l) &= 2^\alpha (l_0 2^{s-1} + \cdots + l_{r-1} 2^{s-r}) \\ &= l_0 2^{s-1+\alpha} + \cdots + l_{r-1} 2^{s-r+\alpha}. \end{aligned}$$

When $\alpha \in r-1 : \nu - 1$, the left side of (4.13.15) equals to unity. The right side of (4.13.15) also equals to unity both when $\alpha \in r : \nu - 1$ and $\alpha = r - 1$. When $\alpha \in 0 : r - 2$, we have to repeat the manipulations from the proof of the former lemma replacing ν by r.

Now we take $l \in \Delta_{r+1} : \Delta_{\nu+1} - 1$ and represent l in form (4.13.17). When $\alpha \in 0 : p - r$, we have $p - (\alpha + 1) \geq r - 1$, therefore

$$\lfloor l/2^{\alpha+1} \rfloor = (1, l_{p-1}, \ldots, l_{\alpha+1})_2 \geq 2^{r+1} = \Delta_r.$$

According to (4.13.11) we write

$$t^{(r)}(\lfloor l/2^{\alpha+1} \rfloor) = 1, \quad \alpha \in 0 : p - r. \tag{4.13.19}$$

We will show that, when $\alpha \in p - r + 1 : \nu - 1$, there holds

$$t^{(r)}(\lfloor l/2^{\alpha+1} \rfloor) = (-1)^{-l_\alpha} \omega_N^{2^\alpha \mathrm{rev}_s(l)}. \tag{4.13.20}$$

When $\alpha \in p : \nu - 1$, the left side of (4.13.20) equals to unity. Since

$$\begin{aligned} 2^{\alpha}\operatorname{rev}_s(l) &= 2^{\alpha}(l_0 2^{s-1} + \cdots + l_{p-1}2^{s-p} + 2^{s-p-1}) \\ &= l_0 2^{s-1+\alpha} + \cdots + l_{p-1}2^{s-p+\alpha} + 2^{s-p-1+\alpha}, \end{aligned}$$

the right side of (4.13.20) also equals to unity both when $\alpha \in p+1 : \nu - 1$ and $\alpha = p$. When $\alpha \in p - r + 1 : p - 1$, an inequality $p - (\alpha + 1) \le r - 2$ holds, therefore $\lfloor l/2^{\alpha+1} \rfloor \in 0 : \Delta_r - 1$ and

$$\operatorname{rev}_s(2\lfloor l/2^{\alpha+1}\rfloor) = 2^{\alpha}\operatorname{rev}_s(l) - (l_0 2^{s-1+\alpha} + \cdots + l_{\alpha}2^{s-1}).$$

From here and from (4.13.14) follows formula (4.13.20).

Let us turn to a calculation of $T_{\nu}^{(r)}[l, q]$ for $l \in \Delta_{r+1} : \Delta_{\nu+1} - 1$. On the basis of (4.13.19) and (4.13.20) we write

$$\begin{aligned} \prod_{\alpha=0}^{\nu-1} \left[t^{(r)}(\lfloor l/2^{\alpha+1}\rfloor)\right]^{q_{\alpha}} &= \prod_{\alpha=p-r+1}^{\nu-1} (-1)^{-l_{\alpha}q_{\alpha}}\, \omega_N^{2^{\alpha}\operatorname{rev}_s(l)q_{\alpha}} \\ &= (-1)^{-\sum_{\alpha=p-r+1}^{\nu-1} l_{\alpha}q_{\alpha}}\, \omega_N^{\operatorname{rev}_s(l)\sum_{\alpha=p-r+1}^{\nu-1} q_{\alpha}2^{\alpha}}. \end{aligned}$$

Substituting this expression into (4.13.7) and taking into account that

$$\sum_{\alpha=p-r+1}^{\nu-1} q_{\alpha}2^{\alpha} = [q]_{p-r+1},$$

$$\{l, q\}_{\nu} - \sum_{\alpha=p-r+1}^{\nu-1} l_{\alpha}q_{\alpha} = \{l, q\}_{p-r+1},$$

we obtain the required formula for $T_{\nu}^{(r)}[l, q]$ when $l \in \Delta_{r+1} : \Delta_{\nu+1} - 1$.

The lemma is proved. □

According to Theorem 4.13.1 signals (4.13.12) can be represented in form (4.13.5) with $T_{\nu} = T_{\nu}^{(r)}$. Elements of the matrix $T_{\nu}^{(r)}$ for $r \in 2 : s - 1$ are defined by formula (4.13.13) or (4.13.18), depending on the relation between the values of r, ν and l. Thereby the question of the explicit representation of the signals $g_{\nu}^{(r)}(k;\, j)$, $k, j \in 0 : N - 1$, is resolved.

Theorem 4.13.4 *Ahmed–Rao functions $g_s^{(r)}(k;\, j)$ for $r \in 2 : s - 1$ can be represented by the formula*

$$g_s^{(r)}(k;\, j) = \begin{cases} \omega_N^{\operatorname{rev}_s(k)j}, & k \in 0 : \Delta_{r+1} - 1; \\ (-1)^{\{k,j\}_{p-r+1}}\omega_N^{\operatorname{rev}_s(k)[j]_{p-r+1}}, & \\ \qquad k = (1, k_{p-1}, \ldots, k_0)_2,\ p \in r : s - 1. & \end{cases}$$

The proof immediately follows from (4.13.6) and (4.13.18).

4.13.6 Below we present a characteristic property of Ahmed–Rao functions.

Theorem 4.13.5 *A function* $g_s^{(r)}(k;\ j)$ *for each* j *takes one of* 2^r *values* $\omega_{2^r}^q$, $q \in 0 : 2^r - 1$.

Proof For $k \in 0 : \Delta_{r+1} - 1$, $k = (k_{r-1}, \ldots, k_0)_2$, we have

$$\mathrm{rev}_s(k) = k_0 2^{s-1} + \cdots + k_{r-1} 2^{s-r} = 2^{s-r}\,\mathrm{rev}_r(k),$$

so that Theorem 4.13.4 yields

$$g_s^{(r)}(k;\ j) = \omega_N^{\mathrm{rev}_s(k)j} = \omega_{2^r}^{\langle \mathrm{rev}_r(k)j\rangle_{2^r}}.$$

Let $k = (1, k_{p-1}, \ldots, k_0)_2$, $p \in r : s - 1$. Then

$$\mathrm{rev}_s(k) = k_0 2^{s-1} + \cdots + k_{p-1} 2^{s-p} + 2^{s-p-1} = 2^{s-p-1} k',$$

where $k' = \mathrm{rev}_{p+1}(k)$. At the same time,

$$[j]_{p-r+1} = \sum_{\alpha=p-r+1}^{s-1} j_\alpha 2^\alpha = 2^{p-r+1} j',$$

where $j' = \lfloor j/2^{p-r+1} \rfloor$. Hence

$$\mathrm{rev}_s(k)[j]_{p-r+1} = 2^{s-r} k' j'.$$

On the basis of Theorem 4.13.4 we gain

$$g_s^{(r)}(k;\ j) = (-1)^{\{k,j\}_{p-r+1}} \omega_N^{\mathrm{rev}_s(k)[j]_{p-r+1}} = \omega_{2^r}^{\langle 2^{r-1}\{k,j\}_{p-r+1} + k'j'\rangle_{2^r}}.$$

The theorem is proved. □

4.13.7 In a conclusion we will consider a question of the frequency of discrete Ahmed–Rao functions.

Theorem 4.13.6 *The frequency of the function* $g_s^{(r)}(k;\ j)$ *is equal to* $\mathrm{rev}_s(k)$ *for each* $r \in 2 : s - 1$.

Proof When $k \in 0 : \Delta_{r+1} - 1$, the assertion is evident.

Let $k = (1, k_{p-1}, \ldots, k_0)_2$, $p \in r : s - 1$. Note that $1 \le p - r + 1 \le s - r \le s - 2$. We take $j \in 0 : N - 1$ and represent it in a form $j = j' 2^{p-r+1} + j''$, where $j'' \in 0 : 2^{p-r+1} - 1$ and $j' \in 0 : 2^{s-p+r-1} - 1$. Then

$$g_s^{(r)}(k;\ j) = (-1)^{\{k,j''\}_{p-r+1}} \omega_N^{\mathrm{rev}_s(k) j' 2^{p-r+1}}.$$

We have

$$(-1)^{\{k,j''\}_{p-r+1}} = (-1)^{\sum_{\alpha=0}^{p-r} k_\alpha (j''_\alpha + j''_{\alpha+1} 2 + \cdots + j''_{p-r} 2^{p-r-\alpha})}$$
$$= (-1)^{\sum_{\alpha=0}^{p-r} k_\alpha \lfloor j''/\Delta_{\alpha+1} \rfloor}.$$

Denote

$$\theta_k(j'') = \sum_{\alpha=0}^{p-r} k_\alpha \lfloor j''/\Delta_{\alpha+1} \rfloor. \tag{4.13.21}$$

In this notation

$$g_s^{(r)}(k;\ j) = \exp\bigl(i\pi[\theta_k(j'') + 2^{-s+p-r+2}\,\mathrm{rev}_s(k)\, j']\bigr).$$

By putting

$$\zeta_k(j) = \theta_k(j'') + 2^{-s+p-r+2}\,\mathrm{rev}_s(k)\, j'$$

we come to a representation

$$g_s^{(r)}(k;\ j) = \exp\bigl(i\pi\zeta_k(j)\bigr), \quad j \in 0 : N-1. \tag{4.13.22}$$

One can see that formula (4.13.22) is valid for $j = N$ as well. Indeed, on the strength of N-periodicity we have $g_s^{(r)}(k;\ N) = g_s^{(r)}(k;\ 0) = 1$. At the same time $j' = 2^{s-p+r-1}$ and $j'' = 0$ when $j = N$, so $\zeta_k(N) = 2\mathrm{rev}_s(k)$ and $\exp\bigl(i\pi\zeta_k(N)\bigr) = 1$.

Since $\zeta_k(0) = 0$ and $\zeta_k(N) = 2\mathrm{rev}_s(k)$, to complete the proof of the theorem we only need to verify that the function $\zeta_k(j)$ monotonically nondecreases while j varies from 0 to N.

We take $j, l \in 0 : N$ and represent them in a form $j = j' 2^{p-r+1} + j''$, $l = l' 2^{p-r+1} + l''$. Presuppose that $j > l$. Then $j' \geq l'$ because $|j'' - l''| \leq 2^{p-r+1} - 1$. When $j' = l'$, the inequality $\zeta_k(j) \geq \zeta_k(l)$ follows from monotonic nondecrease of the function $\theta_k(j'')$.

Assume that $j' > l'$. Let us estimate $\theta_k(j'')$. According to (4.13.21) we gain

$$\theta_k(2^{p-r+1} - 1) \leq \sum_{\alpha=0}^{p-r} k_\alpha \lfloor 2^{p-r+1}/\Delta_{\alpha+1} \rfloor = \sum_{\alpha=0}^{p-r} k_\alpha 2^{p-r-\alpha+1}$$
$$= 2^{-s+p-r+2} \sum_{\alpha=0}^{p-r} k_\alpha 2^{s-1-\alpha} \leq 2^{-s+p-r+2}\,\mathrm{rev}_s(k).$$

On the strength of monotonic nondecrease of $\theta_k(j'')$ we have

$$0 \leq \theta_k(j'') \leq 2^{-s+p-r+2}\,\mathrm{rev}_s(k)$$

and

$$|\theta_k(j'') - \theta_k(l'')| \le 2^{-s+p-r+2}\,\mathrm{rev}_s(k).$$

Now we write

$$\begin{aligned}\zeta_k(j) - \zeta_k(l) &\ge 2^{-s+p-r+2}\,\mathrm{rev}_s(k)(j'-l') - |\theta_k(j'') - \theta_k(l'')| \\ &\ge 2^{-s+p-r+2}\,\mathrm{rev}_s(k)[(j'-l') - 1] \ge 0.\end{aligned}$$

The theorem is proved. □

4.14 Calculation of DFT of Any Order

4.14.1 We will show that discrete Fourier transform of any order can be reduced to DFT whose order is a power of two.

We take Fourier matrix F_N of order $N \ge 3$ with the elements

$$F_N[k, j] = \omega_N^{kj}, \quad k, j \in 0 : N-1.$$

Fourier spectrum of a signal $x \in \mathbb{C}_N$ can be represented in a form $X = \overline{F}_N\, x$, where $\overline{F}_N[k, j] = \omega_N^{-kj}$.

We introduce two more matrices of order N: a diagonal matrix D_N with diagonal elements

$$D_N[k, k] = \omega_{2N}^{-k^2}, \quad k \in 0 : N-1,$$

and a Toeplitz matrix G_N with elements

$$G_N[k, j] = \omega_{2N}^{(k-j)^2}, \quad k, j \in 0 : N-1.$$

Theorem 4.14.1 *The following factorization holds:*

$$\overline{F}_N = D_N G_N D_N. \tag{4.14.1}$$

Proof We have

$$\overline{F}_N[k, j] = \omega_N^{-kj} = \omega_{2N}^{-k^2}\,\omega_{2N}^{(k-j)^2}\,\omega_{2N}^{-j^2}.$$

At the same time,

$$\begin{aligned}\big(D_N G_N D_N\big)[k, j] &= \sum_{l=0}^{N-1} D_N[k, l] \times \big(G_N D_N\big)[l, j] \\ &= \sum_{l=0}^{N-1} D_N[k, l] \sum_{l'=0}^{N-1} G_N[l, l'] D_N[l', j]\end{aligned}$$

$$= \sum_{l=0}^{N-1} D_N[k,l] G_N[l,j] \omega_{2N}^{-j^2}$$

$$= \omega_{2N}^{-k^2} G_N[k,j]\, \omega_{N2}^{-j^2} = \omega_{2N}^{-k^2} \omega_{2N}^{(k-j)^2} \omega_{2N}^{-j^2}.$$

Comparing the obtained formulae we come to (4.14.1). □

4.14.2 We denote $a_k = \omega_{2N}^{k^2}$. Then a row of the matrix G_N with an index k can be represented in a form

$$G_N[k, 0:N-1] = \big(a_k, a_{k-1}, \ldots, a_1, a_0, a_1, \ldots, a_{N-k-1}\big), \tag{4.14.2}$$

$$k \in 0:N-1.$$

Let m be the minimal natural number such that

$$M := 2^m > 2N-1.$$

We introduce a signal $\widehat{h} \in \mathbb{C}_M$ with the following values on the main period:

$$\widehat{h}[0:M-1] = \big(a_0, a_1, \ldots, a_{N-1}, \underbrace{0, \ldots, 0}_{(M-(2N-1))\text{ times}}, a_{N-1}, a_{N-2}, \ldots, a_1\big).$$

We denote by $\widehat{G}_M$ a Toeplitz matrix of order M with elements

$$\widehat{G}_M[k,j] = \widehat{h}(k-j), \quad k, j \in 0:M-1. \tag{4.14.3}$$

Note that

$$\widehat{G}_M[k,j] = G_N[k,j] \quad \text{for } k, j \in 0:N-1. \tag{4.14.4}$$

Indeed, by the definition

$$\widehat{h}(-j) = \widehat{h}(M-j) = a_j \quad \text{for } j \in 1:N-1.$$

For $k \in 0:N-1$ we gain

$$\begin{aligned}\widehat{G}_M[k, 0:N-1] &= \big(\widehat{h}(k), \widehat{h}(k-1), \ldots, \widehat{h}(1), \widehat{h}(0), \widehat{h}(-1), \ldots, \widehat{h}(-N+k+1)\big) \\ &= \big(a_k, a_{k-1}, \ldots, a_1, a_0, a_1, \ldots, a_{N-k-1}\big). \end{aligned} \tag{4.14.5}$$

Now (4.14.4) follows from (4.14.2) and (4.14.5).

Equality (4.14.4) can be written in a matrix form

$$G_N = \left[I_N, \mathbb{O}^\top\right] \widehat{G}_M \begin{bmatrix} I_N \\ \mathbb{O} \end{bmatrix}, \tag{4.14.6}$$

where I_N is an identity matrix of order N and $\mathbb{O}$ is a zero matrix of order $(M - N) \times N$.

4.14.3 According to (4.14.1) and (4.14.6) we have

$$\overline{F}_N x = D_N \left[I_N, \mathbb{O}^\top\right] \widehat{G}_M \begin{bmatrix} I_N \\ \mathbb{O} \end{bmatrix} D_N x. \tag{4.14.7}$$

We denote $z = D_N x$ and augment the vector z with zeroes to obtain an M-dimensional vector $\widehat{z}$. Now formula (4.14.7) can be rewritten as follows:

$$\overline{F}_N x = \left[D_N, \mathbb{O}^\top\right] \widehat{G}_M \widehat{z}. \tag{4.14.8}$$

By a definition (4.14.3) of the matrix $\widehat{G}_M$ we gain

$$\big(\widehat{G}_M \widehat{z}\big)[k] = \sum_{j=0}^{M-1} \widehat{z}(j)\, \widehat{h}(k - j), \quad k \in 0 : M - 1. \tag{4.14.9}$$

The right side of this equality contains a convolution $\widehat{z} * \widehat{h}$. The convolution theorem yields

$$\widehat{z} * \widehat{h} = \mathcal{F}_M^{-1}\big(\mathcal{F}_M(\widehat{z})\, \mathcal{F}_M(\widehat{h})\big). \tag{4.14.10}$$

The spectrum $\widehat{H} = \mathcal{F}_M(\widehat{h})$ does not depend on x. It can be calculated in advance:

$$\begin{aligned} \widehat{H}(k) &= 1 + \sum_{j=1}^{N-1} \omega_{2N}^{j^2} \omega_M^{-kj} + \sum_{j=1}^{N-1} \omega_{2N}^{j^2} \omega_M^{-k(M-j)} \\ &= 1 + 2 \sum_{j=1}^{N-1} \omega_{2N}^{j^2} \cos\big(\tfrac{2\pi k}{M} j\big), \quad k \in 0 : M - 1. \end{aligned} \tag{4.14.11}$$

On the basis of (4.14.8)–(4.14.11) we come to the following scheme of calculation of a spectrum $X = \overline{F}_N x$.

1. Form a vector z with the components $z(j) = \omega_{2N}^{-j^2} x(j)$, $j \in 0 : N - 1$, and augment it with zeroes up to an M-dimensional vector $\widehat{z}$.
2. Calculate $\widehat{Z} = \mathcal{F}_M(\widehat{z})$.
3. Component-wise multiply the vector $\widehat{Z}$ by the vector $\widehat{H}$ of form (4.14.11). Denote the resulting vector by $\widehat{Y}$.
4. Calculate $\widehat{X} = \mathcal{F}_M^{-1}(\widehat{Y})$.

5. Find the components of the spectrum X by the formula

$$X(k) = \omega_{2N}^{-k^2} \widehat{X}(k), \quad k \in 0 : N-1.$$

The worst case for this scheme is $N = 2^s + 1$ when

$$2N - 1 = 2^{s+1} + 1, \qquad M = 2^{s+2} = 4(N-1).$$

Exercises

4.1 Let $N = 2^s$ and x be a signal such that $x(j) = 0$ for $j \in \Delta_{\nu+1} : N-1$. Consider a signal $y(j) = x\big(\mathrm{rev}_s(j)\big)$, $j \in 0 : N-1$. Prove that the samples $y(j)$ are equal to zero when j is not divisible by N_ν.

4.2 Consider a signal

$$h(j) = \sum_{q=0}^{\Delta_{\nu+1}-1} \delta_N(j-q).$$

Prove that $h\big(\mathrm{rev}_s(j)\big) = \delta_{N_\nu}(j)$, $j \in 0 : N-1$.

4.3 Let $\psi_\nu(j)$ and $\varphi_\nu(j)$ be the signals introduced in Sects. 4.8.1 and 4.8.3, respectively. Prove that

$$\varphi_\nu\big(\mathrm{rev}_s(j)\big) = \psi_\nu(j), \quad j \in 0 : N-1.$$

4.4 Prove that the signal $\psi_\nu(j)$ satisfies to a recurrent relation

$$\psi_1(j) = \delta_N(j) - \delta_N(j - N/2),$$

$$\psi_\nu(j) = \psi_{\nu-1}(2j), \quad \nu = 2, \dots, s.$$

4.5 Expand the unit pulse δ_N over the Haar basis related to decimation in time.

4.6 Expand the unit pulse δ_N over the Haar basis related to decimation in frequency.

4.7 Let $q = (q_{s-1}, q_{s-2}, \dots, q_0)_2$. Prove that

$$\delta_N(j-q) = 2^{-s} + \sum_{\nu=1}^{s} (-1)^{q_{\nu-1}} 2^{-\nu} \varphi_\nu(j - \lfloor q/\Delta_{\nu+1} \rfloor \Delta_{\nu+1}).$$

4.8 Let $q = (q_{s-1}, q_{s-2}, \dots, q_0)_2$. Prove that

$$\delta_N(j-q) = 2^{-s} + \sum_{\nu=1}^{s} (-1)^{q_{s-\nu}} 2^{-\nu} \psi_\nu(j - \langle q \rangle_{N_\nu}).$$

4.9 We introduce a signal

$$x(j) = 2^{-s} + \sum_{\nu=1}^{s} 2^{-\nu} \sum_{p=0}^{N_\nu - 1} \psi_\nu(j-p).$$

Prove that $x(j) = 1 - 2j/N$ for $j \in 0 : N-1$.

4.10 We introduce a signal

$$y(j) = 2^{-s} + \sum_{\nu=1}^{s} 2^{-\nu} \sum_{p=0}^{N_\nu - 1} \varphi_\nu(j - p\Delta_{\nu+1}).$$

Prove that $y(j) = 1 - 2\,\mathrm{rev}_s(j)/N$ for $j \in 0{:}N-1$.

4.11 Consider unit steps of a form

$$h_k(j) = \sum_{q=0}^{\Delta_{k+1}-1} \delta_N(j-q), \quad k \in 1 : s-1.$$

Prove that there holds

$$h_k(j) = 2^{-s+k} + \sum_{\nu=k+1}^{s} 2^{-\nu+k} \varphi_\nu(j).$$

4.12 Prove that the following expansion is true for the unit steps from the previous exercise:

$$h_k(j) = 2^{-s+k} + \sum_{\nu=1}^{s-k} 2^{-\nu} \sum_{p=0}^{\Delta_{k+1}-1} \psi_\nu(j-p).$$

4.13 We take expansion (4.8.12) of a signal $x(j)$ and suppose that $q = (q_{s-1}, q_{s-2}, \ldots, q_0)_2$. Prove that

$$x(j \oplus q) = 2^{-s}\alpha + \sum_{\nu=1}^{s} (-1)^{q_{\nu-1}} 2^{-\nu} \sum_{p=0}^{N_\nu - 1} \widehat{\xi}_\nu(p \oplus \lfloor q/\Delta_{\nu+1} \rfloor)\, \varphi_\nu(j - p\Delta_{\nu+1})$$

holds for $j \in 0 : N-1$.

4.14 We take expansion (4.8.2) of a signal $y(j)$. Prove that for all $q \in \mathbb{Z}$ there holds

$$y(j-q) = 2^{-s}\beta + \sum_{\nu=1}^{s} 2^{-\nu} \sum_{p=0}^{N_\nu - 1} (-1)^{\lfloor (p-q)/N_\nu \rfloor}\, \widehat{y}_\nu\big(\langle p-q \rangle_{N_\nu}\big)\, \psi_\nu(j-p).$$

4.15 We denote by V_ν a linear hull spanned by the signals $g_\nu(k;\ j)$, $k = 0, 1, \ldots, N_\nu - 1$. Prove that $V_\nu = \mathbb{C}_{N_\nu}$.

4.16 We denote by W_ν a linear hull spanned by the signals $\psi_\nu(j-k)$, $k = 0, 1, \ldots, N_\nu - 1$. Prove that

$$W_\nu = \big\{w \in \mathbb{C}_{N_{\nu-1}} \;\big|\; w(j - N_\nu) = -w(j),\ \ j \in \mathbb{Z}\big\}.$$

4.17 Prove that Walsh functions $v_k(j)$ have the following properties:

$$1/v_k(j) = v_k(j),$$

$$v_k(j)\, v_{k'}(j) = v_m(j),$$

where $m = k \oplus k'$.

4.18 Prove that

$$v_{2k+1}(j) = v_{2k}(j)\, v_1(j).$$

4.19 We denote by $v_k^{(1)}(j)$ Walsh functions with the main period $0 : N_1 - 1$, i.e.

$$v_k^{(1)}(j) = (-1)^{\{k,\, j\}_{s-1}}, \quad k, j \in 0 : N_1 - 1.$$

Prove that the formulae

$$v_k(j) = v_k^{(1)}(j), \qquad v_k(N_1 + j) = v_k^{(1)}(j),$$

$$v_{N_1+k}(j) = v_k^{(1)}(j), \qquad v_{N_1+k}(N_1 + j) = -v_k^{(1)}(j)$$

are valid for $k, j \in 0 : N_1 - 1$.

4.20 Prove that for $N = 2^s$ the equality

$$\{N - 1 - j,\ j\}_s = 0$$

holds for all $j \in 0 : N - 1$.

4.21 Let $v_k(j)$ be one of Walsh functions with the property $v_k(N-1) = 1$. Prove that

$$v_k(N - 1 - j) = v_k(j), \quad j \in 0 : N - 1.$$

4.22 Let $v_k(j)$ be one of Walsh functions with the property $v_k(N-1) = -1$. Prove that

$$v_k(N - 1 - j) = -v_k(j), \quad j \in 0 : N - 1.$$

4.23 Prove that for $N \geq 8$ and $p \in 0 : N_2 - 1$ there holds

$$v_{3N_2+p}(j) = v_{N_1+p}\big(\langle j + N_2\rangle_N\big), \quad j \in 0 : N-1.$$

4.24 Functions $r_\nu(j) = v_{N_\nu}(j)$, $\nu \in 1 : s$, are referred to as *Rademacher functions*. Plot the graphs of Rademacher functions on the main period for $N = 8$.

4.25 Express Walsh functions via Rademacher functions.

4.26 Prove that

$$\frac{1}{N}\sum_{k=0}^{N-1} v_k(j) = \delta_N(j), \quad j \in 0 : N-1.$$

4.27 We turn to discrete Walsh transform $\mathcal{W}_N$ (see par. 4.10.1). Prove the Parseval equality

$$\|x\|^2 = N^{-1}\,\|\mathcal{W}_N(x)\|^2.$$

4.28 Let z be a dyadic convolution of signals x and y from $\mathbb{C}_N$ with $N = 2^s$ (see par. 4.8.4). Prove that

$$\mathcal{W}_N(z) = \mathcal{W}_N(x)\,\mathcal{W}_N(y).$$

4.29 Calculate the Fourier spectrum of Walsh functions v_1, v_2, and v_3.

4.30 We denote by R_ν the Fourier spectrum of the Rademacher function r_ν (see Exercise 4.24). Prove that

$$R_\nu(k) = \begin{cases} 2^\nu\Big(1 - i\cot\dfrac{(2l+1)\pi}{N_{\nu-1}}\Big) & \text{for } k = (2l+1)\Delta_\nu,\ l \in 0 : N_\nu - 1; \\ 0 & \text{for others } k \in 0 : N-1. \end{cases}$$

4.31 Let $V_p = \mathcal{F}_N(v_p)$. Prove that for $p \in 0 : N_1 - 1$ there holds

$$V_{2p+1}(k) = V_{2p}(k - N_1), \quad k \in 0 : N-1.$$

4.32 Along with the Fourier spectrum V_p of the Walsh function v_p with the main period $0 : N-1$ we consider the Fourier spectrum $V_p^{(1)}$ of the Walsh function $v_p^{(1)}$ with the main period $0 : N_1 - 1$. Prove that the spectra V_p and $V_p^{(1)}$ are bound with the relations

$$V_p(2k) = 2V_p^{(1)}(k), \qquad V_p(2k+1) = 0,$$

$$V_{N_1+p}(2k) = 0, \quad V_{N_1+p}(2k+1) = \frac{1}{N_1}\sum_{l=0}^{N_1-1} V_p^{(1)}(l)\,h(k-l)$$

for all $p, k \in 0 : N_1 - 1$, where

$$h(j) = 2\Big(1 - i\cot\frac{\pi(2j+1)}{N}\Big).$$

4.33 Prove that for $N \geq 8$ and $p \in 0 : N_2 - 1$ there holds

$$V_{3N_2+p}(k) = i^k\, V_{N_1+p}(k), \quad k \in 0 : N - 1.$$

4.34 Prove that for Walsh functions $\widehat{v}_k$ ordered by frequency the formula

$$\widehat{v}_{2k}(j) = \widehat{v}_k(2j)$$

is true for all $k, j \in 0 : N_1 - 1$.

4.35 Let $\widetilde{v}_k$ be a Walsh function ordered by the number of sign changes (see par. 4.11.3). Prove that $\widetilde{v}_{N-1} = v_1$.

4.36 We take a step-function (4.12.17) and expand it over the Walsh basis ordered by frequency. What are the coefficients $\xi(k)$ of this expansion for $k \in 0 : \Delta_{\nu+1} - 1$?

4.37 Calculate the inverse permutation $\mathrm{wal}_3^{-1}(k)$ values for $k = 0, 1, \ldots, 7$.

4.38 We introduce a Frank–Walsh signal

$$f(j_1 N + j_0) = v_{j_1}(j_0), \quad j_1, j_0 \in 0 : N - 1,$$

and put $F = \mathcal{W}_{N^2}(f)$. Prove that

$$F(k_1 N + k_0) = N v_{k_1}(k_0), \quad k_1, k_0 \in 0 : N - 1.$$

Comments

Goertzel's algorithm of fast calculation of a single component of Fourier spectrum is published in [10].

Introduction of the recurrent sequence of orthogonal bases (4.2.1) is a crucial point. Each signal of the ν-th basis in sequence (4.2.1) depends only on two signals of the $(\nu - 1)$-th basis. This produces a simple recurrent scheme of calculation of coefficients of expansion of a signal over all bases in the sequence. The last step gives us a complete Fourier spectrum. Moreover, sequence (4.2.1) lets us form wavelet bases; coefficients of expansion of a signal over these bases are contained in the table of coefficients of expansion of a signal over the bases of the initial recurrent sequence.

We briefly reminded the contents of Sects. 4.2–4.4. They are written on the basis of the papers [35–37].

The simplest wavelet basis is Haar basis. We pay a lot of attention to it in our book (Sects. 4.5–4.8, Exercises 4.3–4.14). The fact is, there are two Haar bases. One of them is related to decimation in time, it is well known (e.g. see [51] where an extensive bibliography is collected). Another one is related to decimation in frequency, it is introduced in [30]. The paper [25] is devoted to a comparative study of the two Haar bases.

Among the results concerning Haar bases we lay emphasis on a convolution theorem; to be more accurate, on a cyclic convolution theorem in case of Haar basis related to decimation in frequency, and on a dyadic convolution theorem in case of decimation in time. This topic has a long history. The final results were obtained in [24].

The recurrent sequence of orthogonal bases (4.6.2) can be defined explicitly, as it was done in [35], or can be deduced from (4.2.1) with the aid of *reverse* permutation. In our book we use the latter approach as being more didactic. This approach is published for the first time ever.

The recurrent sequence of orthogonal bases (4.9.1) that leads to fast Walsh transform was studied in [37]. A question of ordering of discrete Walsh functions and their generalizations was considered in [23, 49].

Note that the recurrent relations (4.2.1) that play a central role in construction of fast algorithms have just been *presented*. However they could be *deduced* on the basis of a factorization of a discrete Fourier transform matrix into a product of sparse matrices. This requires more sophisticated technical tools, in particular, a Kronecker product of matrices. On the subject of details and generalizations, refer to [12, 22, 32, 44, 46, 48, 50].

A sampling theorem in Walsh basis (in a more general form) is published in [39].

Ahmed–Rao bases are introduced in the book [1]. In case of $N = 2^s$ they parametrically consolidate Walsh basis and the exponential basis. Signals of Walsh basis take two values of ± 1; signals of the exponential basis take N values ω_N^q, $q \in 0 : N - 1$. Signals of Ahmed–Rao basis with an index $r \in 1 : s$ take 2^r values $\omega_{2^r}^q$, $q \in 0 : 2^r - 1$. Ahmed–Rao bases are studied in the papers [14–16].

In Sect. 4.14 we ascertain that calculation of DFT of any order can be reduced to calculation of DFT whose order is a power of two. This fact is mentioned in [46, pp. 208–211].

Solutions

To Chapter 1

1.1 The assertion is evident if $N = 1$. Let $N \geq 2$ and $j \in pN + 1 : (p+1)N$ for some $p \in \mathbb{Z}$. Then

$$\left\lfloor \frac{j-1}{N} \right\rfloor = p, \quad \left\lfloor -\frac{j}{N} \right\rfloor = -(p+1),$$

whence the required equality follows.

1.2 By a definition of a residual we write

$$\langle nj \rangle_{nN} = nj - \lfloor (nj)/(nN) \rfloor nN = n(j - \lfloor j/N \rfloor N) = n\langle j \rangle_N.$$

1.3 For $j \in 0 : N - 1$ we have

$$f\big(f(j)\big) = \langle f(j)n \rangle_N = \big\langle \langle jn \rangle_N \, n \big\rangle_N = \langle jn^2 \rangle_N = \big\langle j \langle n^2 \rangle_N \big\rangle_N = \langle j \rangle_N = j.$$

1.4 Since $f(j) \in 0 : N - 1$, it is sufficient to ascertain that $f(j) \neq f(j')$ for $j \neq j'$, $j, j' \in 0 : N - 1$. Let us assume the contrary: $f(j) = f(j')$ for some j, j' with mentioned properties. Then

$$\begin{aligned} \langle (j - j')n \rangle_N &= \langle (jn + l) - (j'n + l) \rangle_N = \big\langle \langle jn + l \rangle_N - \langle j'n + l \rangle_N \big\rangle_N \\ &= \langle f(j) - f(j') \rangle_N = 0. \end{aligned}$$

Taking into account relative primality of n and N we conclude that $j - j'$ is divisible by N. But $|j - j'| \leq N - 1$, therefore $j = j'$. This contradicts with our assumption.

1.5 We have $n = pd$ and $N = qd$, herein $\gcd(p, q) = 1$. According to the result of Exercise 1.2 we have

V. N. Malozemov and S. M. Masharsky, *Foundations of Discrete Harmonic Analysis*, Applied and Numerical Harmonic Analysis,
https://doi.org/10.1007/978-3-030-47048-7

$$f(j) = \langle jpd \rangle_{qd} = d\langle jp \rangle_q = d\big\langle \langle j \rangle_q\, p \big\rangle_q.$$

Hence

$$f(j) \in \{0,\, d,\, 2d,\, \ldots,\, (q-1)d\}.$$

1.6 Use the equality of sets

$$\{aj + bk \mid a \in \mathbb{Z},\ b \in \mathbb{Z}\} = \{p(j-k) + qk \mid p \in \mathbb{Z},\ q \in \mathbb{Z}\}$$

and Theorem 1.2.1.

1.7 According to (1.3.1) we have $a_1 n_2 + a_2 n_1 = 1$ for some integer a_1 and a_2. We put $j_1 = \langle a_1 j \rangle_{n_1}$ and $j_2 = \langle a_2 j \rangle_{n_2}$. Taking into account the result of Exercise 1.2 we gain

$$\begin{aligned}\langle j_1 n_2 + j_2 n_1 \rangle_N &= \big\langle \langle a_1 j \rangle_{n_1} n_2 + \langle a_2 j \rangle_{n_2} n_1 \big\rangle_N = \big\langle \langle a_1 j n_2 \rangle_{n_1 n_2} + \langle a_2 j n_1 \rangle_{n_1 n_2} \big\rangle_N \\ &= \langle j(a_1 n_2 + a_2 n_1) \rangle_N = j.\end{aligned}$$

Let us verify that this representation of a number j is unique. Assume that there is another representation $j = \langle j_1' n_2 + j_2' n_1 \rangle_N$ with $j_1' \in 0 : n_1 - 1$, $j_2' \in 0 : n_2 - 1$. Then $\langle (j_1 - j_1')n_2 + (j_2 - j_2')n_1 \rangle_N = 0$ holds. It means that

$$(j_1 - j_1')n_2 + (j_2 - j_2')n_1 = pN$$

for some integer p. Taking modulo n_1 residuals we come to the equality $\langle (j_1 - j_1')n_2 \rangle_{n_1} = 0$. On the strength of relative primality of n_1 and n_2 the difference $j_1 - j_1'$ is divisible by n_1. Since at the same time $|j_1 - j_1'| \le n_1 - 1$, it is necessary that $j_1' = j_1$. Similarly one can show that $j_2' = j_2$.

1.8 According to (1.3.1) we have $a_\alpha n_\alpha + b_\alpha m = 1$ for some integer a_α and b_α, $\alpha \in 1 : s$. Multiplying out these equalities we gain

$$(a_1 a_2 \cdots a_s)(n_1 n_2 \cdots n_s) + pm = 1.$$

Hence relative primality of the product $N = n_1 n_2 \cdots n_s$ and the number m follows.

1.9 At first we consider the case of $s = 2$. Let $j = p_1 n_1$ and $j = p_2 n_2$. Then $0 = \langle j \rangle_{n_2} = \langle p_1 n_1 \rangle_{n_2}$. Since gcd $(n_1, n_2) = 1$, Theorem 1.3.1 yields $\langle p_1 \rangle_{n_2} = 0$. Taking into account the result of Exercise 1.2 we gain

$$\langle j \rangle_{n_1 n_2} = \langle p_1 n_1 \rangle_{n_1 n_2} = n_1 \langle p_1 \rangle_{n_2} = 0.$$

We perform an induction step from s to $s+1$. By virtue of the inductive hypothesis j is divisible by the product $n_1 n_2 \cdots n_s$. Furthermore j is divisible by n_{s+1}. According to the result of the previous exercise the number $n_1 n_2 \cdots n_s$ is relatively prime with n_{s+1}. Therefore, j is divisible by the product $N = (n_1 n_2 \cdots n_s) n_{s+1}$.

1.10 For $s = 2$ we have $N = n_1 n_2$, $\widehat{N}_1 = n_2$, and $\widehat{N}_2 = n_1$. The statement of the exercise corresponds to formula (1.3.1). We perform an induction step from s to $s + 1$.

Let $N = n_1 \cdots n_s n_{s+1}$. By virtue of the inductive hypothesis there exist integer numbers $a_1, a_2, \ldots, a_s$ such that

$$\sum_{\alpha=1}^{s} a_\alpha \frac{\widehat{N}_\alpha}{n_{s+1}} = 1.$$

Furthermore, by virtue of relative primality of $\widehat{N}_{s+1}$ and n_{s+1} there exist integer numbers a_{s+1} and b_{s+1} with the property

$$a_{s+1}\widehat{N}_{s+1} + b_{s+1}n_{s+1} = 1.$$

Combining two presented equalities we gain

$$b_{s+1}\sum_{\alpha=1}^{s} a_\alpha \widehat{N}_\alpha = b_{s+1}n_{s+1} = 1 - a_{s+1}\widehat{N}_{s+1},$$

which is equivalent to what was required.

1.11 According to the result of the previous exercise there exist integer numbers $a_1, a_2, \ldots, a_s$ such that

$$\sum_{\alpha=1}^{s} a_\alpha \widehat{N}_\alpha = 1. \tag{S.1}$$

We put $j_\alpha = \langle a_\alpha j\rangle_{n_\alpha}$. Taking into account the result of Problem 1.2 we gain

$$\Big\langle \sum_{\alpha=1}^{s} j_\alpha \widehat{N}_\alpha \Big\rangle_N = \Big\langle \sum_{\alpha=1}^{s} \langle a_\alpha j\rangle_{n_\alpha} \widehat{N}_\alpha \Big\rangle_N = \Big\langle \sum_{\alpha=1}^{s} \langle a_\alpha j \widehat{N}_\alpha\rangle_{n_\alpha \widehat{N}_\alpha} \Big\rangle_N$$
$$= \Big\langle j \sum_{\alpha=1}^{s} a_\alpha \widehat{N}_\alpha \Big\rangle_N = j.$$

Uniqueness of this representation is verified in the same way as in the solution of Problem 1.7.

It follows from (S.1) that $\big\langle \langle a_\alpha\rangle_{n_\alpha} \widehat{N}_\alpha \big\rangle_{n_\alpha} = 1$. It means that the number $p_\alpha = \langle a_\alpha\rangle_{n_\alpha}$ is a unique solution of the equation $\langle x\widehat{N}_\alpha\rangle_{n_\alpha} = 1$ on the set $0 : n_\alpha - 1$. Taking into account this fact we come to a final expression for the coefficients j_α:

$$j_\alpha = \big\langle \langle a_\alpha\rangle_{n_\alpha} j \big\rangle_{n_\alpha} = \langle p_\alpha j\rangle_{n_\alpha}, \quad \alpha \in 1 : s.$$

1.12 We will show that the formula

$$k = \left\langle \sum_{\alpha=1}^{s} k_\alpha p_\alpha \widehat{N}_\alpha \right\rangle_N , \tag{S.2}$$

where $k_\alpha \in 0 : n_\alpha - 1$, establishes a bijective mapping between assemblies of coefficients $(k_1, k_2, \ldots, k_s)$ and numbers $k \in 0 : N - 1$. Since both considered sets have the same number of elements, it is sufficient to verify that different integers k correspond to different assemblies of coefficients. Assume that along with (S.2) there holds

$$k = \left\langle \sum_{\alpha=1}^{s} k'_\alpha p_\alpha \widehat{N}_\alpha \right\rangle_N ,$$

where $k'_\alpha \in 0 : n_\alpha - 1$. Then

$$\left\langle \sum_{\alpha=1}^{s} (k_\alpha - k'_\alpha) p_\alpha \widehat{N}_\alpha \right\rangle_N = 0,$$

so that

$$\sum_{\nu=1}^{s} (k_\nu - k'_\nu) p_\nu \widehat{N}_\nu = pN$$

for some integer p. Taking modulo n_α residuals we gain $\langle (k_\alpha - k'_\alpha) p_\alpha \widehat{N}_\alpha \rangle_{n_\alpha} = 0$. By virtue of relative primality of the numbers $p_\alpha \widehat{N}_\alpha$ and n_α the difference $k_\alpha - k'_\alpha$ is divisible by n_α. Since at the same time $|k_\alpha - k'_\alpha| \le n_\alpha - 1$, it is necessary that $k_\alpha = k'_\alpha$ for all $\alpha \in 1 : s$.

It is ascertained that any number $k \in 0 : N - 1$ allows representation (S.2) where $k_\alpha \in 0 : n_\alpha - 1$. Let us rewrite formula (S.2) in a form

$$\sum_{\nu=1}^{s} k_\nu p_\nu \widehat{N}_\nu = qN + k, \quad q \in \mathbb{Z}.$$

Taking modulo n_α residuals and bearing in mind the definition of p_α we gain

$$\langle k \rangle_{n_\alpha} = \left\langle k_\alpha \langle p_\alpha \widehat{N}_\alpha \rangle_{n_\alpha} \right\rangle_{n_\alpha} = k_\alpha,$$

i.e. $k_\alpha = \langle k \rangle_{n_\alpha}$.

1.13 Let us show that

$$\mathrm{grey}_\nu \Big(j_{\nu-1} 2^{\nu-1} + \sum_{k=2}^{\nu} j_{\nu-k} 2^{\nu-k} \Big) = j_{\nu-1} 2^{\nu-1} + \mathrm{grey}_{\nu-1} \Big(\sum_{k=2}^{\nu} \langle j_{\nu-1} + j_{\nu-k} \rangle_2 \, 2^{\nu-k} \Big). \tag{S.3}$$

When $j_{\nu-1} = 0$, this is another notation of the second line of relations (1.4.5). Let $j_{\nu-1} = 1$. Equality (1.4.6) yields

$$\text{grey}_\nu\Big(2^{\nu-1} + \sum_{k=2}^{\nu} j_{\nu-k}2^{\nu-k}\Big) = 2^{\nu-1} + \text{grey}_{\nu-1}\Big(\sum_{k=2}^{\nu}(1 - j_{\nu-k})\,2^{\nu-k}\Big).$$

As long as $1 - j_{\nu-k} = \langle 1 + j_{\nu-k}\rangle_2$, the latter formula corresponds to (S.3) for $j_{\nu-1} = 1$.

Further, the same considerations yield

$$\text{grey}_{\nu-1}\Big(\sum_{k=2}^{\nu}\langle j_{\nu-1} + j_{\nu-k}\rangle_2\,2^{\nu-k}\Big) = \langle j_{\nu-1} + j_{\nu-2}\rangle_2\,2^{\nu-2}$$
$$+ \text{grey}_{\nu-2}\Big(\sum_{k=3}^{\nu}\langle j_{\nu-2} + j_{\nu-k}\rangle_2\,2^{\nu-k}\Big).$$

We have used the equality

$$\big\langle\langle j_{\nu-1} + j_{\nu-2}\rangle_2 + \langle j_{\nu-1} + j_{\nu-k}\rangle_2\big\rangle_2 = \langle j_{\nu-2} + j_{\nu-k}\rangle_2.$$

Continuing this process we gain

$$\begin{aligned}\text{grey}_\nu(j) = {} & j_{\nu-1}2^{\nu-1} + \langle j_{\nu-1} + j_{\nu-2}\rangle_2\,2^{\nu-2} + \langle j_{\nu-2} + j_{\nu-3}\rangle_2\,2^{\nu-3}\\ & + \cdots + \langle j_1 + j_0\rangle_2.\end{aligned}$$

We could use more compact representation $\text{grey}_\nu(j) = j \oplus \lfloor j/2\rfloor$.

1.14 By virtue of the result of the previous exercise it is sufficient to solve a system of equations

$$j_{\nu-1} = p_{\nu-1},$$
$$\langle j_{\nu-k+1} + j_{\nu-k}\rangle_2 = p_{\nu-k}, \quad k = 2, \ldots, \nu.$$

This system is solved in an elementary way. Indeed, for $k \in 2 : \nu$ we have

$$\langle p_{\nu-1} + p_{\nu-2} + \cdots + p_{\nu-k}\rangle_2 = \langle j_{\nu-1} + (j_{\nu-1} + j_{\nu-2}) + (j_{\nu-2} + j_{\nu-3})$$
$$+ \cdots + (j_{\nu-k+1} + j_{\nu-k})\rangle_2 = j_{\nu-k}.$$

1.15 We have

$$\sum_{k=1}^{n}\big[k^3 - (k-1)^3\big] = \sum_{k=1}^{n}k^3 - \sum_{k=0}^{n-1}k^3 = n^3.$$

At the same time

$$\sum_{k=1}^{n}\left[k^3-(k-1)^3\right]=\sum_{k=1}^{n}(3k^2-3k+1).$$

Therefore

$$3\sum_{k=1}^{n}k^2=n^3+\tfrac{3}{2}n(n+1)-n=\tfrac{1}{2}n(2n^2+3n+1)=\tfrac{1}{2}n(n+1)(2n+1),$$

which is equivalent to what was required.

1.16 We denote

$$S_n=\sum_{k=1}^{n}k\binom{n}{k}.$$

Taking into account the formula $\binom{n}{k}=\binom{n}{n-k}$ we gain

$$S_n=\sum_{k=0}^{n}[n-(n-k)]\binom{n}{n-k}=\sum_{k'=0}^{n}(n-k')\binom{n}{k'}=n\,2^n-S_n.$$

Hence the required equality follows in an obvious way.

1.17 We have

$$\varepsilon_n^k=\omega_N^{kn}=\omega_N^{\langle kn\rangle_N}.$$

It is remaining to take into account that by virtue of relative primality of n and N the set of powers $\left\{\langle kn\rangle_N\right\}_{k=0}^{N-1}$ is a permutation of the set $\{0,\,1,\,\ldots,\,N-1\}$.

1.18 We have $a_0m+b_0n=1$ for some integer a_0 and b_0. We write

$$\omega_m^{b_0}\,\omega_n^{a_0}=\omega_{mn}^{b_0n}\,\omega_{nm}^{a_0m}=\omega_{mn}^{a_0m+b_0n}=\omega_{mn}.$$

Hence

$$\omega_{mn}=\omega_m^{\langle b_0\rangle_m}\omega_n^{\langle a_0\rangle_n}.$$

Putting $p=\langle b_0\rangle_m$ and $q=\langle a_0\rangle_n$ we obtain the required decomposition $\omega_{mn}=\omega_m^p\,\omega_n^q$. Let us prove its uniqueness. Assume that $\omega_{mn}=\omega_m^{p'}\,\omega_n^{q'}$ for some $p'\in 0:m-1$ and $q'\in 0:n-1$. Then there holds $\omega_m^{p-p'}=\omega_n^{q'-q}$. Raising to the power n we come to the equality $\omega_m^{\langle(p-p')n\rangle_m}=1$. Hence $\langle(p-p')n\rangle_m=0$. We know that $\gcd(m,n)=1$, therefore the difference $p-p'$ is divisible by m. Since at the same time $|p-p'|\le m-1$, we gain $p'=p$. Similarly one can show that $q'=q$.

Relative primality of p and m and q and n follows from the relations

$$(a_0+\lfloor b_0/m\rfloor n)\,m+\langle b_0\rangle_m\,n=1,$$

$$\langle a_0 \rangle_n\, m + (\lfloor a_0/n \rfloor m + b_0)\, n = 1.$$

1.19 Denote

$$P_{N-1}(z) = \sum_{k=0}^{N-1} z^k.$$

For $z \neq 1$ we have

$$P_{N-1}(z) = \frac{1 - z^N}{1 - z}.$$

It is clear that $P_{N-1}(\omega_N^j) = 0$ for $j = 1, 2, \ldots, N-1$. Thus, we know $N-1$ different roots $\omega_N, \omega_N^2, \ldots, \omega_N^{N-1}$ of the polynomial $P_{N-1}(z)$ of degree $N-1$. This lets us write down the representation

$$P_{N-1}(z) = \prod_{j=1}^{N-1} (z - \omega_N^j).$$

1.20 Let

$$P_r(z) = \sum_{k=0}^{r} a_k z^k.$$

Then

$$\Delta P_r(j) = a_r\big[(j+1)^r - j^r\big] + \sum_{k=0}^{r-1} a_k\big[(j+1)^k - j^k\big] =: P_{r-1}(j).$$

Therefore the finite difference of the first order of a polynomial of degree r is a polynomial of degree $r-1$. The same considerations yield that the finite difference of the second order of a polynomial of degree r is a polynomial of degree $r-2$ and so on. The finite difference of the r-th order of a polynomial of degree r identically equals to a constant, so the finite difference of the $(r+1)$-th order of a polynomial of degree r identically equals to zero.

To Chapter 2

2.1 If x is an even signal then $x(0) = \overline{x}(0)$ holds, so the value $x(0)$ is real. By virtue of N-periodicity we have $x(N-j) = \overline{x}(j)$ for $j \in 1 : N-1$.

Conversely, let $x(0)$ be a real number and $x(N-j) = \overline{x}(j)$ hold for $j \in 1 : N-1$. Then $x(-j) = \overline{x}(j)$ holds for $j \in 0 : N-1$, and N-periodicity yields $x(-j) = \overline{x}(j)$ for all $j \in \mathbb{Z}$.

This exercise shows how to determine an even signal through its values on the main period only.

2.2 The solution is similar to the previous one.

2.3 Assume that there exist an even signal x_0 and an odd signal x_1 such that $x(j) = x_0(j) + x_1(j)$ holds. Then

$$\overline{x}(-j) = \overline{x}_0(-j) + \overline{x}_1(-j) = x_0(j) - x_1(j).$$

Adding and subtracting the given equalities we gain

$$x_0(j) = \tfrac{1}{2}\big(x(j) + \overline{x}(-j)\big), \quad x_1(j) = \tfrac{1}{2}\big(x(j) - \overline{x}(-j)\big). \tag{S.4}$$

Now one can easily verify that the signals x_0 and x_1 of a form (S.4) are even and odd, respectively, and that $x = x_0 + x_1$ holds.

2.4 A signal $\delta_{mn}(mj)$ is n-periodic; hence, it is sufficient to verify the equality $\delta_{mn}(mj) = \delta_n(j)$ for $j \in 0 : n-1$. It is obviously true for $j = 0$. If $1 \le j \le n-1$ then $m \le mj \le mn - m$ hold, so $\delta_{mn}(mj) = 0$. And $\delta_n(j) = 0$ for these j as well.

2.5 A signal $x(j) = \sum_{l=0}^{m-1} \delta_{mn}(j + ln)$ by virtue of Lemma 2.1.3 is n-periodic, so it is sufficient to prove the equality $x(j) = \delta_n(j)$ for $j \in 0 : n-1$. Note that the inequalities $0 \le j + ln \le mn - 1$ hold for given j and $l \in 0 : m-1$; moreover, the left inequality turns into an equality only when $j = 0$ and $l = 0$. Hence it follows that $x(0) = 1 = \delta_n(0)$ and $x(j) = 0 = \delta_n(j)$ for $j \in 1 : n-1$.

2.6 We will use Lemma 2.1.4. Taking into account that $r \le N - 1$ we gain

$$\begin{aligned}
\|\Delta^r(\delta_N)\|^2 &= \Big\langle \sum_{s=0}^{r} (-1)^{r-s}\binom{r}{s}\delta_N(\cdot + s),\ \sum_{s'=0}^{r}(-1)^{r-s'}\binom{r}{s'}\delta_N(\cdot + s')\Big\rangle \\
&= \sum_{s,s'=0}^{r} (-1)^{s+s'}\binom{r}{s}\binom{r}{s'}\delta_N(s - s') = \sum_{s=0}^{r}\binom{r}{s}^2.
\end{aligned}$$

It's not difficult to show that

$$\sum_{s=0}^{r}\binom{r}{s}^2 = \binom{2r}{r}.$$

To do so we should take an identity $(1+z)^r(1+z)^r = (1+z)^{2r}$ and similize the coefficients at z^r. We come to a more compact formula

$$\|\Delta^r(\delta_N)\|^2 = \binom{2r}{r}.$$

2.7 Lemma 2.1.1 yields

$$\sum_{l=0}^{m-1} x(s+ln) = \sum_{l=0}^{m-1}\sum_{j=0}^{N-1} x(j)\,\delta_N(s+ln-j)$$
$$= \sum_{j=0}^{N-1} x(j) \sum_{l=0}^{m-1} \delta_{mn}\big((s-j)+ln\big) = \sum_{j=0}^{N-1} x(j)\,\delta_n(s-j).$$

We used the result of Exercise 2.5 in the last transition.

2.8 It is sufficient to verify that the equality $\langle j\rangle_{kN} = 0$ holds if and only if $\langle j\rangle_k = 0$ and $\langle j\rangle_N = 0$.

Let $\langle j\rangle_{kN} = 0$. It means that $j = pkN$ holds for some integer p. Hence it follows that j is divisible both by k and by N, i.e. that $\langle j\rangle_k = 0$ and $\langle j\rangle_N = 0$.

The converse proposition constitutes the contents of Problem 1.9 (for $s = 2$).

2.9 When k and N are relative primes, the mapping $j \to \langle kj\rangle_N$ is a permutation of the set $\{0, 1, \ldots, N-1\}$, therefore

$$\sum_{j=0}^{N-1} \delta_N(kj+l) = \sum_{j=0}^{N-1} \delta_N\big(\langle kj\rangle_N + l\big) = \sum_{j'=0}^{N-1} \delta_N(j'+l) = \sum_{j=0}^{N-1} \delta_N(j) = 1.$$

2.10 The DFT inversion formula yields that a signal oddity condition $x(-j) = -\overline{x}(j)$ is equivalent to the identity

$$\sum_{k=0}^{N-1} X(k)\,\omega_N^{-kj} = \sum_{k=0}^{N-1} \big(-\overline{X}(k)\big)\,\omega_N^{-kj}, \quad j \in \mathbb{Z},$$

which, in turn, holds if and only if $X(k) = -\overline{X}(k)$ or $\overline{X}(k) = -X(k)$ for all $k \in \mathbb{Z}$. The latter characterizes the spectrum X as pure imaginary.

2.11 Linearity of DFT yields

$$X(k) = A(k) + iB(k),$$

where the spectra A and B are even (see Theorem 2.2.3). Further,

$$\overline{X}(N-k) = \overline{X}(-k) = \overline{A}(-k) - i\overline{B}(-k) = A(k) - iB(k).$$

It is remaining to add and subtract the obtained equalities.

We see that calculation of spectra of two real signals a and b is reduced to calculation of a spectrum of one complex signal $x = a + ib$.

2.12 The DFT inversion formula yields

$$
\begin{aligned}
x_a(j) &= \frac{1}{N}\left[\sum_{k=0}^{N/2} X(k)\,\omega_N^{kj} + \sum_{k=1}^{N/2-1} X(k)\,\omega_N^{kj}\right] \\
&= \frac{1}{N}\left[\sum_{k=0}^{N/2} X(k)\,\omega_N^{kj} + \sum_{k=N/2+1}^{N-1} X(N-k)\,\omega_N^{(N-k)j}\right].
\end{aligned}
$$

The spectrum X of a real signal x is even, therefore

$$
\sum_{k=N/2+1}^{N-1} X(N-k)\,\omega_N^{(N-k)j} = \sum_{k=N/2+1}^{N-1} \overline{X}(k)\,\omega_N^{-kj}.
$$

Taking into account that $\operatorname{Re}\overline{z} = \operatorname{Re} z$ we gain

$$
\begin{aligned}
\operatorname{Re} x_a(j) &= \frac{1}{N}\left[\operatorname{Re}\sum_{k=0}^{N/2} X(k)\,\omega_N^{kj} + \operatorname{Re}\sum_{k=N/2+1}^{N-1} X(k)\,\omega_N^{kj}\right] \\
&= \operatorname{Re} x(j) = x(j).
\end{aligned}
$$

2.13 We take a real signal x and correspond it with a complex signal x_a with a spectrum

$$
X_a(k) = \begin{cases} X(k) & \text{for } k = 0, \\ 2X(k) & \text{for } k \in 1 : (N-1)/2, \\ 0 & \text{for } k \in (N+1)/2 : N-1. \end{cases}
$$

Let us show that $\operatorname{Re} x_a = x$. We have

$$
x_a(j) = \frac{1}{N}\left[\sum_{k=0}^{(N-1)/2} X(k)\,\omega_N^{kj} + \sum_{k=1}^{(N-1)/2} X(k)\,\omega_N^{kj}\right].
$$

Since

$$
\sum_{k=1}^{(N-1)/2} X(k)\,\omega_N^{kj} = \sum_{k=(N+1)/2}^{N-1} X(N-k)\,\omega_N^{(N-k)j} = \sum_{k=(N+1)/2}^{N-1} \overline{X}(k)\,\omega_N^{-kj},
$$

we gain

$$
\operatorname{Re} x_a(j) = \operatorname{Re}\left\{\frac{1}{N}\sum_{k=0}^{N-1} X(k)\,\omega_N^{kj}\right\} = \operatorname{Re} x(j) = x(j).
$$

2.14 Let us verify, for instance, the second equality. We have

$$\begin{aligned} X(N/2+k) &= \sum_{j=0}^{N-1} x(j)\, \omega_N^{-(N/2+k)j} = \sum_{j=0}^{N-1} (-1)^j\, x(j)\, \omega_N^{-kj} \\ &= \sum_{j=0}^{N/2-1} x(2j)\, \omega_N^{-k(2j)} - \sum_{j=0}^{N/2-1} x(2j+1)\, \omega_N^{-k(2j+1)} \\ &= \sum_{j=0}^{N/2-1} \big[x(2j) - \omega_N^{-k}\, x(2j+1)\big]\, \omega_{N/2}^{-kj}. \end{aligned}$$

2.15 Let us verify, for instance, the second equality. We have

$$\begin{aligned} X(2k+1) &= \sum_{j=0}^{N/2-1} x(j)\, \omega_N^{-(2k+1)j} + \sum_{j=0}^{N/2-1} x(N/2+j)\, \omega_N^{-(2k+1)(N/2+j)} \\ &= \sum_{j=0}^{N/2-1} \big[x(j) - x(N/2+j)\big]\, \omega_N^{-j}\, \omega_{N/2}^{-kj}. \end{aligned}$$

2.16 We immediately obtain

$$\begin{aligned} X(k) &= \sum_{j=0}^{N-1} \sin\frac{\pi j}{N}\, \omega_N^{-kj} = \frac{1}{2i} \sum_{j=0}^{N-1} \left(\omega_{2N}^{-(2k-1)j} - \omega_{2N}^{-(2k+1)j}\right) \\ &= \frac{1}{2i} \left(\frac{1-\omega_{2N}^{-(2k-1)N}}{1-\omega_{2N}^{-(2k-1)}} - \frac{1-\omega_{2N}^{-(2k+1)N}}{1-\omega_{2N}^{-(2k+1)}} \right) \\ &= \frac{1}{i} \left(\frac{1}{1-\omega_{2N}^{-(2k-1)}} - \frac{1}{1-\omega_{2N}^{-(2k+1)}} \right) \\ &= \frac{1}{i} \left(\frac{\omega_{2N}^{-(2k-1)} - \omega_{2N}^{-(2k+1)}}{1-\omega_{2N}^{-(2k-1)} - \omega_{2N}^{-(2k+1)} + \omega_{2N}^{-4k}} \right) \\ &= \frac{2\, \omega_{2N}^{-2k} \sin\frac{\pi}{N}}{1-\omega_{2N}^{-(2k-1)} - \omega_{2N}^{-(2k+1)} + \omega_{2N}^{-4k}} \\ &= \frac{2 \sin\frac{\pi}{N}}{\omega_{2N}^{2k} - \omega_{2N}^{1} - \omega_{2N}^{-1} + \omega_{2N}^{-2k}} = \frac{\sin\frac{\pi}{N}}{\cos\frac{2\pi k}{N} - \cos\frac{\pi}{N}}. \end{aligned}$$

2.17 Since $(-1)^j = \omega_2^j = \omega_{2n}^{nj}$, Lemma 2.2.1 for $N = 2n$ yields

$$X(k) = \sum_{j=0}^{N-1} (-1)^j\, \omega_N^{-kj} = \sum_{j=0}^{N-1} \omega_N^{(n-k)j} = N\, \delta_N(k-n).$$

Let $N = 2n + 1$. We will show that $X(k) = 1 + i \tan \frac{\pi k}{N}$ holds for $k \in 0 : N - 1$.
The definition of DFT yields

$$X(k) = \sum_{j=0}^{n} (-1)^{2j}\, \omega_N^{-2kj} + \sum_{j=0}^{n-1} (-1)^{2j+1}\, \omega_N^{-k(2j+1)}$$
$$= \omega_N^{-2kn} + (1 - \omega_N^{-k}) \sum_{j=0}^{n-1} \omega_N^{-2kj}.$$

In particular, $X(0) = 1$. Similar to (2.2.7), for $k \in 1 : N - 1$ we gain

$$X(k) = \omega_N^{-2kn} + (1 - \omega_N^{-k}) \frac{1 - \omega_N^{-2kn}}{1 - \omega_N^{-2k}} = \omega_N^{-2kn} + \frac{1 - \omega_N^{-2kn}}{1 + \omega_N^{-k}}$$
$$= \frac{2}{1 + \omega_N^{-k}} = 1 + i \tan \frac{\pi k}{N}.$$

2.18 We write

$$X(k) = \sum_{j=0}^{n} j\, \omega_N^{-kj} + \sum_{j=n+1}^{2n} (j - N)\, \omega_N^{-k(j-N)} = \sum_{j=-n}^{n} j\, \omega_N^{-kj}.$$

It is obvious that $X(0) = 0$. Let $k \in 1 : N - 1$. We denote $z = \omega_N^{-k}$. Taking into account that $z^{-n} = z^{n+1}$ we gain

$$(1 - z)\, X(k) = \sum_{j=-n}^{n} j z^j - \sum_{j=-n+1}^{n+1} (j - 1) z^j = -nz^{-n} - nz^{n+1} + \sum_{j=-n+1}^{n} z^j$$
$$= -2nz^{-n} + \frac{z^{-n+1} - z^{n+1}}{1 - z} = -2nz^{-n} + \frac{z^{-n}(z - 1)}{1 - z} = -Nz^{-n}.$$

Let us perform some transformations:

$$z^{-n} = \omega_N^{kn} = \cos \frac{2\pi kn}{N} + i \sin \frac{2\pi kn}{N} = \cos \frac{(N-1)\pi k}{N} + i \sin \frac{(N-1)\pi k}{N}$$
$$= (-1)^k \Big(\cos \frac{\pi k}{N} - i \sin \frac{\pi k}{N} \Big),$$

$$1 - z = 2 \sin \frac{\pi k}{N} \Big(\sin \frac{\pi k}{N} + i \cos \frac{\pi k}{N} \Big) = 2i \sin \frac{\pi k}{N} \Big(\cos \frac{\pi k}{N} - i \sin \frac{\pi k}{N} \Big).$$

We come to the final formula

$$X(k) = \frac{1}{2}\,Ni\,\frac{(-1)^k}{\sin\dfrac{\pi k}{N}}, \quad k \in 1:N-1.$$

2.19 The definition of DFT yields

$$X(k) = \sum_{j=0}^{n} j\,\omega_N^{-kj} + \sum_{j=n+1}^{N-1} (N-j)\,\omega_N^{k(N-j)} = n\,\omega_N^{-kn} + \sum_{j=1}^{n-1} j\,\omega_N^{-kj} + \sum_{j=1}^{n-1} j\,\omega_N^{kj}$$

$$= n\,(-1)^k + 2\,\mathrm{Re}\left\{\sum_{j=1}^{n-1} j\,\omega_N^{-kj}\right\}.$$

It is clear that $X(0) = n^2$. Let $k \in 1:N-1$. We will use formula

$$\sum_{j=1}^{n-1} jz^j = \frac{z}{(1-z)^2}\big[1 - nz^{n-1} + (n-1)z^n\big], \quad z \neq 1 \tag{S.5}$$

(see Sect. 1.6). Since the number $z = \omega_N^{-k}$ is other than unity and $z^n = (-1)^k$, formula (S.5) yields

$$\sum_{j=1}^{n-1} j\,\omega_N^{-kj} = \frac{\omega_N^{-k}}{(1-\omega_N^{-k})^2}\big[1 - n(-1)^k\,\omega_N^k + (n-1)(-1)^k\big]. \tag{S.6}$$

It is not difficult to verify that

$$\frac{\omega_N^{-k}}{(1-\omega_N^{-k})^2} = -\frac{1}{4\sin^2\dfrac{\pi k}{N}}, \quad k \in 1:N-1. \tag{S.7}$$

Indeed,

$$\frac{\omega_N^{-k}}{(1-\omega_N^{-k})^2} = \frac{\omega_N^{-k}}{1 - 2\,\omega_N^{-k} + \omega_N^{-2k}} = \frac{1}{\omega_N^k - 2 + \omega_N^{-k}}$$

$$= -\frac{1}{2\left(1 - \cos\dfrac{2\pi k}{N}\right)} = -\frac{1}{4\sin^2\dfrac{\pi k}{N}}.$$

On the basis of (S.6) and (S.7) we gain

$$\operatorname{Re}\left\{\sum_{j=1}^{n-1} j\,\omega_N^{-kj}\right\} = -\frac{1}{4\sin^2\dfrac{\pi k}{N}}\left[\left(1-(-1)^k\right)+n(-1)^k\,2\sin^2\frac{\pi k}{N}\right]$$
$$= -\frac{1}{2}\,n(-1)^k - \frac{1-(-1)^k}{4\sin^2\dfrac{\pi k}{N}},$$

so that

$$X(k) = -\frac{1-(-1)^k}{2\sin^2\dfrac{\pi k}{N}}, \quad k \in 1:N-1.$$

2.20 According to Lemmas 2.1.2 and 2.2.1 we have

$$X(k)\,\overline{X}(k) = \sum_{j=0}^{N-1}\omega_N^{j^2-kj}\sum_{l=0}^{N-1}\omega_N^{-l^2+kl} = \sum_{j,\,l=0}^{N-1}\omega_N^{(j^2-l^2)-k(j-l)}$$
$$= \sum_{l=0}^{N-1}\sum_{j=0}^{N-1}\omega_N^{(j-l)((j-l)+2l-k)} = \sum_{l=0}^{N-1}\sum_{j=0}^{N-1}\omega_N^{j(j+2l-k)}$$
$$= \sum_{j=0}^{N-1}\omega_N^{j(j-k)}\sum_{l=0}^{N-1}\omega_N^{2jl} = N\sum_{j=0}^{N-1}\omega_N^{j(j-k)}\,\delta_N(2j).$$

It is clear that provided N is odd the equality $|X(k)| = \sqrt{N}$ holds for all $k \in \mathbb{Z}$. As for $N = 2n$, there holds $|X(k)|^2 = N\left(1+\omega_N^{n(n-k)}\right) = N\left(1+(-1)^{n-k}\right)$, so in this case we have $|X(k)| = \sqrt{N\left(1+(-1)^{n-k}\right)}$ for all $k \in \mathbb{Z}$.

2.21 The definition of DFT and Lemma 2.1.2 yield

$$X_l(k) = \sum_{j=0}^{N-1} x(j+l)\,\omega_N^{-k(j+l)+kl} = \omega_N^{kl}\sum_{j=0}^{N-1} x(j)\,\omega_N^{-kj} = \omega_N^{kl}\,X(k), \quad k \in \mathbb{Z}.$$

2.22 We have

$$X_l(k) = \frac{1}{2}\sum_{j=0}^{N-1}\left(\omega_N^{lj}+\omega_N^{-lj}\right)x(j)\,\omega_N^{-kj}$$
$$= \frac{1}{2}\left(\sum_{j=0}^{N-1} x(j)\,\omega_N^{-(k-l)j} + \sum_{j=0}^{N-1} x(j)\,\omega_N^{-(k+l)j}\right)$$
$$= \tfrac{1}{2}\left(X(k-l)+X(k+l)\right), \quad k \in \mathbb{Z}.$$

2.23 Since $\gcd(p, N) = 1$, there exist integers r and q such that $rp + qN = 1$, herein we can suppose that $r \in 1 : N - 1$ and $\gcd(r, N) = 1$. Taking into account that the mapping $j \to \langle pj \rangle_N$ is a permutation of the set $\{0, 1, \ldots, N-1\}$, we write

$$Y_p(k) = \sum_{j=0}^{N-1} x\big(\langle pj \rangle_N\big)\, \omega_N^{-kj(rp+qN)} = \sum_{j=0}^{N-1} x\big(\langle pj \rangle_N\big)\, \omega_N^{-rk\langle pj \rangle_N}$$
$$= \sum_{j=0}^{N-1} x(j)\, \omega_N^{-\langle rk \rangle_N j} = X\big(\langle rk \rangle_N\big), \quad k \in 0 : N-1.$$

We see that Euler permutation $j \to \langle pj \rangle_N$ of samples of a signal leads to Euler permutation $k \to \langle rk \rangle_N$ of components of its spectrum. Here $\langle rp \rangle_N = 1$.

2.24 On the basis of the definition of DFT and the inversion formula we gain

$$X_n(k) = \sum_{j=0}^{N-1} x(j)\, \omega_{nN}^{-kj} = \sum_{j=0}^{N-1} \left\{ \frac{1}{N} \sum_{l=0}^{N-1} X(l)\, \omega_N^{lj} \right\} \omega_{nN}^{-kj}$$
$$= \sum_{l=0}^{N-1} X(l) \left\{ \frac{1}{N} \sum_{j=0}^{N-1} \omega_{nN}^{-j(k-ln)} \right\} = \sum_{l=0}^{N-1} X(l)\, \overline{h}(k - ln),$$

where $h(k) = \dfrac{1}{N} \sum_{j=0}^{N-1} \omega_{nN}^{kj}$.

2.25 By virtue of N-periodicity we have $x(j) = x\big(\langle j \rangle_N\big)$, hence

$$X_n(k) = \sum_{j=0}^{nN-1} x(j)\, \omega_{nN}^{-kj} = \sum_{l=0}^{n-1} \sum_{p=0}^{N-1} x(lN + p)\, \omega_{nN}^{-k(lN+p)}$$
$$= \sum_{p=0}^{N-1} x(p)\, \omega_{nN}^{-kp} \sum_{l=0}^{n-1} \omega_n^{-kl} = \left(\sum_{p=0}^{N-1} x(p)\, \omega_{nN}^{-kp} \right) n\, \delta_n(k).$$

We come to the following formula:

$$X_n(k) = \begin{cases} nX(k/n) & \text{if } \langle k \rangle_n = 0, \\ 0 & \text{otherwise.} \end{cases}$$

2.26 The definition of DFT yields

$$X_n(k) = \sum_{l=0}^{N-1} x(l)\, \omega_{nN}^{-kln} = X\big(\langle k \rangle_N\big), \quad k \in 0 : nN - 1.$$

2.27 Let us represent a number $j \in 0 : nN - 1$ in a form $j = ln + p$, where $p \in 0 : n - 1$ and $l \in 0 : N - 1$. Since $\lfloor j/n \rfloor = l$, we gain

$$X_n(k) = \sum_{l=0}^{N-1} \sum_{p=0}^{n-1} x(l)\, \omega_{nN}^{-k(ln+p)} = \sum_{l=0}^{N-1} x(l)\, \omega_N^{-kl} \sum_{p=0}^{n-1} \omega_{nN}^{-kp}.$$

We denote $h(k) = \frac{1}{n} \sum_{p=0}^{n-1} \omega_{nN}^{kp}$. Then

$$X_n(k) = nX\big(\langle k \rangle_N\big)\, \overline{h}(k), \quad k \in 0 : nN - 1.$$

2.28 According to (2.2.1) we have

$$\begin{aligned} Y_n(k) &= \sum_{j=0}^{n-1} x(jm)\, \omega_{nm}^{-kjm} = \sum_{j'=0}^{N-1} x(j')\, \omega_N^{-kj'}\, \delta_m(j') \\ &= \sum_{j=0}^{N-1} x(j)\, \omega_N^{-kj} \left\{ \frac{1}{m} \sum_{p=0}^{m-1} \omega_{mn}^{-pjn} \right\} \\ &= \frac{1}{m} \sum_{p=0}^{m-1} \sum_{j=0}^{N-1} x(j)\, \omega_N^{-j(k+pn)} = \frac{1}{m} \sum_{p=0}^{m-1} X(k + pn). \end{aligned}$$

2.29 The inclusion $y_n \in \mathbb{C}_n$ follows from Lemma 2.1.3. Let us calculate the spectrum of the signal y_n. We have

$$Y_n(k) = \sum_{j=0}^{n-1} \sum_{p=0}^{m-1} x(j + pn)\, \omega_n^{-k(j+pn)} = \sum_{l=0}^{N-1} x(l)\, \omega_N^{-kml} = X(km).$$

The inversion formula

$$\sum_{p=0}^{m-1} x(j + pn) = \frac{1}{n} \sum_{k=0}^{n-1} X(km)\, \omega_n^{kj}, \quad j \in 0 : n - 1$$

is referred to as a Poisson summation formula. Here $x \in \mathbb{C}_N$ and $N = mn$.

2.30 The DFT inversion formula yields

$$y_n(j) = \sum_{p=0}^{m-1} x(p + jm) = \frac{1}{N} \sum_{p=0}^{m-1} \sum_{l=0}^{N-1} X(l)\, \omega_N^{l(p+jm)}$$

$$= \frac{1}{N} \sum_{p=0}^{m-1} \sum_{q=0}^{m-1} \sum_{k=0}^{n-1} X(qn+k)\, \omega_N^{(qn+k)(p+jm)}$$

$$= \frac{1}{n} \sum_{k=0}^{n-1} \left\{ \sum_{q=0}^{m-1} X(qn+k) \left[\frac{1}{m} \sum_{p=0}^{m-1} \omega_N^{(qn+k)p} \right] \right\} \omega_n^{kj}.$$

On the strength of uniqueness of the expansion over an orthogonal basis we gain

$$Y_n(k) = \sum_{q=0}^{m-1} X(k+qn)\, h(k+qn),$$

where $h(j) = \frac{1}{m} \sum_{p=0}^{m-1} \omega_N^{pj}$.

2.31 Let us introduce a signal $y_n(j) = \sum_{p=0}^{m-1} x(p+jm)$, $y_n \in \mathbb{C}_n$. Then $y(j) = y_n(\lfloor j/m \rfloor)$. We denote $X = \mathcal{F}_N(x)$, $Y = \mathcal{F}_N(y)$, and $Y_n = \mathcal{F}_n(y_n)$. Taking into consideration solutions of Exercises 2.27 (changing n by m and N by n) and 2.30, we write

$$Y(k) = m\, Y_n\big(\langle k \rangle_n\big)\, \overline{h}(k), \quad k \in 0 : N-1,$$

$$Y_n(k) = \sum_{q=0}^{m-1} X(k+qn)\, h(k+qn), \quad k \in 0 : n-1,$$

where $h(j) = \frac{1}{m} \sum_{p=0}^{m-1} \omega_N^{pj}$. Hence

$$Y(k) = m\, \overline{h}(k) \sum_{q=0}^{m-1} X\big(\langle k \rangle_n + qn\big)\, h\big(\langle k \rangle_n + qn\big), \quad k \in 0 : N-1.$$

2.32 We have

$$X_1(k) = \sum_{j=0}^{N-1} c_{j+1}\, \omega_N^{-k(j+1)+k} = \omega_N^k \sum_{j=1}^{N} c_j\, \omega_N^{-kj}$$

$$= \omega_N^k \left(\sum_{j=0}^{N-1} c_j\, \omega_N^{-kj} + c_N - c_0 \right) = \omega_N^k \big(X_0(k) + (c_N - c_0) \big).$$

2.33 A setting of the exercise states that $y(k) = X(N-k)$ holds for all $k \in \mathbb{Z}$. According to (2.1.4) we have

$$[\mathcal{F}_N(y)](j) = \sum_{k=0}^{N-1} X(N-k)\,\omega_N^{(N-k)j} = \sum_{k=0}^{N-1} X(k)\,\omega_N^{kj} = Nx(j).$$

Thereby we ascertained that calculation of inverse DFT of the spectrum X is reduced to calculation of direct DFT of the signal y.

2.34 Let $X = \mathcal{F}_N(x)$. Then the DFT inversion formula yields

$$[\mathcal{F}_N^2(x)](j) = \sum_{k=0}^{N-1} X(k)\,\omega_N^{k(N-j)} = N\,x(N-j).$$

Applying the latter formula to a signal $\mathcal{F}_N^2(x)$ we gain

$$[\mathcal{F}_N^4(x)](j) = N[\mathcal{F}_N^2(x)](N-j) = N^2\,x\big(N-(N-j)\big) = N^2\,x(j).$$

This result can be restated as follows: the mapping $N^{-2}\mathcal{F}_N^4 \colon \mathbb{C}_N \to \mathbb{C}_N$ is the identity one.

2.35 We have

$$\begin{aligned}
\frac{1}{N}[\mathcal{F}_N^{-1}(X * Y)](j) &= \frac{1}{N^2} \sum_{k=0}^{N-1}\sum_{l=0}^{N-1} X(l)\,Y(k-l)\,\omega_N^{(k-l)j+lj} \\
&= \frac{1}{N^2} \sum_{l=0}^{N-1} X(l)\,\omega_N^{lj} \sum_{k=0}^{N-1} Y(k-l)\,\omega_N^{(k-l)j} \\
&= \frac{1}{N^2} \sum_{l=0}^{N-1} X(l)\,\omega_N^{lj} \sum_{k=0}^{N-1} Y(k)\,\omega_N^{kj} = x(j)\,y(j),
\end{aligned}$$

which is equivalent to what was required.

2.36 According to the result of Exercise 2.35 we have

$$X * Y = N\mathcal{F}_N(xy) = N\mathcal{F}_N(\mathbb{1}) = N^2\delta_N.$$

2.37 Provided the signals x and y are even, the corresponding spectra X and Y are real (see Theorem 2.2.3). The spectrum of the convolution $x * y$, which is equal to XY, is also real; hence, the convolution $x * y$ itself is even.

2.38 Auto-correlation is even because its Fourier transform is real. The immediate proof is also possible:

$$\overline{R_{xx}}(-j) = \sum_{k=0}^{N-1} \overline{x}(k)\, x(k+j) = \sum_{k=0}^{N-1} \overline{x}(k-j)\, x(k) = R_{xx}(j).$$

2.39 The correlation theorem yields

$$\sum_{j=0}^{N-1} R_{xx}(j) = \big[\mathcal{F}_N(R_{xx})\big](0) = |X(0)|^2.$$

2.40 The correlation theorem and the convolution theorem yield

$$\mathcal{F}_N(R_{uu}) = |\mathcal{F}_N(u)|^2 = |\mathcal{F}_N(x * y)|^2 = |X|^2\, |Y|^2,$$

$$\mathcal{F}_N(R_{xx} * R_{yy}) = \mathcal{F}_N(R_{xx})\,\mathcal{F}_N(R_{yy}) = |X|^2\, |Y|^2.$$

Hence it follows that $\mathcal{F}_N(R_{uu}) = \mathcal{F}_N(R_{xx} * R_{yy})$. The inversion formula yields $R_{uu} = R_{xx} * R_{yy}$.

The obtained result can be stated this way: auto-correlation of convolution equals to convolution of auto-correlations.

2.41 As long as x and y are delta-correlated signals, there holds $|X(k)| \equiv \sqrt{R_{xx}(0)}$ and $|Y(k)| \equiv \sqrt{R_{yy}(0)}$. On the basis of the Parseval equality and the convolution theorem we gain

$$E(u) = N^{-1}\, E(U) = N^{-1} \sum_{k=0}^{N-1} |X(k)\, Y(k)|^2 = R_{xx}(0)\, R_{yy}(0) = E(x)\, E(y).$$

2.42 Let $u = x * y$ and $R_{xx} = E(x)\,\delta_N$, $R_{yy} = E(y)\,\delta_N$. According to the results of Exercises 2.40 and 2.41 we have

$$R_{uu} = R_{xx} * R_{yy} = E(x)\, E(y)\, \delta_N * \delta_N = E(u)\, \delta_N,$$

which was to be ascertained.

2.43 It is evident that $E(v) = N^2$; therefore, it is sufficient to prove that $|V(k)| \equiv N$. The definition of DFT yields

$$V(k) = \sum_{j=0}^{N^2-1} v(j)\, \omega_{N^2}^{-kj} = \sum_{j_1, j_0=0}^{N-1} \omega_N^{j_1 j_0}\, \omega_{N^2}^{-k(j_1 N + j_0)} = \sum_{j_0=0}^{N-1} \omega_{N^2}^{-kj_0} \sum_{j_1=0}^{N-1} \omega_N^{j_1(j_0-k)}$$

$$= N \sum_{j_0=0}^{N-1} \omega_{N^2}^{-kj_0}\, \delta_N\big(j_0 - \langle k\rangle_N\big) = N \omega_{N^2}^{-k\langle k\rangle_N}.$$

As a consequence we gain the required identity.

2.44 Let us investigate, for example, a case of an odd N. Since the number $s(sN+1)$ is even for any $s \in \mathbb{Z}$, we have

$$a(j+sN) = \omega_{2N}^{(j+sN)(j+sN+1)+2q(j+sN)} = a(j)\,\omega_{2N}^{sN(sN+1)} = a(j).$$

It means that $a \in \mathbb{C}_N$. Further

$$\begin{aligned} R_{aa}(j) &= \sum_{k=0}^{N-1} a(k)\,\overline{a}(k-j) = \sum_{k=0}^{N-1} a(k+j)\,\overline{a}(k) \\ &= \sum_{k=0}^{N-1} \omega_{2N}^{(k+j)(k+j+1)+2q(k+j)-k(k+1)-2qk} \\ &= \omega_{2N}^{j(j+1)+2qj} \sum_{k=0}^{N-1} \omega_{2N}^{2kj} = N\,a(j)\,\delta_N(j) = N\,\delta_N(j). \end{aligned}$$

We ascertained that the signal a is delta-correlated.

2.45 If a binary signal $x \in \mathbb{C}_N$ is delta-correlated then, in particular, $R_{xx}(1) = 0$ holds. It means that

$$\sum_{k=0}^{N-1} x(k)\,x(k-1) = 0.$$

Each product $x(k)\,x(k-1)$ equals either to $+1$ or to -1. Their sum can equal to zero only when N is even, i.e. when $N = 2n$.

Further, a binary delta-correlated signal satisfies to a relation $|X(0)| = \sqrt{R_{xx}(0)} = \sqrt{N}$ or, in more detail,

$$\bigl|x(0) + x(1) + \cdots + x(N-1)\bigr| = \sqrt{2n}.$$

The left side of the latter equality is an integer number, so the square root of $2n$ must be integer as well. This is possible only when $n = 2p^2$. Thus, binary delta-correlated signals can exist only for $N = 4p^2$.

If $p = 1$, a binary delta-correlated signal exists. It is, for instance, $x = (1,\ 1,\ 1,\ -1)$. There is a hypothesis that no binary delta-correlated signals exist for $p > 1$. This hypothesis is not entirely proved yet.

2.46 Let us show that

$$R_{x_n x_n}(j) = \begin{cases} R_{xx}(j/n) \text{ if } \langle j\rangle_n = 0, \\ 0 \text{ for others } j \in 0:nN-1. \end{cases} \tag{S.8}$$

In the solution of Exercise 2.26 it was ascertained that $X_n(k) = X\bigl(\langle k\rangle_N\bigr)$, therefore

$$\left[\mathcal{F}_{nN}(R_{x_n x_n})\right](k) = |X_n(k)|^2 = \left|X\left(\langle k\rangle_N\right)\right|^2, \quad k \in 0 : nN - 1.$$

According to the same Exercise 2.26, the discrete Fourier transform of the signal from the right side of the formula (S.8) looks this way: $\left[\mathcal{F}_N(R_{xx})\right]\left(\langle k\rangle_N\right) = \left|X\left(\langle k\rangle_N\right)\right|^2$. We gained that the DFTs of the signals from the left and from the right sides of formula (S.8) are equal. Therefore, equal are the signals themselves.

2.47 The correlation theorem yields

$$R_{u_1 v_1} = U_1 \overline{V_1} = (X\,\overline{Y})(\overline{W}\,Z),$$

$$R_{u_2 v_2} = U_2 \overline{V_2} = (X\,\overline{W})(\overline{Y}\,Z).$$

The right sides of these relations are equal. Hence the left sides are equal too.

2.48 We have $R_{xy} = \mathbb{O}$. Keeping the notations of the previous exercise we consecutively gain $u_1 = R_{xy} = \mathbb{O}$, $U_1 = \mathbb{O}$, $R_{u_1 v_1} = \mathbb{O}$, $R_{u_2 v_2} = R_{u_1 v_1} = \mathbb{O}$. It is remaining to recall that $u_2 = R_{xw}$ and $v_2 = R_{yz}$.

2.49 Let $k,\ k' \in 0 : N - 1$, $k \neq k'$, and $x = \delta_N(\cdot - k)$, $y = \delta_N(\cdot - k')$. According to Lemma 2.1.4 we have

$$\begin{aligned} R_{xy}(j) &= \left\langle \delta_N(\cdot - k),\ \delta_N(\cdot - k' - j)\right\rangle = \delta_N(k - k' - j) \\ &= \delta_N\left(j - \langle k - k'\rangle_N\right). \end{aligned}$$

For $j' = \langle k - k'\rangle_N$ we gain $R_{xy}(j') = 1$. This characterizes the signals as being correlated.

2.50 We take a linear combination of shifts of a signal x and equate it to zero:

$$\sum_{k=0}^{N-1} c(k)\, x(j - k) = 0 \text{ for all } j \in \mathbb{Z}.$$

This condition can be rewritten in a form $c * x = \mathbb{O}$, which by virtue of the convolution theorem is equivalent to a relation $CX = \mathbb{O}$. Now it is clear that $C = \mathbb{O}$ (and therefore $c = \mathbb{O}$) if and only if all the components of the spectrum X are nonzero.

2.51 The proof is similar to the proof of Lemma 2.6.1.

2.52 By virtue of the result of the previous exercise, we need to construct a signal y such that $R_{xy} = \delta_N$. Using the correlation theorem we write the equivalent condition $X\overline{Y} = \mathbb{I}$. Here each component of the spectrum X is nonzero (see Exercise 2.50), therefore $Y = (\overline{X})^{-1}$. A desired signal is obtained with the aid of the DFT inversion formula.

2.53 The following inequalities are true:

$$\frac{1}{N}\sum_{j=0}^{N-1}|x(j)|^2 \le \max_{j\in 0:N-1}|x(j)|^2,$$

$$\sum_{j=0}^{N-1}|x(j)|^2 \ge \max_{j\in 0:N-1}|x(j)|^2.$$

Hence it follows that $1 \le p(x) \le N$. The left inequality is fulfilled as an equality only if $|x(j)| \equiv \text{const}$, and the right one only if a signal x has a single nonzero sample on the main period.

2.54 Consider the equivalent equation in a spectral domain

$$-\sum_{j=0}^{N-1}\Delta^2 x(j-1)\,\omega_N^{-kj} + c\sum_{j=0}^{N-1}x(j)\,\omega_N^{-kj} = \sum_{j=0}^{N-1}g(j)\,\omega_N^{-kj}.$$

Since

$$\begin{aligned}\sum_{j=0}^{N-1}\Delta^2 x(j-1)\,\omega_N^{-kj} &= \sum_{j=0}^{N-1}[x(j+1)-2x(j)+x(j-1)]\,\omega_N^{-kj}\\ &= (\omega_N^k - 2 + \omega_N^{-k})\sum_{j=0}^{N-1}x(j)\,\omega_N^{-kj}\\ &= -4\sin^2\left(\tfrac{\pi k}{N}\right)X(k),\end{aligned}$$

the equation for the spectra can be rewritten as

$$\left(4\sin^2\left(\tfrac{\pi k}{N}\right)+c\right)X(k) = G(k).$$

Hence

$$X(k) = G(k)/\left(4\sin^2(\pi k/N)+c\right),\quad k\in\mathbb{Z}.$$

A desired signal $x(j)$ is obtained with the aid of the DFT inversion formula.

2.55 We denote $N = 2n+1$. By virtue of 1-periodicity of the function $f(t)$, the signal $x(j) = f(t_j)$ is N-periodic. Note that

$$\exp(2\pi i k t_j) = \omega_N^{kj},$$

so the interpolation conditions take a form

$$x(j) = \sum_{k=-n}^{n} a(k)\, \omega_N^{kj}, \quad j \in \mathbb{Z}. \tag{S.9}$$

Let us extend the vector of coefficients $a(k)$ periodically with a period of N on all integer indices and rewrite formula (S.9) this way:

$$x(j) = \frac{1}{N} \sum_{k=0}^{N-1} \big(N a(k)\big)\, \omega_N^{kj}, \quad j \in \mathbb{Z}.$$

The latter equality has a form of a DFT inversion formula. Therefore,

$$N a(k) = \sum_{j=0}^{N-1} x(j)\, \omega_N^{-kj}$$

or

$$a(k) = \frac{1}{N} \sum_{j=0}^{N-1} f(t_j) \exp(-2\pi i k t_j), \quad k \in \mathbb{Z}.$$

To Chapter 3

3.1 We have

$$\begin{aligned} \overline{b_r}(j) &= \frac{1}{N} \sum_{k=1}^{N-1} (\omega_N^{N-k} - 1)^{-r}\, \omega_N^{(N-k)j} \\ &= \frac{1}{N} \sum_{k=1}^{N-1} (\omega_N^{k} - 1)^{-r}\, \omega_N^{kj} = b_r(j). \end{aligned}$$

This guarantees reality of $b_r(j)$.

3.2 According to (3.1.4) and (3.1.3) for $j \in 1 : N-1$ we have

$$b_1(j+1) - b_1(1) = \sum_{k=1}^{j} [b_1(k+1) - b_1(k)] = \sum_{k=1}^{j} b_0(k) = -j/N.$$

Denoting $c = b_1(1)$ we gain

$$b_1(j) = c - (j-1)/N, \quad j \in 2 : N.$$

The last equality is true for $j = 1$ as well.

The constant c can be determined from a condition $\sum_{j=1}^{N} b_1(j) = 0$ equivalent to (3.1.2) for $r = 1$:

$$c = \frac{1}{N^2}\sum_{j=1}^{N}(j-1) = \frac{N-1}{2N}.$$

We come to the formula

$$b_1(j) = \frac{1}{N}\left(\frac{N+1}{2} - j\right), \quad j \in 1:N.$$

3.3 According to (3.1.4) and the result of the previous exercise, for $j \in 1 : N-1$ we have

$$\begin{aligned} b_2(j+1) - b_2(1) &= \sum_{k=1}^{j}[b_2(k+1) - b_2(k)] = \sum_{k=1}^{j} b_1(k) = \frac{1}{N}\sum_{k=1}^{j}\left(\frac{N+1}{2} - k\right) \\ &= \frac{1}{N}\left[\frac{N+1}{2}j - \frac{(j+1)j}{2}\right] = \frac{j(N-j)}{2N}. \end{aligned}$$

Denoting $c = b_2(1)$ we gain

$$b_2(j) = c + \frac{(j-1)(N-j+1)}{2N}, \quad j \in 1:N.$$

Like in the previous exercise, the constant c can be determined from a condition $\sum_{j=1}^{N} b_2(j) = 0$ equivalent to (3.1.2) for $r = 2$:

$$\begin{aligned} c &= -\frac{1}{2N^2}\sum_{j=1}^{N}(j-1)(N-j+1) = -\frac{1}{2N^2}\sum_{j=1}^{N-1} j(N-j) \\ &= -\frac{1}{2N^2}\left[\frac{N^2(N-1)}{2} - \frac{(N-1)N(2N-1)}{6}\right] = -\frac{N^2-1}{12N}. \end{aligned}$$

We come to the formula

$$b_2(j) = -\frac{N^2-1}{12N} + \frac{(j-1)(N-j+1)}{2N}, \quad j \in 1:N.$$

3.4 According to (3.2.6) and (3.2.2), for $n = 2$ we have $N = 2m$ and

$$T_r(l) = \tfrac{1}{2}[X_1^r(l) + X_1^r(m+l)], \quad l \in 0 : m-1,$$

where $X_1(k) = \left(2\cos\frac{\pi k}{2m}\right)^2$ for $k \in 0 : N-1$. Taking into account that $X_1(m+l) = \left(2\sin\frac{\pi l}{2m}\right)^2$ for $l \in 0 : m-1$ we come to the required formula.

3.5 We note that

$$Q_r(j - pn) = \frac{1}{N}\sum_{k=0}^{N-1} X_1^r(k)\,\omega_N^{k(j-pn)} = \frac{1}{N}\sum_{k=0}^{N-1}\big[X_1^r(k)\,\omega_N^{-kpn}\big]\omega_N^{kj}.$$

Now we use the generalized Parseval equality (2.3.1). We gain

$$\begin{aligned}\big\langle Q_r(\cdot - pn),\, Q_r(\cdot - p'n)\big\rangle &= \frac{1}{N}\sum_{k=0}^{N-1}\big[X_1^r(k)\,\omega_N^{-kpn}\big]\big[X_1^r(k)\,\omega_N^{kp'n}\big]\\ &= \frac{1}{N}\sum_{k=0}^{N-1} X_1^{2r}(k)\,\omega_N^{k(p'-p)n} = Q_{2r}\big((p'-p)n\big).\end{aligned}$$

3.6 On the strength of (3.2.9) and (3.1.4) we have

$$\begin{aligned}\Delta^{2r} Q_r(j) &= \sum_{l=-r}^{r}(-1)^{r-l}\binom{2r}{r-l}\Delta^{2r} b_{2r}(j + r - ln)\\ &= \sum_{l=-r}^{r}(-1)^{r-l}\binom{2r}{r-l} b_0(j + r - ln).\end{aligned}$$

It is remaining to use formula (3.1.3).

3.7 When $r = 1$, the assertion follows from the definition of $Q_1(j)$. We perform an induction step from $r - 1$ to r, $r \geq 2$. According to (3.2.4) we have

$$\begin{aligned}Q_r(j) &= \left(\sum_{k=0}^{n-1} + \sum_{k=N-n+1}^{N-1}\right) Q_1(k)\, Q_{r-1}(j-k)\\ &= \sum_{k=-n+1}^{n-1} Q_1(k)\, Q_{r-1}(j-k). \qquad \text{(S.10)}\end{aligned}$$

Given $j \in 0 : (r-1)(n-1)$, we take the right side of (S.10) and consider the summand corresponding to $k = 0$. This summand equals to $Q_1(0)\, Q_{r-1}(j)$. By virtue of the inductive hypothesis it is positive. As far as the other summands are nonnegative, we gain $Q_r(j) > 0$.

If $j \in (r-1)(n-1) + 1 : r(n-1)$, we will consider the summand corresponding to $k = n - 1$. It is equal to $Q_1(n-1)\, Q_{r-1}(j - n + 1)$. Since $j - n + 1$ belongs to the set $(r-2)(n-1) + 1 : (r-1)(n-1)$, we have $Q_{r-1}(j - n + 1) > 0$. Therefore, in this case the right side of (S.10) also contains a positive summand which guarantees positivity of $Q_r(j)$.

For $j = r(n-1)$ we have

$$(r-1)(n-1) \le j - k \le (r+1)(n-1),$$

so that the right side of (S.10) contains the only nonzero summand corresponding to $k = n-1$. We gain

$$Q_r\big(r(n-1)\big) = Q_1(n-1)\, Q_{r-1}\big((r-1)(n-1)\big) = 1.$$

At last, let $j \in r(n-1)+1 : N - r(n-1) - 1$. In this case the difference $j - k$ for all $k \in -n+1 : n-1$ belongs to the set

$$(r-1)(n-1)+1 : N-(r-1)(n-1)-1$$

on which $Q_{r-1}(j-k) = 0$ holds. We gain that $Q_r(j) = 0$ holds for the given j.

The fact that $Q_r(j) > 0$ for $j \in N - r(n-1)+1 : N-1$ follows from evenness of a B-spline $Q_r(j)$.

3.8 Let us calculate the DFT of the signal $x(j)$ that stands in the right side of the equality being proved. We have

$$X(k) = \sum_{j=0}^{(n-1)/2} \omega_N^{-kj} + \sum_{j=N-(n-1)/2}^{N-1} \omega_N^{k(N-j)} = \sum_{j=-(n-1)/2}^{(n-1)/2} \omega_N^{kj}.$$

It is evident that $X(0) = n$. Suppose that $k \in 1 : N-1$. Using formula (2.4.4) for $\mu = (n+1)/2$ we gain

$$X(k) = \frac{\sin(\pi k/m)}{\sin(\pi k/N)}, \quad k \in 1 : N-1.$$

We see that the DFTs of the signals x and $Q_{1/2}$ are equal. Therefore, the signals themselves are also equal.

3.9 Denote $G = \mathcal{F}_N(Q_{1/2}^0)$. We have

$$G(0) = 0, \quad G(k) = \frac{\sin(\pi k/m)}{\sin(\pi k/N)} \quad \text{for} \quad k \in 1 : N-1.$$

Note that by virtue of parity of n for $k \in 1 : N-1$ there holds

$$G(N-k) = \frac{\sin\big(\pi(N-k)/m\big)}{\sin\big(\pi(N-k)/N\big)} = \frac{\sin(\pi n - \pi k/m)}{\sin(\pi - \pi k/N)} = -\frac{\sin(\pi k/m)}{\sin(\pi k/N)} = -G(k).$$

Therefore

$$\overline{Q^0_{1/2}(j)} = \frac{1}{N}\sum_{k=1}^{N-1} G(k)\,\omega_N^{-kj} = -\frac{1}{N}\sum_{k=1}^{N-1} G(N-k)\,\omega_N^{(N-k)j}$$
$$= -\frac{1}{N}\sum_{k=1}^{N-1} G(k)\,\omega_N^{kj} = -Q^0_{1/2}(j).$$

Hence follows the equality $\operatorname{Re} Q^0_{1/2} = \mathbb{O}$ which means by a definition that the signal $Q^0_{1/2}$ is pure imaginary.

3.10 Let us use the equality $\overline{c}(p) = c(-p)$, evenness of a B-spline $Q_r(j)$, and formula (2.1.4). We gain

$$\overline{S}(-j) = \sum_{p=0}^{m-1} c(-p)\,Q_r\big(-j + (-p)n\big) = \sum_{p=0}^{m-1} c(p)\,Q_r(j - pn) = S(j),$$

which is equivalent to what was required.

3.11 We have

$$\|\Delta^r(x - S)\|^2 = \|\Delta^r(S_* - S) + \Delta^r(x - S_*)\|^2$$
$$= \|\Delta^r(S_* - S)\|^2 + \|\Delta^r(x - S_*)\|^2$$
$$+ 2\operatorname{Re}\sum_{j=0}^{N-1} \Delta^r\big(S_*(j) - S(j)\big)\Delta^r\big(\overline{x}(j) - \overline{S}_*(j)\big).$$

Theorem 3.3.2 yields

$$\sum_{j=0}^{N-1} \Delta^r\big(S_*(j) - S(j)\big)\Delta^r\big(\overline{x}(j) - \overline{S}_*(j)\big) = (-1)^r \sum_{l=0}^{m-1} d(l)\,\big(\overline{x}(ln) - \overline{S}_*(ln)\big) = 0.$$

Here $d(l)$ are the coefficients of the expansion of the discrete periodic spline $S_* - S$ over the shifts of the Bernoulli function. Combining the given equalities we gain

$$\|\Delta^r(x - S)\|^2 = \|\Delta^r(S_* - S)\|^2 + \|\Delta^r(x - S_*)\|^2.$$

Hence follows optimality of S_*.

Let $S_1 \in \mathcal{S}_r^m$ be another solution of the original problem. Then

$$\|\Delta^r(S_* - S_1)\|^2 = 0.$$

Theorem 3.1.2 yields $S_*(j) - S_1(j) \equiv \text{const}$, i.e. S_1 differs from S_* by an additive constant.

3.12 Basic functions of the spline S_α presented in a form (3.3.4) are real. Its coefficients are determined from the system of linear equations (3.5.6) with the real matrix. Provided the right side of the system is real its solution is real as well.

3.13 The assertion follows from formula (3.8.5).

3.14 A conclusion about evenness of $\mu_k(j)$ with respect to k follows from formulae (3.8.7) and (3.8.8) and from the result of Exercise 2.1.

3.15 In this case an orthogonal basis is formed by two splines $\mu_0(j)$ and $\mu_1(j)$. According to (3.8.7) and (3.8.10) we have

$$\mu_0(j) \equiv 1, \qquad \mu_1(j) = \tfrac{1}{2}\big[Q_1(j) - Q_1(j-2)\big].$$

Since $Q_1(j)$ for $j = 0,\ 1,\ 2,\ 3$ takes the values $2,\ 1,\ 0,\ 1$, we gain that $\mu_1(j)$ for the same arguments j takes the values $1,\ 0, -1,\ 0$.

3.16 We put $c = \mathcal{F}_m^{-1}(\xi)$. Then

$$S(j) = \sum_{p=0}^{m-1} c(p)\, Q_r(j - pn).$$

Coefficients $c(p)$ in this representation are real (see Theorem 2.2.2). It is remaining to take into account that the values of a B-spline Q_r are also real.

3.17 According to Theorems 2.2.2 and 2.2.3 the signal $c = \mathcal{F}_m^{-1}(\xi)$ is real and even. The representation

$$S(j) = \sum_{p=0}^{m-1} c(p)\, Q_r(j - pn)$$

lets us draw a conclusion that S is real (which is obvious) and even (see Exercise 3.10).

3.18 The coefficients $\xi(k) = [T_{2r}(k)]^{-1/2}$ in the expansion of $\varphi_r(j)$ over the orthogonal basis are real and comprise an even sequence. It is remaining to use the result of Exercise 3.17.

3.19 The solution is similar to the previous one.

3.20 According to (3.9.7) and (3.8.9) we have

$$\sum_{q=0}^{m-1} R_r(j - qn) = \sum_{q=0}^{m-1}\sum_{k=0}^{m-1} \frac{\mu_k(j)\,\omega_m^{-kq}}{T_{2r}(k)} = \sum_{k=0}^{m-1} \frac{\mu_k(j)}{T_{2r}(k)} \sum_{q=0}^{m-1} \omega_m^{-kq}$$
$$= m \sum_{k=0}^{m-1} \frac{\mu_k(j)}{T_{2r}(k)}\, \delta_m(k) = m\, \frac{\mu_0(j)}{T_{2r}(0)}.$$

It is remaining to take into account that $\mu_0(j) \equiv \frac{1}{N} n^{2r}$ and $T_{2r}(0) = n^{4r-1}$.

3.21 The solution is similar to the proof of Theorem 3.10.1.

3.22 Since

$$\omega_{m_\nu}^{-l}\,(\omega_{m_\nu}^{l}+1)^2 = \omega_{m_\nu}^{l} + \omega_{m_\nu}^{-l} + 2 = \big(2\cos(\pi l/m_\nu)\big)^2,$$

we have

$$c_\nu(l) = \omega_{m_\nu}^{-lr}\,(\omega_{m_\nu}^{l}+1)^{2r} = \sum_{p=0}^{2r}\binom{2r}{p}\,\omega_{m_\nu}^{-l(r-p)} = \sum_{p=-r}^{r}\binom{2r}{r-p}\,\omega_{m_\nu}^{-lp}.$$

3.23 Taking into account (3.10.2) and the result of the previous exercise we gain

$$\begin{aligned} Q_r^{\nu+1}(j) &= \frac{1}{N}\sum_{l=0}^{N-1} y_{\nu+1}(l)\,\omega_N^{lj} = \frac{1}{N}\sum_{l=0}^{N-1} c_\nu(l)\,y_\nu(l)\,\omega_N^{lj} \\ &= \frac{1}{N}\sum_{p=-r}^{r}\binom{2r}{r-p}\sum_{l=0}^{N-1} y_\nu(l)\,\omega_N^{l(j-pn_\nu)} = \sum_{p=-r}^{r}\binom{2r}{r-p}\,Q_r^{\nu}(j-pn_\nu). \end{aligned}$$

3.24 We have

$$a_\nu(-k) = \omega_{m_\nu}^{-k}\,c_\nu(m_{\nu+1}-k)\,\|\mu_{m_{\nu+1}-k}^{\nu}\|^2.$$

Let us use the fact that the sequences $\{c_\nu(k)\}$ and $\{\mu_k^\nu\}$ are even with respect to k. (The former one is even by a definition. As for evenness of the latter one, see Exercise 3.14.) Taking into account that $m_{\nu+1} - k = m_\nu - (m_{\nu+1} + k)$ we gain

$$a_\nu(-k) = \omega_{m_\nu}^{-k}\,c_\nu(m_{\nu+1}+k)\,\|\mu_{m_{\nu+1}+k}^{\nu}\|^2 = \overline{a}_\nu(k).$$

3.25 In the same way as in the solution of the previous exercise we have

$$w_{-k}^{\nu+1}(j) = \overline{a}_\nu(k)\,\overline{\mu_k^\nu}(j) + \overline{a}_\nu(m_{\nu+1}+k)\,\overline{\mu_{m_{\nu+1}+k}^{\nu}}(j) = \overline{w_k^{\nu+1}}(j).$$

3.26 On the basis of (3.10.8) we gain

$$\begin{aligned} \sum_{j=0}^{N-1}\big[w_k^{\nu+1}(j)\big]^2 &= \sum_{l=0}^{m_{\nu+1}-1}\sum_{p=0}^{n_{\nu+1}-1}\big[w_k^{\nu+1}(p+ln_{\nu+1})\big]^2 \\ &= \sum_{l=0}^{m_{\nu+1}-1}\omega_{m_{\nu+1}}^{2kl}\sum_{p=0}^{n_{\nu+1}-1}\big[w_k^{\nu+1}(p)\big]^2 \\ &= m_{\nu+1}\,\delta_{m_{\nu+1}}(2k)\sum_{p=0}^{n_{\nu+1}-1}\big[w_k^{\nu+1}(p)\big]^2. \end{aligned}$$

Hence the required equality follows obviously.

3.27 According to evenness of the signal a_ν (see Exercise 3.24) we have

$$a_\nu(m_{\nu+1} + m_{\nu+2}) = a_\nu(m_\nu - m_{\nu+2}) = \overline{a}_\nu(m_{\nu+2}).$$

Similarly, with a reference to Exercise 3.14, one can deduce the equality

$$\mu^\nu_{m_{\nu+1}+m_{\nu+2}}(j) = \overline{\mu^\nu_{m_{\nu+2}}}(j).$$

On the basis of (3.10.7) we gain

$$w^{\nu+1}_{m_{\nu+2}}(j) = 2\ \mathrm{Re}\big[a_\nu(m_{\nu+2})\,\mu^\nu_{m_{\nu+2}}(j)\big].$$

3.28 If

$$\varphi(j) = \sum_{k=0}^{m_\nu-1} \beta_\nu(k)\, w^\nu_k(j)$$

then (3.10.8) yields

$$\varphi(j - ln_\nu) = \sum_{k=0}^{m_\nu-1} \beta_\nu(k)\, w^\nu_k(j)\, \omega^{-lk}_{m_\nu}.$$

By virtue of the DFT inversion formula we write

$$\beta_\nu(k)\, w^\nu_k(j) = \frac{1}{m_\nu} \sum_{k=0}^{m_\nu-1} \omega^{lk}_{m_\nu}\, \varphi(j - ln_\nu).$$

Now the solution finishes in the same way as the proof of Theorem 3.9.1.

3.29 We note that

$$\big\langle \varphi(\cdot - ln_\nu),\ \psi(\cdot - l'n_\nu)\big\rangle = \sum_{k=0}^{m_\nu-1} \beta_\nu(k)\, \overline{\gamma}_\nu(k)\, \|w^\nu_k\|^2\, \omega^{k(l-l')}_{m_\nu}.$$

Therefore the equality $\big\langle \varphi(\cdot - ln_\nu),\ \psi(\cdot - l'n_\nu)\big\rangle = \delta_{m_\nu}(l - l')$ holds if and only if

$$m_\nu\, \beta_\nu(k)\, \overline{\gamma}_\nu(k)\, \|w^\nu_k\|^2 \equiv 1$$

holds (see the proof of Theorem 3.9.2).

3.30 Let us use formula (3.10.3). We gain

$$P_r^{\nu+1}(j) = \sum_{k=0}^{m_\nu - 1} a_\nu(k)\, \mu_k^\nu(j). \tag{S.11}$$

Since

$$\mu_k^\nu(j) = \frac{1}{m_\nu} \sum_{p=0}^{m_\nu - 1} \omega_{m_\nu}^{kp}\, Q_r^\nu(j - pn_\nu),$$

we have

$$P_r^{\nu+1}(j) = \sum_{p=0}^{m_\nu - 1} Q_r^\nu(j - pn_\nu) \left\{ \frac{1}{m_\nu} \sum_{k=0}^{m_\nu - 1} a_\nu(k)\, \omega_{m_\nu}^{kp} \right\}.$$

It is remaining to put $d_\nu = \mathcal{F}_{m_\nu}^{-1}(a_\nu)$.

3.31 Recall that

$$a_\nu(k) = \omega_{m_\nu}^k\, c_\nu(m_{\nu+1} + k)\, \|\mu_{m_{\nu+1}+k}^\nu\|^2.$$

According to the result of Exercise 3.22 we have

$$c_\nu(m_{\nu+1} + k) = \sum_{l=-r}^{r} (-1)^l \binom{2r}{r-l} \omega_{m_\nu}^{kl}.$$

Equality (3.8.6) yields

$$\|\mu_{m_{\nu+1}+k}^\nu\|^2 = \frac{1}{m_\nu}\, T_{2r}^\nu(m_{\nu+1} + k).$$

Therefore,

$$a_\nu(k) = \frac{1}{m_\nu}\, T_{2r}^\nu(m_{\nu+1} + k) \sum_{l=-r}^{r} (-1)^l \binom{2r}{r-l} \omega_{m_\nu}^{k(l+1)}.$$

Taking into consideration formula (3.2.7) we gain

$$\begin{aligned}
d_\nu(p) &= \frac{1}{m_\nu} \sum_{k=0}^{m_\nu - 1} a_\nu(k)\, \omega_{m_\nu}^{kp} \\
&= \frac{1}{m_\nu^2} \sum_{l=-r}^{r} (-1)^l \binom{2r}{r-l} \sum_{k=0}^{m_\nu - 1} T_{2r}^\nu(m_{\nu+1} + k)\, \omega_{m_\nu}^{k(p+l+1)} \\
&= \frac{1}{m_\nu} \sum_{l=-r}^{r} (-1)^l \binom{2r}{r-l} \left\{ \frac{1}{m_\nu} \sum_{k=0}^{m_\nu - 1} T_{2r}^\nu(k)\, \omega_{m_\nu}^{(k-m_{\nu+1})(p+l+1)} \right\} \\
&= (-1)^{p+1} \frac{1}{m_\nu} \sum_{l=-r}^{r} \binom{2r}{r-l} Q_{2r}^\nu\big((p + l + 1)\, n_\nu\big).
\end{aligned}$$

3.32 Let us use formula (S.11) and the fact that

$$\mu_k^\nu(j) = \frac{1}{N} \sum_{q=0}^{n_\nu - 1} y_\nu(qm_\nu + k)\, \omega_N^{(qm_\nu + k)j}$$

(see par. 3.10.1). Bearing in mind m_ν-periodicity of the sequence $\{a_\nu(k)\}$ we gain

$$\begin{aligned} P_r^{\nu+1}(j) &= \frac{1}{N} \sum_{k=0}^{m_\nu - 1} \sum_{q=0}^{n_\nu - 1} a_\nu(qm_\nu + k)\, y_\nu(qm_\nu + k)\, \omega_N^{(qm_\nu + k)j} \\ &= \frac{1}{N} \sum_{l=0}^{N-1} a_\nu(l)\, y_\nu(l)\, \omega_N^{lj}. \end{aligned}$$

Hence

$$\big[\mathcal{F}_N(P_r^{\nu+1})\big](l) = a_\nu(l)\, y_\nu(l), \quad l \in 0 : N-1.$$

3.33 According to (3.10.8) and Theorem 3.10.2 we have

$$\begin{aligned} \big\langle P_r^{\nu+1}(\cdot - l n_{\nu+1}),\, P_r^{\nu+1}(\cdot - l' n_{\nu+1}) \big\rangle &= \Big\langle \sum_{k=0}^{m_{\nu+1}-1} \omega_{m_{\nu+1}}^{-kl}\, w_k^{\nu+1},\, \sum_{k'=0}^{m_{\nu+1}-1} \omega_{m_{\nu+1}}^{-k'l'}\, w_{k'}^{\nu+1} \Big\rangle \\ &= \sum_{k=0}^{m_{\nu+1}-1} \|w_k^{\nu+1}\|^2\, \omega_{m_{\nu+1}}^{-k(l-l')}, \end{aligned}$$

which is equivalent to what was required.

3.34 According to the results of Exercises 3.30 and 3.31 we have

$$\begin{aligned} P_r^{\nu+1}(j - n_\nu) &= \sum_{p=0}^{m_\nu - 1} d_\nu(p)\, Q_r^\nu\big(j - (p+1)n_\nu\big) \\ &= \sum_{p=0}^{m_\nu - 1} d_\nu(p-1)\, Q_r^\nu(j - pn_\nu). \end{aligned}$$

Here

$$d_\nu(p-1) = (-1)^p \frac{1}{m_\nu} \sum_{l=-r}^{r} \binom{2r}{r-l} Q_{2r}^\nu\big((p+l)n_\nu\big).$$

At the same time

$$P_r^{\nu+1}(-j-n_\nu) = \sum_{p=0}^{m_\nu-1} d_\nu(p-1)\, Q_r^\nu(j+pn_\nu)$$
$$= \sum_{p=0}^{m_\nu-1} d_\nu(-p-1)\, Q_r^\nu(j-pn_\nu).$$

Since

$$d_\nu(-p-1) = (-1)^p \frac{1}{m_\nu} \sum_{l=-r}^{r} \binom{2r}{r+l} Q_{2r}^\nu\big((p-l)n_\nu\big)$$
$$= (-1)^p \frac{1}{m_\nu} \sum_{l=-r}^{r} \binom{2r}{r-l} Q_{2r}^\nu\big((p+l)n_\nu\big) = d_\nu(p-1),$$

we gain

$$P_r^{\nu+1}(-j-n_\nu) = P_r^{\nu+1}(j-n_\nu).$$

This means that the real spline $P_r^{\nu+1}(j-n_\nu)$ is even with respect to j.

3.35 B-spline $B_1(x)$ is even, it follows from its definition. Assume that $B_{\nu-1}(-x) = B_{\nu-1}(x)$ holds for some $\nu \geq 2$. In this case

$$B_\nu(-x) = \int_0^m B_{\nu-1}(t)\, B_1(x+t)\, dt = \int_0^m B_{\nu-1}(m-t)\, B_1\big(x-(m-t)\big)\, dt$$
$$= \int_0^m B_{\nu-1}(t)\, B_1(x-t)\, dt = B_\nu(x).$$

To Chapter 4

4.1 Let $j = (j_{s-1}, j_{s-2}, \ldots, j_0)_2$. A condition $j \neq pN_\nu$ implies that not all components $j_{s-\nu-1}, \ldots, j_0$ are equal to zero. And then

$$\mathrm{rev}_s(j) \geq j_0 2^{s-1} + \cdots + j_{s-\nu-1} 2^\nu \geq 2^\nu = \Delta_{\nu+1}.$$

Hence it follows that $y(j) := x\big(\mathrm{rev}_s(j)\big) = 0$.

4.2 Use the solution of the previous exercise. Bear in mind that the equality $\mathrm{rev}_s(pN_\nu) = \mathrm{rev}_\nu(p)$ holds for $p \in 0 : \Delta_{\nu+1} - 1$.

4.3 We have

$$\varphi_\nu(j) = \varphi_\nu(N_\nu;\ j) = f_\nu(\Delta_\nu;\ j),$$

$$\psi_\nu(j) = g_\nu(N_\nu;\ j) = f_\nu\big(\mathrm{rev}_s(N_\nu);\ \mathrm{rev}_s(j)\big).$$

It remains to take into consideration that $\text{rev}_s(N_\nu) = \Delta_\nu$.

4.4 The first equality follows from formula (4.8.1). To prove the second equality, let us use the result of Exercise 2.4. We gain

$$\begin{aligned}\psi_{\nu-1}(2j) &= \delta_{N_{\nu-2}}(2j) - \delta_{N_{\nu-2}}(2j - N_{\nu-1})\\ &= \delta_{2N_{\nu-1}}(2j) - \delta_{2N_{\nu-1}}\big(2(j - N_\nu)\big)\\ &= \delta_{N_{\nu-1}}(j) - \delta_{N_{\nu-1}}(j - N_\nu) = \psi_\nu(j).\end{aligned}$$

4.5 The required expansion has a form

$$\delta_N(j) = 2^{-s} + \sum_{\nu=1}^{s} 2^{-\nu}\,\varphi_\nu(j).$$

It can be obtained in the same way as in the example from par. 4.5.2, but it also can be deduced analytically. Indeed, according to (4.5.3) we have

$$2\,\varphi_{\nu-1}(0;\ j) = \varphi_\nu(0;\ j) + \varphi_\nu(N_\nu;\ j).$$

Hence

$$\begin{aligned}\sum_{\nu=1}^{s} 2^{-\nu}\,\varphi_\nu(j) &= 2\sum_{\nu=1}^{s} 2^{-\nu}\,\varphi_{\nu-1}(0;\ j) - \sum_{\nu=1}^{s} 2^{-\nu}\,\varphi_\nu(0;\ j)\\ &= \sum_{\nu=0}^{s-1} 2^{-\nu}\,\varphi_\nu(0;\ j) - \sum_{\nu=1}^{s} 2^{-\nu}\,\varphi_\nu(0;\ j)\\ &= \varphi_0(0;\ j) - 2^{-s}\,\varphi_s(0;\ j).\end{aligned}$$

It is remaining to take into account that $\varphi_0(0;\ j) = \delta_N(j)$ and $\varphi_s(0;\ j) \equiv 1$.

4.6 The required expansion has a form

$$\delta_N(j) = 2^{-s} + \sum_{\nu=1}^{s} 2^{-\nu}\,\psi_\nu(j).$$

It can be obtained in the same way as in the example from par. 4.6.5, but it also can be deduced analytically. Indeed, according to (4.6.10) we have

$$2\,g_{\nu-1}(0;\ j) = g_\nu(0;\ j) + g_\nu(N_\nu;\ j).$$

Hence

$$\begin{aligned}\sum_{\nu=1}^{s} 2^{-\nu}\,\psi_\nu(j) &= 2\sum_{\nu=1}^{s} 2^{-\nu}\,g_{\nu-1}(0;\ j) - \sum_{\nu=1}^{s} 2^{-\nu}\,g_\nu(0;\ j) \\ &= \sum_{\nu=0}^{s-1} 2^{-\nu}\,g_\nu(0;\ j) - \sum_{\nu=1}^{s} 2^{-\nu}\,g_\nu(0;\ j) \\ &= g_0(0;\ j) - 2^{-s}\,g_s(0;\ j).\end{aligned}$$

It is remaining to take into account that $g_0(0;\ j) = \delta_N(j)$ and $g_s(0;\ j) \equiv 1$.

4.7 According to (4.8.12) we have

$$\delta_N(j-q) = 2^{-s}\alpha + \sum_{\nu=1}^{s} 2^{-\nu} \sum_{p=0}^{N_\nu - 1} \widehat{\xi}_\nu(p)\,\varphi_\nu(j - p\Delta_{\nu+1}).$$

Here

$$\alpha = \langle \delta_N(\cdot - q),\ \varphi_s(0)\rangle = \sum_{j=0}^{N-1} \delta_N(j-q) = 1,$$

$$\begin{aligned}\widehat{\xi}_\nu(p) &= \langle \delta_N(\cdot - q),\ \varphi_\nu(p+N_\nu)\rangle = \langle \delta_N(\cdot - q),\ f_\nu(\Delta_\nu + p\Delta_{\nu+1})\rangle \\ &= f_\nu(\Delta_\nu + p\Delta_{\nu+1};\ q) = f_\nu(\Delta_\nu;\ q - p\Delta_{\nu+1}).\end{aligned}$$

Since $-N + \Delta_{\nu+1} \le q - p\Delta_{\nu+1} \le N-1$ for $p \in 0 : N_\nu - 1$, formula (4.4.6) yields that a coefficient $\widehat{\xi}_\nu(p)$ is nonzero if and only if

$$q - p\Delta_{\nu+1} \in 0 : \Delta_{\nu+1} - 1. \tag{S.12}$$

Note that

$$q - p\Delta_{\nu+1} = (\lfloor q/\Delta_{\nu+1}\rfloor - p)\Delta_{\nu+1} + q_{\nu-1}\Delta_\nu + \cdots + q_0,$$

therefore condition (S.12) holds only for $p = \lfloor q/\Delta_{\nu+1}\rfloor$. Referring to formula (4.4.6) again and bearing in mind that $q_{\nu-1} \in 0 : 1$ we gain

$$\widehat{\xi}_\nu(\lfloor q/\Delta_{\nu+1}\rfloor) = f_\nu(\Delta_\nu;\ q_{\nu-1}\Delta_\nu + \cdots + q_0) = (-1)^{q_{\nu-1}}.$$

Thus,

$$\delta_N(j-q) = 2^{-s} + \sum_{\nu=1}^{s} 2^{-\nu}(-1)^{q_{\nu-1}}\varphi_\nu(j - \lfloor q/\Delta_{\nu+1}\rfloor\,\Delta_{\nu+1}).$$

4.8 According to (4.8.2) we have

$$\delta_N(j-q) = 2^{-s}\beta + \sum_{\nu=1}^{s} 2^{-\nu} \sum_{p=0}^{N_\nu - 1} \widehat{y}_\nu(p)\, \psi_\nu(j-p).$$

Here

$$\beta = \langle \delta_N(\cdot - q),\, g_s(0)\rangle = \sum_{j=0}^{N-1} \delta_N(j-q) = 1,$$

$$\begin{aligned}\widehat{y}_\nu(p) &= \langle \delta_N(\cdot - q),\, g_\nu(p+N_\nu)\rangle = g_\nu(p+N_\nu;\, q) = g_\nu(N_\nu;\, q-p) \\ &= \psi_\nu(q-p) = \delta_{N_{\nu-1}}(q-p) - \delta_{N_{\nu-1}}(q-p-N_\nu).\end{aligned}$$

Since $q = lN_{\nu-1} + q_{s-\nu}N_\nu + \langle q\rangle_{N_\nu}$, there holds

$$\widehat{y}_\nu(p) = \delta_{N_{\nu-1}}(q_{s-\nu}N_\nu + \langle q\rangle_{N_\nu} - p) - \delta_{N_{\nu-1}}\big((q_{s-\nu}-1)N_\nu + \langle q\rangle_{N_\nu} - p\big).$$

If $q_{s-\nu} = 0$ then

$$\widehat{y}_\nu(p) = \delta_{N_{\nu-1}}(\langle q\rangle_{N_\nu} - p) - \delta_{N_{\nu-1}}(\langle q\rangle_{N_\nu} - p - N_\nu).$$

Taking into account the inequalities $|\langle q\rangle_{N_\nu} - p| \le N_\nu - 1$ and

$$-N_{\nu-1} + 1 \le \langle q\rangle_{N_\nu} - p - N_\nu \le -1,$$

we conclude that

$$\widehat{y}_\nu(p) = \begin{cases} 1 \text{ for } p = \langle q\rangle_{N_\nu}, \\ 0 \text{ for } p \ne \langle q\rangle_{N_\nu}. \end{cases}$$

If $q_{s-\nu} = 1$ then

$$\widehat{y}_\nu(p) = \delta_{N_{\nu-1}}(\langle q\rangle_{N_\nu} - p + N_\nu) - \delta_{N_{\nu-1}}(\langle q\rangle_{N_\nu} - p).$$

Taking into account that $1 \le \langle q\rangle_{N_\nu} - p + N_\nu \le N_{\nu-1} - 1$ we gain

$$\widehat{y}_\nu(p) = \begin{cases} -1 \text{ for } p = \langle q\rangle_{N_\nu}, \\ 0 \text{ for } p \ne \langle q\rangle_{N_\nu}. \end{cases}$$

Moreover, in both cases $\widehat{y}_\nu(\langle q\rangle_{N_\nu}) = (-1)^{q_{s-\nu}}$ holds.

We come to the formula

$$\delta_N(j-q) = 2^{-s} + \sum_{\nu=1}^{s} 2^{-\nu}\,(-1)^{q_{s-\nu}}\,\psi_\nu(j - \langle q\rangle_{N_\nu}).$$

4.9 Formula (4.6.10) yields

$$2\,g_{\nu-1}(p;\;j) = g_\nu(p;\;j) + \psi_\nu(j-p), \quad p \in 0 : N_\nu - 1.$$

Therefore

$$\begin{aligned}
&\sum_{\nu=1}^{s} 2^{-\nu} \sum_{p=0}^{N_\nu - 1} \psi_\nu(j-p) \\
&= \sum_{\nu=1}^{s} 2^{-\nu} \sum_{p=0}^{N_\nu - 1} [2\,g_{\nu-1}(p;\;j) - g_\nu(p;\;j)] \\
&= \sum_{\nu=0}^{s-1} 2^{-\nu} \sum_{p=0}^{N_{\nu+1}-1} g_\nu(p;\;j) - \sum_{\nu=0}^{s-1} 2^{-\nu} \sum_{p=0}^{N_\nu - 1} g_\nu(p;\;j) - 2^{-s}\,g_s(0;\;j) + \sum_{p=0}^{N-1} g_0(0;\;j) \\
&= \sum_{p=0}^{N-1} \delta_N(j-p) - 2^{-s} - \sum_{\nu=0}^{s-1} 2^{-\nu} \sum_{p=N_{\nu+1}}^{N_\nu - 1} \delta_{N_\nu}(j-p). \qquad \text{(S.13)}
\end{aligned}$$

We used formula (4.7.9) while performing the last transition.

Let $j = (j_{s-1}, j_{s-2}, \ldots, j_0)_2$. We will show that

$$\sum_{p=N_{\nu+1}}^{N_\nu - 1} \delta_{N_\nu}(j-p) = j_{s-\nu-1}. \qquad \text{(S.14)}$$

As far as $j = lN_\nu + j_{s-\nu-1}2^{s-\nu-1} + \cdots + j_0$, we have

$$\sum_{p=N_{\nu+1}}^{N_\nu - 1} \delta_{N_\nu}(j-p) = \sum_{p=N_{\nu+1}}^{N_\nu - 1} \delta_{N_\nu}\big((j_{s-\nu-1}2^{s-\nu-1} + \cdots + j_0) - p\big). \qquad \text{(S.15)}$$

If $j_{s-\nu-1} = 0$ then the right side of (S.15) equals to zero. In this case (S.15) corresponds to (S.14). Let $j_{s-\nu-1} = 1$. Then the right side of (S.15) equals to unity. In this case (S.15) also corresponds to (S.14).

Substituting (S.14) into (S.13) and taking into account that

$$\sum_{p=0}^{N-1} \delta_N(j-p) \equiv 1$$

we gain

$$x(j) = 1 - \sum_{\nu=0}^{s-1} 2^{-\nu} j_{s-\nu-1} = 1 - 2j/N, \quad j \in 0 : N-1.$$

4.10 A definition yields

$$\varphi_\nu(j - p\Delta_{\nu+1}) = \varphi_\nu(N_\nu;\ j - p\Delta_{\nu+1}) = f_\nu(\Delta_\nu + p\Delta_{\nu+1};\ j).$$

We have

$$\begin{aligned} y\big(\mathrm{rev}_s(j)\big) &= 2^{-s} + \sum_{\nu=1}^{s} 2^{-\nu} \sum_{p=0}^{N_\nu - 1} f_\nu\big(\Delta_\nu + p\Delta_{\nu+1};\ \mathrm{rev}_s(j)\big) \\ &= 2^{-s} + \sum_{\nu=1}^{s} 2^{-\nu} \sum_{p=0}^{N_\nu - 1} f_\nu\big(\Delta_\nu + \Delta_{\nu+1}\,\mathrm{rev}_{s-\nu}(p);\ \mathrm{rev}_s(j)\big). \end{aligned}$$

According to (4.6.3) there holds $\Delta_\nu + \Delta_{\nu+1}\,\mathrm{rev}_{s-\nu}(p) = \mathrm{rev}_s(N_\nu + p)$, therefore

$$\begin{aligned} f_\nu\big(\Delta_\nu + \Delta_{\nu+1}\,\mathrm{rev}_{s-\nu}(p);\ \mathrm{rev}_s(j)\big) &= f_\nu\big(\mathrm{rev}_s(p + N_\nu);\ \mathrm{rev}_s(j)\big) \\ &= g_\nu(p + N_\nu;\ j) = \psi_\nu(j - p). \end{aligned}$$

Taking into consideration the result of the previous exercise we gain

$$y\big(\mathrm{rev}_s(j)\big) = 2^{-s} + \sum_{\nu=1}^{s} 2^{-\nu} \sum_{p=0}^{N_\nu - 1} \psi_\nu(j - p) = 1 - 2j/N.$$

Hence

$$y(j) = 1 - 2\,\mathrm{rev}_s(j)/N, \quad j \in 0 : N-1.$$

4.11 In the same way as in the solution of Exercise 4.5 we write

$$\begin{aligned} \sum_{\nu=k+1}^{s} 2^{-\nu} \varphi_\nu(j) &= \sum_{\nu=k}^{s-1} 2^{-\nu} \varphi_\nu(0;\ j) - \sum_{\nu=k+1}^{s} 2^{-\nu} \varphi_\nu(0;\ j) \\ &= 2^{-k}\, \varphi_k(0;\ j) - 2^{-s}\, \varphi_s(0;\ j). \end{aligned} \tag{S.16}$$

It was noted in par. 4.7.1 that

$$\varphi_k(0;\ j) = \sum_{q=0}^{\Delta_{k+1}-1} \delta_N(j - q) =: h_k(j).$$

Now the required expansion follows from (S.16).

4.12 Note that for $\nu \in 1 : s-k$ there holds an inequality

$$N_\nu \geq 2^k = \Delta_{k+1}.$$

According to (4.6.10) we have

$$2\, g_{\nu-1}(p;\ j) = g_\nu(p;\ j) + \psi_\nu(j-p) \quad \text{for} \quad p \in 0 : \Delta_{k+1} - 1.$$

Hence

$$\begin{aligned}\sum_{\nu=1}^{s-k} 2^{-\nu} \psi_\nu(j-p) &= \sum_{\nu=0}^{s-k-1} 2^{-\nu} g_\nu(p;\ j) - \sum_{\nu=1}^{s-k} 2^{-\nu} g_\nu(p;\ j) \\ &= g_0(p;\ j) - 2^{-s+k} g_{s-k}(p;\ j).\end{aligned}$$

Summing up the last equations on p from 0 to $\Delta_{k+1} - 1$ we gain

$$\sum_{\nu=1}^{s-k} 2^{-\nu} \sum_{p=0}^{\Delta_{k+1}-1} \psi_\nu(j-p) = h_k(j) - 2^{-s+k} \sum_{p=0}^{\Delta_{k+1}-1} g_{s-k}(p;\ j). \tag{S.17}$$

Equality (4.7.9) yields $g_{s-k}(p;\ j) = \delta_{\Delta_{k+1}}(j-p)$, so

$$\sum_{p=0}^{\Delta_{k+1}-1} g_{s-k}(p;\ j) = \sum_{p=0}^{\Delta_{k+1}-1} \delta_{\Delta_{k+1}}(j-p) \equiv 1.$$

Now the required expansion follows from (S.17).

4.13 On the basis of (4.8.12) and (4.8.19) we have

$$x(j \oplus q) = 2^{-s}\alpha + \sum_{\nu=1}^{s} 2^{-\nu} \sum_{p=0}^{N_\nu - 1} \widehat{\xi}_\nu(p)\, \varphi_\nu\big((j \oplus q) \oplus p\Delta_{\nu+1}\big).$$

Since

$$\begin{aligned}(j \oplus q) \oplus p\Delta_{\nu+1} &= j \oplus (q \oplus p\Delta_{\nu+1}) \\ &= j \oplus \big((\lfloor q/\Delta_{\nu+1} \rfloor \oplus p)\Delta_{\nu+1} + q_{\nu-1}\Delta_\nu + \langle q \rangle_{\Delta_\nu}\big),\end{aligned}$$

equality (4.8.16) yields

$$x(j \oplus q) = 2^{-s}\alpha + \sum_{\nu=1}^{s} (-1)^{q_{\nu-1}} 2^{-\nu} \sum_{p=0}^{N_\nu - 1} \widehat{\xi}_\nu(p)\, \varphi_\nu\big(j - (\lfloor q/\Delta_{\nu+1} \rfloor \oplus p)\Delta_{\nu+1}\big).$$

Performing a change of variables $p' = \lfloor q/\Delta_{\nu+1} \rfloor \oplus p$ we gain the required expansion.

4.14 On the basis of (4.8.2) and (4.8.6) we have

$$y(j-q) = 2^{-s}\beta + \sum_{\nu=1}^{s} 2^{-\nu} \sum_{p=0}^{N_\nu - 1} \widehat{y}_\nu(p)\,(-1)^{\lfloor (q+p)/N_\nu \rfloor}\,\psi_\nu\big(j - \langle q+p \rangle_{N_\nu}\big).$$

We change the variables: $p' = \langle q+p \rangle_{N_\nu}$. In this case $p = \langle p' - q \rangle_{N_\nu}$ and

$$\Big\lfloor \frac{q+p}{N_\nu} \Big\rfloor = \Big\lfloor \frac{q + \langle p' - q \rangle_{N_\nu}}{N_\nu} \Big\rfloor = \Big\lfloor \frac{p' - \big((p'-q) - \langle p'-q \rangle_{N_\nu}\big)}{N_\nu} \Big\rfloor$$
$$= \Big\lfloor \frac{p' - \lfloor (p'-q)/N_\nu \rfloor N_\nu}{N_\nu} \Big\rfloor = -\Big\lfloor \frac{p'-q}{N_\nu} \Big\rfloor.$$

Finally we gain

$$y(j-q) = 2^{-s}\beta + \sum_{\nu=1}^{s} 2^{-\nu} \sum_{p'=0}^{N_\nu - 1} (-1)^{\lfloor (p'-q)/N_\nu \rfloor}\,\widehat{y}_\nu\big(\langle p'-q \rangle_{N_\nu}\big)\,\psi_\nu(j-p').$$

4.15 According to (4.7.9) the identity

$$g_\nu(k;\, j) = \delta_{N_\nu}(j-k)$$

is true for $k \in 0 : N_\nu - 1$. Hence the required equality follows obviously.

4.16 We denote

$$G_\nu = \big\{ w \in \mathbb{C}_{N_{\nu-1}} \;\big|\; w(j - N_\nu) = -w(j),\ \ j \in \mathbb{Z} \big\}.$$

It is required to verify that $W_\nu = G_\nu$.

Let us take $w \in W_\nu$. Then

$$w(j) = \sum_{k=0}^{N_\nu - 1} a(k)\,\psi_\nu(j-k).$$

By virtue of formula (4.8.1) we have $\psi_\nu(\cdot - k) \in \mathbb{C}_{N_{\nu-1}}$, therefore $w \in \mathbb{C}_{N_{\nu-1}}$. Further, the same formula (4.8.1) yields $\psi_\nu(j - N_\nu) = -\psi_\nu(j)$, whence it follows that $w(j - N_\nu) = -\,w(j)$. So, $w \in G_\nu$. We have ascertained that $W_\nu \subset G_\nu$.

Now let $w \in G_\nu$. Since $w \in \mathbb{C}_{N_{\nu-1}}$, there holds

$$w(j) = \sum_{k=0}^{2N_\nu - 1} w(k)\,\delta_{N_{\nu-1}}(j-k). \tag{S.18}$$

We rewrite the equality $w(j - N_\nu) = -w(j)$ in a form

$$w(j) = -\sum_{k=0}^{2N_\nu - 1} w(k)\,\delta_{N_{\nu-1}}(j - k - N_\nu). \tag{S.19}$$

Summing (S.18) and (S.19) and taking into account (4.8.1) we gain

$$\begin{aligned} 2\,w(j) &= \sum_{k=0}^{2N_\nu - 1} w(k)\,\psi_\nu(j - k) \\ &= \sum_{k=0}^{N_\nu - 1} w(k)\,\psi_\nu(j - k) + \sum_{k=0}^{N_\nu - 1} w(k + N_\nu)\,\psi_\nu(j - k - N_\nu) \\ &= \sum_{k=0}^{N_\nu - 1} [w(k) - w(k + N_\nu)]\,\psi_\nu(j - k). \end{aligned}$$

Hence $w \in W_\nu$. The inclusion $G_\nu \subset W_\nu$ is ascertained, so is the equality $G_\nu = W_\nu$.

4.17 Since discrete Walsh functions $v_k(j)$ take only two values $+1$ and -1, there holds $[v_k(j)]^2 \equiv 1$. An equivalent notation is $1/v_k(j) = v_k(j)$.

Further,

$$v_k(j)\,v_{k'}(j) = \prod_{\alpha=0}^{s-1} (-1)^{\langle k_\alpha + k'_\alpha \rangle_2\, j_\alpha} = v_m(j),$$

where $m = k \oplus k'$.

4.18 Take into account that $2k + 1 = 2k \oplus 1$ and use the result of the previous exercise.

4.19 We remind that the numbers $v_k(0),\ v_k(1),\ \ldots,\ v_k(N-1)$ form the row of the Hadamard matrix A_s with the index k. The required formulae follow from the recurrent relation

$$A_s = \begin{bmatrix} A_{s-1} & A_{s-1} \\ A_{s-1} & -A_{s-1} \end{bmatrix}.$$

4.20 Let $j \in 0 : N - 1$. The numbers $N - 1 - j$ and j belong to the set $0 : N - 1$, and their sum equals to $N - 1 = (1,\ 1,\ \ldots,\ 1)_2$. This is possible only when the binary codes of these numbers satisfy to the following condition: if a binary digit of one of these numbers equals to zero then the same binary digit of another number equals to unity.

4.21 Under the conditions of the exercise we have

$$v_k(j) = v_k(N - 1)v_k(j).$$

Since $N-1=(1,\ \ldots,\ 1)_2$, there holds

$$v_k(j)=(-1)^{\sum_{\alpha=0}^{s-1}k_\alpha(1+j_\alpha)}=(-1)^{\sum_{\alpha=0}^{s-1}k_\alpha(1-j_\alpha)}.$$

Note that for $j\in 0:N-1$ the following equality is true:

$$(N-1-j)_\alpha=1-j_\alpha,\quad \alpha\in 0:s-1.$$

Hence

$$v_k(j)=(-1)^{\sum_{\alpha=0}^{s-1}k_\alpha(N-1-j)_\alpha}=v_k(N-1-j).$$

4.22 The solution is similar to one of the previous exercise.

4.23 We fix $p=(p_{s-3},\ \ldots,\ p_0)_2$ and write

$$3N_2+p=N_1+N_2+p=(1,1,p_{s-3},\ \ldots,\ p_0)_2,$$

$$N_1+p=(1,0,p_{s-3},\ \ldots,\ p_0)_2.$$

Let us take $j=(j_{s-1},\ j_{s-2},\ \ldots,\ j_0)_2$. If $j_{s-2}=0$, i. e. $j=(j_{s-1},\ 0,\ j_{s-3},\ \ldots,\ j_0)_2$, then $\langle j+N_2\rangle_N=(j_{s-1},\ 1,\ j_{s-3},\ \ldots,\ j_0)_2$. We gain

$$\{3N_2+p,\ j\}_s=\{N_1+p,\langle j+N_2\rangle_N\}_s=j_{s-1}+\sum_{\alpha=0}^{s-3}p_\alpha j_\alpha.$$

This guarantees validity of the equality

$$v_{3N_2+p}(j)=v_{N_1+p}\big(\langle j+N_2\rangle_N\big). \tag{S.20}$$

Let $j_{s-2}=1$, i. e. $j=(j_{s-1},\ 1,\ j_{s-3},\ \ldots,\ j_0)_2$. Then

$$\langle j+N_2\rangle_N=(\langle j_{s-1}+1\rangle_2,\ 0,\ j_{s-3},\ \ldots,\ j_0)_2.$$

We have

$$\{3N_2+p,\ j\}_s=j_{s-1}+1+\sum_{\alpha=0}^{s-3}p_\alpha j_\alpha,$$

$$\{N_1+p,\langle j+N_2\rangle_N\}_s=\langle j_{s-1}+1\rangle_2+\sum_{\alpha=0}^{s-3}p_\alpha j_\alpha.$$

Since

$$(-1)^{\langle j_{s-1}+1\rangle_2}=(-1)^{j_{s-1}+1},$$

equality (S.20) holds in this case as well.

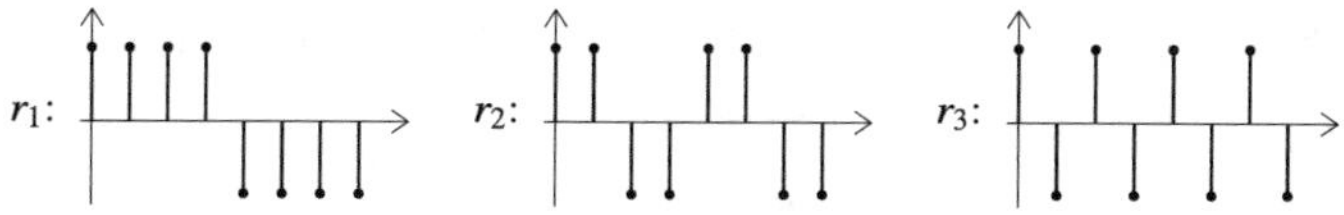

Fig. S.1 Graphs of the Rademacher functions for $N = 8$

4.24 The definition of Rademacher functions yields

$$r_\nu(j) = v_{N_\nu}(j) = (-1)^{j_{s-\nu}} = (-1)^{\lfloor j/N_\nu \rfloor}.$$

For $N = 8$ we have

$$r_1(j) = (-1)^{\lfloor j/4 \rfloor}, \quad r_2(j) = (-1)^{\lfloor j/2 \rfloor}, \quad r_3(j) = (-1)^j,$$

$$j \in 0 : 7.$$

Figure S.1 depicts the graphs of the functions r_1, r_2 and r_3.

4.25 As it was noted in the solution of the previous exercise, $r_\nu(j) = (-1)^{j_{s-\nu}}$ holds. Bearing this in mind we gain

$$v_k(j) = \prod_{\nu=1}^{s} (-1)^{k_{s-\nu} j_{s-\nu}} = \prod_{\nu=1}^{s} [r_\nu(j)]^{k_{s-\nu}}, \quad k \in 0 : N-1.$$

4.26 Let us use the identity $v_0(k) \equiv 1$ and the fact that $v_k(j) = v_j(k)$. We write

$$\sum_{k=0}^{N-1} v_k(j) = \sum_{k=0}^{N-1} v_0(k)\, v_j(k).$$

Now the required equality follows from orthogonality of Walsh functions and the equality $\langle v_0,\, v_0 \rangle = N$.

4.27 We denote $x_s = \mathcal{W}_N(x)$. Equalities (4.10.2) and (4.10.1) yield

$$\begin{aligned} \|x_s\|^2 &= \sum_{k=0}^{N-1} \Bigl(\sum_{j=0}^{N-1} x(j)\, v_k(j) \Bigr) \overline{x}_s(k) \\ &= \sum_{j=0}^{N-1} x(j) \Bigl\{ \sum_{k=0}^{N-1} \overline{x}_s(k)\, v_k(j) \Bigr\} = N \|x\|^2, \end{aligned}$$

which is equivalent to what was required.

4.28 On the basis of the definitions of the discrete Walsh transform and the dyadic convolution we write

$$\begin{aligned}\big[\mathcal{W}_N(z)\big](k) &= \sum_{j=0}^{N-1}\Big(\sum_{l=0}^{N-1} x(l)\, y(j\oplus l)\Big) v_k\big((j\oplus l)\oplus l\big)\\ &= \sum_{l=0}^{N-1} x(l) \sum_{j=0}^{N-1} y(j)\, v_k(j\oplus l).\end{aligned}$$

According to the result of Exercise 4.17 we have

$$v_k(j\oplus l) = v_{j\oplus l}(k) = v_j(k)\, v_l(k) = v_k(j)\, v_k(l).$$

Hence

$$\big[\mathcal{W}_N(z)\big](k) = \sum_{l=0}^{N-1} x(l)\, v_k(l) \sum_{j=0}^{N-1} y(j)\, v_k(j) = \big[\mathcal{W}_N(x)\big](k)\,\big[\mathcal{W}_N(y)\big](k)$$

as was to be proved.

4.29 We denote $V_p = \mathcal{F}_N(v_p)$. Taking into account that $v_1(j) = (-1)^{j_0} = (-1)^j$ for $j \in 0 : N-1$ we gain

$$V_1(k) = \sum_{j=0}^{N-1}(-1)^j\, \omega_N^{-kj} = \sum_{j=0}^{N-1} \omega_2^j\, \omega_N^{-kj} = \sum_{j=0}^{N-1} \omega_N^{-j(k-N_1)} = N\delta_N(k-N_1).$$

Further, $v_2(j) = (-1)^{j_1} = (-1)^{\lfloor j/2\rfloor}$ for $j \in 0 : N-1$. We put $j = 2l+q$, $q \in 0:1$, $l \in 0 : N_1 - 1$. Then $\lfloor j/2\rfloor = l$ and

$$\begin{aligned}V_2(k) &= \sum_{j=0}^{N-1}(-1)^{\lfloor j/2\rfloor}\, \omega_N^{-kj} = \sum_{q=0}^{1}\sum_{l=0}^{N_1-1}(-1)^l\, \omega_N^{-k(2l+q)}\\ &= \sum_{q=0}^{1} \omega_N^{-kq} \sum_{l=0}^{N_1-1} \omega_{N_1}^{-l(k-N_2)} = N_1(1+\omega_N^{-k})\delta_{N_1}(k-N_2).\end{aligned}$$

Since $v_3(j) = (-1)^{j_1+j_0} = (-1)^{\lfloor j/2\rfloor + j} = (-1)^{3l+q}$, we have

$$V_3(k) = N_1(1-\omega_N^{-k})\delta_{N_1}(k-N_2).$$

The Fourier spectra $V_2(k)$ and $V_3(k)$ on the main period are not equal to zero only for $k = N_2 = N/4$ and $k = N_2 + N_1 = 3N/4$. Herein

$$V_2(N/4) = N_1(1+\omega_4^{-1}) = N_1(1-i),$$

$$V_2(3N/4) = N_1(1 + \omega_4^{-3}) = N_1(1 + i);$$

$$V_3(N/4) = N_1(1 + i), \quad V_3(3N/4) = N_1(1 - i).$$

Further results on Fourier spectrum of Walsh functions can be found in [17].

4.30 We will use the fact that $r_\nu(j) = (-1)^{\lfloor j/N_\nu \rfloor}$ for $j \in 0 : N - 1$ (see the solution of Exercise 4.24). We put $j = pN_\nu + q,\ q \in 0 : N_\nu - 1,\ p \in 0 : \Delta_{\nu+1} - 1$. Then $\lfloor j/N_\nu \rfloor = p$ and

$$R_\nu(k) = \sum_{p=0}^{\Delta_{\nu+1}-1} (-1)^p \sum_{q=0}^{N_\nu-1} \omega_N^{-k(pN_\nu+q)} = \sum_{p=0}^{\Delta_{\nu+1}-1} (-1)^p\, \omega_{\Delta_{\nu+1}}^{-kp} \sum_{q=0}^{N_\nu-1} \omega_N^{-kq}$$

$$= \sum_{p=0}^{\Delta_{\nu+1}-1} \omega_{\Delta_{\nu+1}}^{-p(k-\Delta_\nu)} \sum_{q=0}^{N_\nu-1} \omega_N^{-kq} = 2^\nu \delta_{\Delta_{\nu+1}}(k - \Delta_\nu) \sum_{q=0}^{N_\nu-1} \omega_N^{-kq}.$$

It is clear that the Fourier spectrum $R_\nu(k)$ on the main period is not equal to zero only for $k = \Delta_\nu + l\Delta_{\nu+1} = (2l + 1)\Delta_\nu,\ l \in 0 : N_\nu - 1$. Given these k, according to (2.2.7) we gain

$$R_\nu(k) = 2^\nu \sum_{q=0}^{N_\nu-1} \omega_N^{-(2l+1)\Delta_\nu q} = 2^\nu \sum_{q=0}^{N_\nu-1} \omega_{N_{\nu-1}}^{-(2l+1)q}$$

$$= 2^\nu \frac{1 - \omega_{N_{\nu-1}}^{-(2l+1)N_\nu}}{1 - \omega_{N_{\nu-1}}^{-(2l+1)}} = 2^\nu \Big(1 - i \cot \frac{(2l + 1)\pi}{N_{\nu-1}}\Big).$$

4.31 It is known that $v_{2p+1}(j) = v_{2p}(j)\, v_1(j)$ (see Exercise 4.18). Let us use the result of Exercise 2.35 which yields

$$V_{2p+1} = N^{-1}(V_{2p} * V_1).$$

Since $V_1(k) = N\delta_N(k - N_1)$ (see Exercise 4.29), we have

$$V_{2p+1}(k) = \sum_{l=0}^{N-1} V_{2p}(l)\, \delta_N(k - l - N_1) = V_{2p}(k - N_1).$$

4.32 According to the result of Exercise 4.19 for $p \in 0 : N_1 - 1$ we have

$$V_p(k) = \sum_{j=0}^{N_1-1} v_p(j)\, \omega_N^{-kj} + \sum_{j=0}^{N_1-1} v_p(N_1 + j)\, \omega_N^{-k(N_1+j)}$$

$$= \big(1 + (-1)^k\big) \sum_{j=0}^{N_1-1} v_p^{(1)}(j)\, \omega_N^{-kj}.$$

Hence immediately follows that

$$V_p(2k) = 2V_p^{(1)}(k), \quad V_p(2k+1) = 0, \qquad k \in 0 : N_1 - 1. \tag{S.21}$$

Now we note that $N_1 + p = N_1 \oplus p$ for $p \in 0 : N_1 - 1$. Therefore (see Exercise 4.17)

$$v_{N_1+p}(j) = v_{N_1}(j)\, v_p(j).$$

Going over to Fourier transforms we gain (see Exercise 2.35)

$$V_{N_1+p} = N^{-1}(V_p * V_{N_1}). \tag{S.22}$$

The spectrum V_{N_1} is calculated easily. Indeed,

$$v_{N_1}(j) = (-1)^{j_{s-1}} = (-1)^{\lfloor j/N_1 \rfloor} = \begin{cases} 1 \text{ for } j \in 0 : N_1 - 1, \\ -1 \text{ for } j \in N_1 : N - 1. \end{cases}$$

As it is shown in par. 2.2.5,

$$V_{N_1}(k) = \begin{cases} 0 \text{ for even } k, \\ 2(1 - i \cot \frac{\pi k}{N}) \text{ for odd } k. \end{cases}$$

Let us write down formula (S.22) in more detail:

$$V_{N_1+p}(2k) = \frac{1}{N} \sum_{l=0}^{N_1-1} V_p(2l+1)\, V_{N_1}\big(2(k-l)-1\big) + \frac{1}{N} \sum_{l=0}^{N_1-1} V_p(2l)\, V_{N_1}\big(2(k-l)\big). \tag{S.23}$$

Both sums in the right side of (S.23) equal to zero: the former one due to $V_p(2l+1)$, the latter one due to $V_{N_1}\big(2(k-l)\big)$. Thus, $V_{N_1+p}(2k) = 0$ for $k \in 0 : N_1 - 1$. Further, using equality (S.21) we gain, similar to (S.23),

$$V_{N_1+p}(2k+1) = \frac{1}{N_1} \sum_{l=0}^{N_1-1} V_p^{(1)}(l)\, V_{N_1}\big(2(k-l)+1\big),$$

$$k \in 0 : N_1 - 1.$$

It is remaining to take into account that $V_{N_1}(2j+1) = h(j)$.

4.33 By virtue of N-periodicity of Walsh functions and equality (S.20) for $p \in 0 : N_2 - 1$ we gain

$$V_{3N_2+p}(k) = \sum_{j=0}^{N-1} v_{N_1+p}(j+N_2)\,\omega_N^{-kj} = \sum_{j=0}^{N-1} v_{N_1+p}(j)\,\omega_N^{-k(j-N_2)}$$
$$= \omega_4^k\, V_{N_1+p}(k) = i^k\, V_{N_1+p}(k), \quad k \in 0 : N-1.$$

4.34 It is sufficient to verify that

$$\{\mathrm{rev}_s(2k),\, j\}_s = \{\mathrm{rev}_s(k), 2j\}_s, \quad k, j \in 0 : N_1 - 1.$$

Let $k = (0,\, k_{s-2},\, \ldots,\, k_0)_2$ and $j = (0,\, j_{s-2},\, \ldots,\, j_0)_2$. Then

$$\mathrm{rev}_s(k) = (k_0,\, \ldots,\, k_{s-2},\, 0)_2, \quad \mathrm{rev}_s(2k) = (0,\, k_0,\, \ldots,\, k_{s-2})_2.$$

We gain

$$\{\mathrm{rev}_s(2k),\, j\}_s = \{\mathrm{rev}_s(k), 2j\}_s = \sum_{\alpha=0}^{s-2} k_{s-2-\alpha}\, j_\alpha.$$

4.35 It is required to verify that there holds the equality $\mathrm{wal}_s^{-1}(N-1) = 1$ or, which is equivalent, $\mathrm{wal}_s(1) = N-1$. By the definition, $\mathrm{wal}_s(1)$ is the number of sign changes of the Walsh function $v_1(j)$ on the main period. Since $v_1(j) = (-1)^j$ for $j \in 0 : N-1$, we have $\mathrm{wal}_s(1) = N-1$.

4.36 It follows from the proof of Theorem 4.12.2 that the formula

$$\xi(k) = N_\nu \sum_{l=0}^{\Delta_{\nu+1}-1} a(l)\, \widehat{v}_k(lN_\nu)$$

is true for $k \in 0 : \Delta_{\nu+1} - 1$.

4.37 With the aid of Theorem 4.11.1 we consecutively fill out the table of values of the permutations $\mathrm{wal}_1(k)$, $\mathrm{wal}_2(k)$, and $\mathrm{wal}_3(k)$ (Table S.1).

On the basis of the definition of an inverse mapping we gain

$$\mathrm{wal}_3^{-1}(k) = \{0,\ 4,\ 6,\ 2,\ 3,\ 7,\ 5,\ 1\}.$$

Table S.1 Values of $\mathrm{wal}_\nu(k)$ for $\nu = 1, 2, 3$

ν	$\mathrm{wal}_\nu(k)$ for $k = 0, 1, \ldots, 2^\nu - 1$
1	0 1
2	0 3 1 2
3	0 7 3 4 1 6 2 5

4.38 We note that

$$v_{k_1 N+k_0}(j_1 N + j_0) = (-1)^{\{k_1, j_1\}_s+\{k_0, j_0\}_s} = v_{k_1}(j_1)\, v_{k_0}(j_0).$$

Therefore

$$\begin{aligned}
F(k_1 N + k_0) &= \sum_{j_1, j_0=0}^{N-1} f(j_1 N + j_0)\, v_{k_1 N+k_0}(j_1 N + j_0) \\
&= \sum_{j_1, j_0=0}^{N-1} v_{j_1}(j_0)\, v_{k_1}(j_1)\, v_{k_0}(j_0) \\
&= \sum_{j_1=0}^{N-1} v_{k_1}(j_1) \sum_{j_0=0}^{N-1} v_{j_1}(j_0)\, v_{k_0}(j_0) \\
&= N \sum_{j_1=0}^{N-1} v_{k_1}(j_1)\, \delta_N(j_1 - k_0) = N v_{k_1}(k_0).
\end{aligned}$$

References

1. Ahmed, N., Rao, K.R.: Orthogonal Transforms for Digital Signal Processing. Springer, Heidelberg, New York (1975)
2. Ber, M.G., Malozemov, V.N.: On the recovery of discrete periodic data. Vestnik Leningrad. Univ. Math. **23**(3), 8–14 (1990)
3. Ber, M.G., Malozemov, V.N.: Interpolation of discrete periodic data. Probl. Inf. Transm. **28**(4), 351–359 (1992)
4. Ber, M.G., Malozemov, V.N.: Best formulas for the approximate calculation of the discrete Fourier transform. Comput. Math. Math. Phys. **32**(11), 1533–1544 (1992)
5. Blahut, R.E.: Fast Algorithms for Digital Signal Processing. Addison-Wesley, Reading, MA (1984)
6. Chashnikov, N.V.: Hermite spline interpolation in the discrete periodic case. Comput. Math. Math. Phys. **51**(10), 1664–1678 (2011)
7. Chashnikov, N.V.: Discrete Periodic Splines and Coons Surfaces. Lambert Academic Publishing (2010) (in Russian)
8. Cooley, J.W., Tukey, J.W.: An algorithm for the machine calculation of complex Fourier series. Math. Comput. **19**(90), 297–301 (1965)
9. Donoho, D.L., Stark, P.B.: Uncertainty principles and signal recovery. SIAM J. Appl. Math. **49**(3), 906–931 (1989)
10. Goertzel, G.: An algorithm for the evaluation of finite trigonometric series. Am. Math. Monthly **65**, 34–35 (1958)
11. Ipatov, V.P.: Periodic Discrete Signals with Optimal Correlation Properties. Radio i Svyaz, Moscow (1992). (in Russian)
12. Johnson, J., Johnson, R.W., Rodriguez, D., Tolimieri, R.: A methodology for designing, modifying and implementing Fourier transform algorithms on various architectures. Circuits Syst. Signal Process. **9**(4), 449–500 (1990)
13. Kirushev, V.A., Malozemov, V.N., Pevnyi, A.B.: Wavelet decomposition of the space of discrete periodic splines. Mathem. Notes **67**(5), 603–610 (2000)
14. Korovkin, A.V.: Generalized discrete Ahmed-Rao transform. Vestnik Molodyh Uchenyh **2**, 33–41 (2003). (in Russian)
15. Korovkin, A.V., Malozemov, V.N.: Ahmed-Rao bases. Mathem. Notes **75**(5), 780–786 (2004)
16. Korovkin, A.V., Masharsky, S.M.: On the fast Ahmed-Rao transform with subsampling in frequency. Comput. Math. Math. Phys. **44**(6), 934–944 (2004)

V. N. Malozemov and S. M. Masharsky, *Foundations of Discrete Harmonic Analysis*, Applied and Numerical Harmonic Analysis,
https://doi.org/10.1007/978-3-030-47048-7

17. Lvovich, A.A., Kuzmin, B.D.: Analytical expression for spectra of Walsh functions. Radiotekhnika **35**(1), 33–39 (in Russian) (1980)
18. Mallat, S.: A Wavelet Tour of Signal Processing, 2nd edn. Press, Acad (1999)
19. Malozemov, V.N., Chashnikov, N.V.: Limit theorems of the theory of discrete periodic splines. J. Math. Sci. **169**(2), 188–211 (2010)
20. Malozemov, V.N., Chashnikov, N.V.: Limit theorems in the theory of discrete periodic splines. Doklady Math. **83**(1), 39–40 (2011)
21. Malozemov, V.N., Chashnikov, N.V.: Discrete periodic splines with vector coefficients for computer-aided geometric design. Doklady Math. **80**(3), 797–799 (2009)
22. Malozemov, V.N., Masharsky, S.M.: Glassman's formula, fast Fourier transform, and wavelet expansions. Am. Math. Soc. Transl. **209**(2), 93–114 (2003)
23. Malozemov, V.N., Masharsky, S.M.: Generalized wavelet bases related with discrete Vilenkin-Chrestenson transform. St. Petersburg Math. J. **13**(1), 75–106 (2002)
24. Malozemov, V.N., Masharsky, S.M.: Haar spectra of discrete convolutions. Comput. Math. Math. Phys. **40**(6), 914–921 (2000)
25. Malozemov, V.N., Masharsky, S.M.: Comparative study of two wavelet bases. Probl. Inf. Transm. **36**(2), 114–124 (2000)
26. Malozemov, V.N., Masharsky, S.M., Tsvetkov, K.Yu.: Frank signal and its generalizations. Probl. Inf. Transm. **37**(2), 100–107 (2001)
27. Malozemov, V.N., Pevnyi, A.B.: Polynomial Splines. LGU, Leningrad (1986). (in Russian)
28. Malozemov, V.N., Pevnyi, A.B.: Discrete periodic B-splines. Vestnik St. Petersburg Univ. Math. **30**(4), 10–14 (1997)
29. Malozemov, V.N., Pevnyi, A.B.: Discrete periodic splines and their numerical applications. Comput. Math. Math. Phys. **38**(8), 1181–1192 (1998)
30. Malozemov, V.N., Pevnyi, A.B., Tretyakov, A.A.: Fast wavelet transform for discrete periodic signals and patterns. Probl. Inf. Transm. **34**(2), 161–168 (1998)
31. Malozemov, V.N., Prosekov, O.V.: Fast Fourier transform of small orders. Vestnik St. Petersburg Univ. Math. **36**(1), 28–35 (2003)
32. Malozemov, V.N., Prosekov, O.V.: Parametric versions of the fast Fourier transform. Doklady Math. **78**(1), 576–578 (2008)
33. Malozemov, V.N., Solov'eva, N.A.: Parametric lifting schemes of wavelet decompositions. J. Math. Sci. **162**(3), 319–347 (2009)
34. Malozemov, V.N., Solov'eva, N.A.: Wavelets and Frames in Discrete Analysis. Lambert Academic Publishing (2012) (in Russian)
35. Malozemov, V.N., Tret'yakov, A.A.: New approach to the Cooley-Tukey algorithm. Vestnik St. Petersburg Univ. Math. **30**(3), 47–50 (1997)
36. Malozemov, V.N., Tret'yakov, A.A.: The Cooley-Tukey algorithm and discrete Haar transform. Vestnik St. Petersburg Univ. Math. **31**(3), 27–30 (1998)
37. Malozemov, V.N., Tret'yakov, A.A.: Partitioning, orthogonality and permutations. Vestnik St. Petersburg Univ. Math. **32**(1), 14–19 (1999)
38. Malozemov, V.N., Tsvetkov, K.Yu.: On optimal signal-filter pairs. Probl. Inf. Transm. **39**(2), 216–226 (2003)
39. Malozemov, V.N., Tsvetkov, K.Yu.: A sampling theorem in Vilenkin-Chrestenson basis. Commun. Appl. Anal. **10**(2), 201–207 (2006)
40. Kamada, Masaru: Toraichi, Kazuo, Mori, Ryoichi: Periodic spline orthogonal bases. J. Approx. Theory **55**(1), 27–34 (1988)
41. McClellan, J.H., Rader, C.M.: Number Theory in Digital Signal Processing. Prentice-Hall, Englewood Cliffs, NJ (1979)
42. Morozov, V.A.: Regular Methods for Solving Ill-Posed Problems. Nauka, Moscow (1987). (in Russian)
43. Narcowich, F.J., Ward, J.D.: Wavelets associated with periodic basis functions. Appl. Comput. Harmonic Anal. **3**(1), 40–56 (1996)
44. Prosekov, O.V., Malozemov, V.N.: Parametric Variants of the Fast Fourier Transform. Lambert Academic Publishing (2010) (in Russian)

45. Sarwate, D., Pursley, M.: Cross-correlation properties of pseudorandom and related sequences. Proc. IEEE **68**(5), 593–619 (1980)
46. Malozemov, V.N. (ed.): Selected Chapters of Discrete Harmonic Analysis and Geometric Modeling. Part One. VVM, St. Petersburg (2014). (in Russian)
47. Malozemov, V.N. (ed.): Selected Chapters of Discrete Harmonic Analysis and Geometric Modeling. Part Two. VVM, St. Petersburg (2014). (in Russian)
48. Temperton, C.: Self-sorting in-place fast Fourier transform. SIAM J. Sci. Statist. Comput. **12**(4), 808–823 (1991)
49. Trakhtman, A.M., Trakhtman, V.A.: Fundamentals of the Theory of Discrete Signals on Finite Intervals. Sov. Radio, Moscow (1975). (in Russian)
50. Vlasenko, V.A., Lappa, Yu.M., Yaroslavsky, L.P.: Methods of Synthesis of Fast Algorithms for Signal Convolution and Spectral Analysis. Nauka, Moscow (1990). (in Russian)
51. Zalmanzon, L.A.: Fourier, Walsh and Haar Transforms and Their Application to Control. Communications and Other Fields. Nauka, Moscow (1989). (in Russian)
52. Zheludev, V.A.: Wavelets based on periodic splines. Rus. Acad. Sci. Dokl. Math. **49**(2), 216–222 (1994)
53. Zheludev, V.A., Pevnyi, A.B.: Biorthogonal wavelet schemes based on discrete spline interpolation. Comput. Math. Math. Phys. **41**(4), 502–513 (2001)

Index

V. N. Malozemov and S. M. Masharsky, *Foundations of Discrete Harmonic Analysis*, Applied and Numerical Harmonic Analysis,
https://doi.org/10.1007/978-3-030-47048-7

Printed in the United States
By Bookmasters